Advances in Industrial Control

Advances in Industrial Control is a series of monographs and contributed titles focusing on the applications of advanced and novel control methods within applied settings. This series has worldwide distribution to engineers, researchers and libraries.

The series promotes the exchange of information between academia and industry, to which end the books all demonstrate some theoretical aspect of an advanced or new control method and show how it can be applied either in a pilot plant or in some real industrial situation. The books are distinguished by the combination of the type of theory used and the type of application exemplified. Note that "industrial" here has a very broad interpretation; it applies not merely to the processes employed in industrial plants but to systems such as avionics and automotive brakes and drivetrain. This series complements the theoretical and more mathematical approach of Communications and Control Engineering.

Indexed by SCOPUS and Engineering Index.

Publishing Ethics

Researchers should conduct their research from research proposal to publication in line with best practices and codes of conduct of relevant professional bodies and/or national and international regulatory bodies. For more details on individual ethics matters please see:

https://www.springer.com/gp/authors-editors/journal-author/journal-author-helpdesk/publishing-ethics/14214

More information about this series at http://www.springer.com/series/1412

Prashant Mhaskar · Abhinav Garg
Brandon Corbett

Modeling and Control of Batch Processes

Theory and Applications

Springer

Prashant Mhaskar
Department of Chemical Engineering
McMaster University
Hamilton, ON, Canada

Brandon Corbett
Department of Chemical Engineering
McMaster University
Hamilton, ON, Canada

Abhinav Garg
Department of Chemical Engineering
McMaster University
Hamilton, ON, Canada

ISSN 1430-9491 ISSN 2193-1577 (electronic)
Advances in Industrial Control
ISBN 978-3-030-04139-7 ISBN 978-3-030-04140-3 (eBook)
https://doi.org/10.1007/978-3-030-04140-3

Library of Congress Control Number: 2018961228

This Springer imprint is published by the registered company Springer Nature Switzerland AG
The registered company address is: Gewerbestrasse 11, 6330 Cham, Switzerland

Series Editor's Foreword

Control systems engineering is viewed very differently by researchers and those that practice the craft. The former group develops general algorithms with a strong underlying mathematical basis while for the latter concerns over the limits of equipment and plant downtime often dominate. The series *Advances in Industrial Control* attempts to bridge this divide and to promote an understanding of the problems that really need a solution.

The rapid development of new control theory and technology has an impact on all areas of control engineering and applications. There are new control theories, actuators, sensors, communication and computing methods, and of course new application areas. It is important that new control theory and design methods are stimulated and driven by the needs and challenges of applications. A focus on applications is also essential if the different aspects of the control design problem are to receive sufficient attention. In fact, there is a lot of work on the analysis and synthesis problems in control systems engineering but much less on the problems of control design. The path from a control loop specification to a suitable design is often paved with uncertainties and confusion. The series provides an opportunity for researchers to present an exposition of new work on industrial control, raising awareness of the benefits that can accrue, and the challenges that can arise, and dealing with this issue of control design.

The batch control topic covered by this text is important in many industries and particularly in the pharmaceuticals industry's manufacture of healthcare products. These are of high value commercially but also of immense value to society, so it is valuable to explore what advanced control tools can improve in systems that do not have the luxury of long periods of steady-state operation. As the authors explain, fault tolerance in batch processes is a special problem where reliability is mainly concerned with achieving the desired end-point. Disturbances can result in the nonlinear system being in a region where the end-point is not achievable, and this can represent a challenging control problem.

It is not surprising that model predictive control (MPC) is proposed for this type of application since, as shown, it has very relevant features, particularly regarding constraint handling. The use of robust MPC is particularly notable in accounting for uncertainties. The first-principles process models are often presented in physical equation form, but the development of suitable models by online estimation methods is also described including online learning features. A model predictive quality control approach is introduced using an inferential quality model to ensure low variance of the final tracking errors.

The authors mainly cover chemical types of application, but they also cover electric arc furnaces: another batch operation but of a very different nature. A multi-model control approach, which is now very popular, is discussed for this and other applications. The very topical subject of economic MPC is also discussed in terms of its role in the hierarchy for solving batch control problems. The development of models is as important as control techniques, and the use of subspace identification methods for data-driven modeling is described. The chemical engineering application problems are treated in some detail as expected from authors with great expertise in the subject. This is therefore a welcome addition to the series *Advances in Industrial Control*.

Glasgow, UK
September 2018

Michael J. Grimble

Preface

Competitive economic conditions have compelled the manufacturing industries in most industrialized countries to pursue improved economic margins through the production of low-volume, higher-value-added specialty chemicals and materials, such as advanced alloys, polymers, herbicides, insecticides, pharmaceuticals, and biochemicals, that are manufactured predominantly in batch processes. Moreover, startups and shutdowns (that are batch-like processes) are an integral constituent of almost every process operation. The operation of these processes, however, has to grapple with several challenges, such as the lack of online sensors for measuring critical process variables, the finite duration of the process operation, the presence of significant nonlinear dynamics (due to predominantly transient operation), and rejecting raw material variability. Modeling and control of these batch and batch-like processes are therefore essential to ensure their safe and reliable function and to guarantee that they produce consistent and high-quality products or, in the case of startup operation, transit smoothly to continuous operation. Batch process operation, however, differs from operation around equilibrium points, both in the model identification aspects and control design.

Motivated by the above considerations, this book presents methods for the modeling and control of batch and batch-like processes with techniques ranging from mechanistic model to data-driven models. Specifically, the book proposes: (1) a novel batch control design with well-characterized feasibility properties; (2) a modeling approach that unites multi-model techniques and partial least squares; (3) a generalization of the subspace identification approach for batch processes; and (4) application to several detailed case studies, ranging from complex simulation test bed to industrial data. The proposed methodology employs statistical tools such as partial least squares and subspace identification and synergizes them with notions from state-space-based models to provide solutions to the quality control problem for batch processes.

The application of the proposed modeling and control tools is expected to significantly improve the operation of batch and batch-like processes. The book requires basic knowledge of statistical modeling, differential equations, and optimization methods and is intended for researchers, graduate students, and process control engineers. Throughout the book, practical implementation issues are discussed to help engineers and researchers understand the application of the methods in greater depth.

Finally, we would like to thank all the people who contributed in some way to this project. These include former graduate students and our colleagues at McMaster University for creating a pleasant working environment. Last but not least, we would like to express our deepest gratitude to our families for their dedication, encouragement, and support over the course of this project. We dedicate this book to them.

Hamilton, ON, Canada
September 2018

Prashant Mhaskar
Abhinav Garg
Brandon Corbett

Contents

List of Figures

List of Tables

Part I
Motivation

Chapter 1
Introduction

1.1 Motivation

Competitive economic conditions have compelled the manufacturing industries in most industrialized countries to pursue improved economic margins through the production of low volume, higher-value-added specialty chemicals and materials, such as advanced alloys, polymers, herbicides, insecticides, pharmaceuticals, and biochemicals, that are manufactured predominantly in batch processes. Moreover, startups and shutdowns (that are batch-like processes) are an integral constituent of almost every process operation. The operation of these processes, however, has to grapple with several challenges, such as the lack of online sensors for measuring critical process variables, the finite duration of the process operation, the presence of significant nonlinear dynamics (due to a predominantly transient operation), and rejecting raw material variability. Modeling and control of these batch and batch-like processes are therefore essential to ensure their safe and reliable function, and to guarantee that they produce consistent and high-quality products or, in the case of startup operation, transit smoothly to continuous operation. Batch process operation, however, differs from operation around equilibrium points, both in the model identification aspects and in the control design. Motivated by the above, this book presents batch-specific modeling and control approaches along with their application to nonlinear process systems.

1.2 Background

Many batch processes are operated under a simple open-loop 'control' policy where the same, time-indexed input trajectory is applied for each batch [1]. This approach is motivated by the assumption that by repeating historically successful actions, the desired product quality will be obtained. Open-loop policies are beneficial in that they are easy to implement, do not require online measurement or process models,

P. Mhaskar et al., *Modeling and Control of Batch Processes*,
Advances in Industrial Control, https://doi.org/10.1007/978-3-030-04140-3_1

and can be easily adapted from laboratory-scale procedures. However, because there is no feedback mechanism in this approach, it is incapable of rejecting disturbances that affects the process.

One common control approach for batch process is trajectory tracking. In trajectory tracking control, a predefined set-point trajectory for a measurable process variable is tracked. The underlying assumption for this approach is that if the process is tightly controlled to follow the same trajectory as a historically successful batch, the desired quality (often measured only after the batch termination) will be obtained. In its simplest form, trajectory tracking can be attempted using a PI controller. To account for the nonlinear evolution of the process states, techniques such as gain scheduling may be applied [2, 3]. While this approach provides a simple, easily implemented solution, even with perfect tracking the desired quality may not be obtained. This is because relationship between the controlled (and online measured) variable and the final quality may change significantly with changes in feedstock and other variations. Furthermore, this approach is inherently time-indexed which may become a problem when the duration of batches is not consistent.

More advanced control approaches universally rely on the availability of some sort of model for the process. In many studies, the existence of an accurate first-principles dynamic model of the process is assumed. Given such a model, approaches such as linearizing control [2] and direct model predictive control [4–10] can be applied. In these existing approaches, however, the control design does not explicitly recognize the set of states that can be driven to the desired end-point, and does not utilize that information in the control design; this is one of the contributions in the book.

Furthermore, a large part of the reason that batch processes are so ubiquitous is that, compared to their continuous counterparts, batch processes are relatively simple to design and implement. This leads to a short time to market but also has important implications in terms of process control. Because maintaining low development cost is often of utmost importance, it is often impractical to carry out the thorough identification experiments necessary to determine the parameters in first-principles models [11]. This motivates the use instead of simpler process models, where the parameters can be determined using statistical techniques. In these control approaches, models built from historical and limited identification data are used to calculate control actions. Because of the central role of the models in these approaches, the model building methodology is an integral part of the resulting control design.

Statistical batch control methods (and the related model identification) can be broadly classified into two classes: batch-to-batch approaches and within-batch approaches. In batch-to-batch approaches, the key assumption is that there are disturbances influencing the batch quality that are correlated from one batch to the next. Therefore, these approaches build models to predict the quality of the next batch from the qualities of the previous series of batches and their corresponding input trajectories. Between batches, the input trajectory is updated to reject disturbances observed in the preceding batches. A significant number of the literature contributions have been made addressing this approach [12–19]. Most notably related is a proposed batch-to-batch approach where the dynamic evolution of the quality variables between batches is identified using subspace identification [19]. While batch-to-batch approaches are valuable if there is a correlation between the batches, it is not applicable when batches are independent events. Furthermore, even if consecutive batches are correlated, batch-to-batch methods do not address the problem of rejecting disturbances that arise within individual batches. This motivates the need for within-batch control.

Within-batch approaches utilize measurements taken during the batch in conjunction with a model to reject disturbances as they arise throughout the batch. These approaches can be further categorized as trajectory tracking and quality control methods. As previously discussed, the premise of trajectory tracking control is to regulate a process variable to track a historically successful trajectory. Advanced trajectory tracking control is most commonly achieved by replacing traditional PI controllers with model-based control. Specifically, in these contributions, some form of dynamic model is used to predict the evolution of the controlled variable for a candidate input trajectory. The primary difference between these contributions lies in the models they employ. Of particular relevance are contributions in which empirical dynamic models are employed. In [20, 21], neural networks are used to identify a dynamic model. One of the most commonly studied approaches in the literature is the application of latent variable methods and in particular principal component analysis (PCA). In these approaches, the correlation between subsequent measurements is employed to describe the evolution of the process [22–25]. As with more traditional trajectory tracking approaches, the drawback of all trajectory tracking approaches is that the relationship between the controlled variable and quality may change. The solution is to apply direct quality control.

In general, direct quality control is achieved using model-based control. Specifically, a model is developed that relates the past observations, current observations, and the candidate input trajectory to quality. A variety of modeling methodologies have been applied to identify such models. One approach has been to train artificial neural networks on batch data [26]. The most commonly used modeling method, however, is projection to latent spaces (PLSs) [27–31]. In these contributions, trajectories of process measurements and inputs are related to quality in the model building step. Then, to control a new batch, the model is applied to available process measurements to determine an appropriate input. A key challenge in the control step is that fewer measurements may be available than were used in model fitting. Specifically, partway through the batch, only process measurements from the past and current sampling instant are available. A number of approaches have been taken to solve this problem including use of missing data algorithms [27] and independently building dynamic models [29] to estimate the necessary values. Note that the problem of estimating future process measurements is essentially a dynamic modeling problem.

Subspace identification is a class of linear, time-invariant, dynamic model identification techniques in which the first step is to identify a valid state trajectory from training data before fitting model parameters. Early contributions in this field established the broad success of these methods for identifying LTI models of continuous processes [32–35]. However, very little work has been carried out on applying subspace identification to batch processes. In [36], CVA (a class of subspace identification algorithms) is used to build a model for monitoring of batch processes, but the issue of control is not addressed. In [19], subspace identification is used for batch-to-batch control. Neither of these contributions treat the use of subspace identification for within-batch control.

In summary, a close examination of the existing literature indicates a lack of dedicated modeling and control designs for batch process control accounting explicitly for input constraints, variable duration nature of the modeling and control problem, and applications to complex nonlinear systems and to data from industrial processes.

1.3 Objectives and Organization of the Book

The broad objectives of this book are as follows:

1. To develop dedicated control design to account for the presence of input constraints and nonlinearity.
2. To develop modeling techniques to account for process nonlinearity.
3. To develop integrated data-driven modeling techniques with state-space-based models for batch processes.
4. To develop methods for handling issues such as multi-rate measurements and lack of online quality measurements.
5. To illustrate the applications of the developed modeling and control designs to nonlinear process systems.

This book is organized into four parts. Part I presents general background on the batch process control literature and the objectives behind this book. Part II deals with first-principles model-based control of batch processes. Parts III and IV then deal with data-based modeling approaches for batch process control. While the former presents designs integrating multi-model dynamics with partial least squares (PLS)-based methods, the latter presents the theory and applications of batch subspace identification to modeling and control of batch and batch-like processes.

Chapter 2 summarizes the existing tools in the area of batch process modeling and control.

Chapter 3 presents a novel control design that utilizes the characterization of so-called reachability regions for batch operation to determine the control action. The resultant control design is then shown to have superior fault-handling characteristics, hence the focus on safe-steering framework for batch processes. A simulated fed-batch process is used as an example to illustrate efficacy of the proposed framework.

In Chap. 4, the problem of uncertainty in the control design is explicitly addressed. Thus, robust reverse-time reachability region (RTRR)-based MPC is used to develop robust safe-steering framework to address the problem of operating the functioning inputs during the fault repair period. The computational efficiency and control performance of the proposed method are demonstrated using a simulated fed-batch process.

In Chap. 5, a model predictive control-based approach for transition of batch to continuous operation in bioreactors and subsequent control subject to constraints and performance considerations is presented. The key idea in the predictive controller is the formulation of appropriate stability constraints that allow an explicit characterization of the set of initial conditions from where feasibility of the optimization problem and hence closed-loop stability is guaranteed. Additional constraints are incorporated in the predictive control design to expand on the set of initial conditions that can be stabilized by control designs that only require the value of the Lyapunov function to decay. Then, the explicit characterization of the set of stabilizable initial conditions is used in determining the appropriate time for which the reactor must be run in batch mode. In Chaps. 3–5, the approaches are illustrated using first-principles models of the process.

In the next part of this book, in Chap. 6, empirical modeling approaches are united with nonlinear control tools for control of batch systems. The empirical models are in the form of multiple, ARX models that are appropriately 'combined' for the purpose of prediction, and the problem of trajectory tracking is addressed. Applications to fed-batch reactor and a nylon-6,6 batch polymerization process are presented to demonstrate the suitability of the proposed framework.

In Chap. 7, online learning ability is integrated within data-based modeling approach presented in Chap. 6, making it adaptive. Applications of the approach are demonstrated on a nylon-6,6 batch polymerization reactor case study.

In Chap. 8, the previously developed data-driven modeling methodology, which combines multiple local linear models with an appropriate weighting function to describe nonlinearities, is integrated with the inferential model, which relates the process conditions over the entire batch duration to the final quality. The key feature of this approach is that the causality and nonlinear relationships between the future inputs and outputs are accounted for in predicting the final quality and computing the manipulated input trajectory. The efficacy of the proposed predictive control design is illustrated via closed-loop simulations of a nylon-6,6 batch polymerization process with limited measurements.

In Chap. 9, the problem of quality control for the production of polymethyl methacrylate (PMMA) to achieve prescribed number and weight average molecular weights is considered. For this, a dynamic multiple model-based approach integrated with a quality model, to enable predicting the end quality based on initial conditions and candidate control input (jacket temperature) moves, is implemented to capture the process dynamics from past batch data. A data-driven model predictive controller is then designed to achieve the desired product quality while satisfying input constraint, a lower bound on the conversion, as well as additional constraints that enforce the validity of data-driven models for the range of chosen input moves.

In Chap. 10, the problem of multi-rate modeling and economic model predictive control (EMPC) of electric arc furnaces (EAFs) is considered. The two main challenges addressed are (a) the multi-rate nature of the measurement availability and (b) the requirement to achieve final product of a desired characteristic, while minimizing the operation cost. Multi-rate models are identified that include predictions for both the infrequently and frequently measured process variables. The models comprise local linear models and an appropriate weighting scheme to capture the nonlinear nature of the EAF. The resulting model is integrated into a two-tiered predictive controller that enables achieving the target end-point while minimizing the associated cost. The EMPC is implemented in the EAF process, and the closed-loop simulation results subject to the limited availability of process measurements and noise illustrate the improvement in economic performance over existing trajectory tracking approaches.

In the next part of this book, in Chap. 11, subspace identification methods are adapted for use with batch data to identify a state-space model from available process data. The resulting LTI, dynamic, state-space model is shown to describe the transient behavior of (finite duration) batch processes. The theoretical properties of the proposed approach are studied and compared to state-of-the-art latent variable control approaches, and applications to batch polymethyl methacrylate (PMMA) polymerization reactor are discussed to demonstrate the efficacy of the proposed approach.

In Chap. 12, the problem of variable duration economic model predictive control (EMPC) of batch processes subject to multi-rate and missing data is considered. For this, the batch subspace identification approach is generalized to handle multi-rate and missing data by utilizing the incremental singular value decomposition technique. Exploiting the fact that the proposed identification approach is capable of handling inconsistent batch lengths, the resulting dynamic model is integrated into a tiered EMPC formulation that optimizes process economics (including batch duration). Application of the proposed approach is demonstrated on an energy-intensive electric arc furnace process.

In Chap. 13, the problem of direct, data-driven, quality control for batch processes is considered. Specifically, previous results using subspace identification for modeling dynamic evolution and making quality predictions are extended with two key novel contributions: First, a method is proposed to account for mid-batch ingredient additions in both the modeling and the control stages. Second, a novel model predictive control scheme is proposed that includes batch duration as a decision variable. The efficacy of the proposed modeling and control approaches is demonstrated using a simulation study of a polymethyl methacrylate (PMMA) reactor.

In Chap. 14, the problem of subspace-based model identification and predictive control of particulate process subject to uncertainty and time-varying parameters is considered. A linear model predictive controller (MPC) formulation is shown to achieve a particle size distribution with desired characteristics subject to both manipulated input and product quality constraints in comparison with traditional control practices.

In Chap. 15, the problem of determining a data-driven model for the startup of a hydrogen production unit is considered. The proposed approach is demonstrated both on detailed first-principles simulation model and on real plant data.

References

1. Bonvin, D., Srinivasan, B., Hunkeler, D.: Control and optimization of batch processes. IEEE Control Syst. Mag. **26**, 34–45 (2006)
2. Bequette, B.: Nonlinear control of chemical processes - a review. Ind. Eng. Chem. Res. **30**, 1391–1413 (1991)
3. Berber, R.: Control of batch reactors: a review (Reprinted from methods of model based process control, 1995). Chem. Eng. Res. Des. **74**, 3–20 (1996)
4. Shi, D., El-Farra, N.H., Li, M., Mhaskar, P., Christofides, P.D.: Predictive control of particle size distribution in particulate processes. Chem. Eng. Sci. **61**, 268–281 (2006)
5. Nayhouse, M., Tran, A., Kwon, J.S.-I., Crose, M., Orkoulas, G., Christofides, P.D.: Modeling and control of ibuprofen crystal growth and size distribution. Chem. Eng. Sci. **134**, 414–422 (2015)
6. Valappil, J., Georgakis, C.: Accounting for batch reactor uncertainty in the nonlinear MPC of end-use properties. AIChE J. **49**, 1178–1192 (2003)
7. Bonvin, D., Srinivasan, B.: On the role of the necessary conditions of optimality in structuring dynamic real-time optimization schemes. Comput. Chem. Eng. **51**, 172–180 (2013)
8. Rafizadeh, M., Solgi, R., Abbaszadeh, M.: Constrained model predictive control of MMA polymerization reactor based on genetic algorithm optimization. In: CCA 2003: Proceedings of 2003 IEEE Conference on Control Applications, Istanbul, Turkey, 23–25 June 2003, Vols. 1 and 2, 345 E 47TH ST, New York, NY 10017 USA, pp. 464–469
9. Kozub, D., Macgregor, J.: Feedback-control of polymer quality in semibatch copolymerization reactors. Chem. Eng. Sci. **47**, 929–942 (1992)
10. Nayhouse, M., Kwon, J.S.-I., Christofides, P.D., Orkoulas, G.: Crystal shape modeling and control in protein crystal growth. Chem. Eng. Sci. **87**, 216–223 (2013)
11. Bonvin, D.: Optimal operation of batch reactors - a personal view. J. Process Control **8**, 355–368 (1998)
12. Kwon, J.S.-I., Nayhouse, M., Orkoulas, G., Ni, D., Christofides, P.D.: Run-to-run-based model predictive control of protein crystal shape in batch crystallization. Ind. Eng. Chem. Res. **54**, 4293–4302 (2015)
13. Flores-Cerrillo, J., MacGregor, J.: Iterative learning control for final batch product quality using partial least squares models. Ind. Eng. Chem. Res. **44**, 9146–9155 (2005)
14. Clarke-Pringle, T., MacGregor, J.: Optimization of molecular-weight distribution using batch-to-batch adjustments. Ind. Eng. Chem. Res. **37**, 3660–3669 (1998)
15. Chin, I., Lee, K., Lee, J.: A technique for integrated quality control, profile control, and constraint handling for batch processes. Ind. Eng. Chem. Res. **39**, 693–705 (2000)
16. Lee, J.H., Lee, K.S.: Iterative learning control applied to batch processes: an overview. Control Eng. Pract. **15**, 1306–1318 (2007). IFAC International Symposium on Advanced Control of Chemical Processes, Gramado, Brazil, 02–05 April 2006
17. Lee, K., Chin, I., Lee, H., Lee, J.: Model predictive control technique combined with iterative learning for batch processes. AIChE J. **45**, 2175–2187 (1999). This is available in print at the library in storage
18. Camacho, J., Pico, J., Ferrer, A.: Self-tuning run to run optimization of fed-batch processes using unfold-PLS. AIChE J. **53**, 1789–1804 (2007)
19. Dorsey, A., Lee, J.: Building inferential prediction models of batch processes using subspace identification. J. Process Control **13**, 397–406 (2003)
20. Sjoberg, J., Agarwal, M.: Trajectory tracking in batch processes using neural controllers. Eng. Appl. Artif. Intell. **15**, 41–51 (2002)
21. Hosen, M.A., Hussain, M.A., Mjalli, F.S.: Control of polystyrene batch reactors using neural network based model predictive control (NNMPC): an experimental investigation. Control Eng. Pract. **19**, 454–467 (2011)
22. Yu, H., Flores-Cerrillo, J.: Latent variable model predictive control for trajectory tracking in batch processes: internal model control interpretation and design methodology. Ind. Eng. Chem. Res. **52**, 12437–12450 (2013)

23. Golshan, M., MacGregor, J.F., Mhaskar, P.: Latent variable model predictive control for trajectory tracking in batch processes: alternative modeling approaches. J. Process Control **21**, 1345–1358 (2011)
24. Golshan, M., MacGregor, J.F., Bruwer, M.-J., Mhaskar, P.: Latent variable model predictive control (LV-MPC) for trajectory tracking in batch processes. J. Process Control **20**, 538–550 (2010)
25. Flores-Cerrillo, J., MacGregor, J.F.: Latent variable MPC for trajectory tracking in batch processes. J. Process Control **15**, 651–663 (2005)
26. Tsen, A., Jang, S., Wong, D., Joseph, B.: Predictive control of quality in batch polymerization using hybrid ANN models. AIChE J. **42**, 455–465 (1996)
27. Flores-Cerrillo, J., MacGregor, J.F.: Control of batch product quality by trajectory manipulation using latent variable models. J. Process Control **14**, 539–553 (2004)
28. Flores-Cerrillo, J., MacGregor, J.F.: Control of particle size distributions in emulsion semibatch polymerization using mid-course correction policies. Ind. Eng. Chem. Res. **41**(7), 1805–1814 (2002)
29. Corbett, B., Macdonald, B., Mhaskar, P.: Model predictive quality control of polymethyl methacrylate. IEEE Trans. Control Syst. Tech. **23**, 687–692 (2015)
30. Aumi, S., Corbett, B., Mhaskar, P., Clarke-Pringle, T.: Data-based modeling and control of nylon-6, 6 batch polymerization. IEEE Trans. Control Syst. Tech. **21**, 94–106 (2013)
31. Wan, J., Marjanovic, O., Lennox, B.: Disturbance rejection for the control of batch end-product quality using latent variable models. J. Process Control **22**, 643–652 (2012)
32. Overschee, P.v., : Subspace Identification for Linear Systems: Theory, Implementation, Applications. Kluwer Academic Publishers, Boston (1996)
33. VanOverschee, P., DeMoor, B.: N4SID - subspace algorithms for the identification of combined deterministic stochastic-systems. Automatica **30**, 75–93 (1994)
34. Moonen, M., Demoor, B., Vandenberghe, L., Vandewalle, J.: Online and off-line identification of linear state-space models. Int. J. Control **49**, 219–232 (1989)
35. Shi, R., MacGregor, J.: Modeling of dynamic systems using latent variable and subspace methods. J. Chemom. **14**, 423–439 (2000). 6th Scandinavian Symposium on Chemometrics, Porsgrunn, Norway, 15–19 August 1999
36. Yao, Y., Gao, F.: Subspace identification for two-dimensional dynamic batch process statistical monitoring. Chem. Eng. Sci. **63**, 3411–3418 (2008)

Chapter 2
Batch Process Modeling and Control: Background

2.1 Batch Process

Batch processes are an indispensable constituent of chemical process industries and are universally used for manufacturing of high-quality products. The preeminent reason for their popularity can be attributed to their flexibility to control different grades of products by changing the initial conditions and input trajectories. However, a batch process is characterized by absence of operation around equilibrium conditions resulting in highly nonlinear dynamics, which make the classical approaches (for continuous processes) not directly applicable.

Typically, a batch process consists of the following main steps.

1. Reactor is charged with ingredients following some predefined recipe.
2. A transforming process is carried out over a finite duration of time.
3. Finally, the process is terminated upon meeting some predefined criterion.

Apart from conventional batch processes, startups and shutdowns in continuous plants also mimic batch behavior and are termed as batch-like processes in this work. The specific characteristics of batch and batch-like processes impact both the modeling and control problems, as discussed next.

2.2 Data-Driven Process Modeling

We will focus our attention on data-driven modeling approaches and explain the differences between data from continuous operation and data from batch processes, especially with regard to the modeling problem. The data-driven modeling problem, generally speaking, can be stated as follows: Given a set of input–output data, pick a model structure (linear/nonlinear) and determine the associated parameters. As opposed to first-principles models, where the model structure is guided by governing equations, the choice of the model structure in data-driven modeling approaches is

P. Mhaskar et al., *Modeling and Control of Batch Processes*,
Advances in Industrial Control, https://doi.org/10.1007/978-3-030-04140-3_2

often based on simplicity and ease of use. When using data from operation around a nominal equilibrium point, local linear model often suffices to capture the observed dynamics. The emphasis there is on ensuring and checking sufficient excitation to capture the process dynamics around nominal operation.

In contrast, for batch operation, where a process typically transits over a wide dynamical range, the choice of the model structure, and more importantly modeling approach, becomes key so as to not lead to a model with too many parameters (resulting in over fitting), while still having the ability to capture process dynamics over a sufficiently wide range. In the remainder of this chapter, a brief overview of some promising data-driven modeling techniques, namely partial least squares (PLS), classical system identification, and subspace identification, is presented.

2.2.1 PLS

Among data-driven approaches for modeling of batch processes, latent variable-based methods have been widely used over past few decades in various applications [1–6]. Geometrically, in PLS regression, the variables in the regressor and response matrices, **X** and **Y**, are projected onto corresponding orthogonal subspaces of A-pairs of latent variables. Each pair of latent variables accounts for a certain percentage of the variance in the original matrices. Mathematically, PLS regression consists of decomposing **X** and **Y** as the sum of the outer products of a score and loading vector:

$$\mathbf{X} = \sum_{a=1}^{A} \boldsymbol{t}_a \boldsymbol{p}'_a + \mathbf{E} \tag{2.1}$$

$$\mathbf{Y} = \sum_{a=1}^{A} \boldsymbol{d}_a \boldsymbol{r}'_a + \mathbf{F} \tag{2.2}$$

where $\boldsymbol{t}_a$ and $\boldsymbol{d}_a$ are the input and output scores representing the projections of the variables in **X** and **Y** on the subspaces, $\boldsymbol{p}_a$ and $\boldsymbol{r}_a$ define the orientation of the corresponding subspaces, and **E** and **F** denote residual matrices. Because it is desired to obtain a useful relationship between the original data matrices, the two matrices are linked by an inner relation between their scores of the form:

$$\boldsymbol{d}_a = \boldsymbol{b}_a \boldsymbol{t}_a + \boldsymbol{e}_a \quad \forall a \in [1, A] \tag{2.3}$$

where $\boldsymbol{b}_a$ are the coefficients of the inner relationship and $\boldsymbol{e}_a$ are the residuals. In PLS algorithms, such as Nonlinear Iterative Partial Least Squares (NIPALS) [7], the subspace orientation and scores for both matrices are determined simultaneously to maximize the correlation between **X** and **Y** and therefore obtain the optimal fit for the inner relationship. The final result from PLS regression is a linear model between **X** and **Y** where the coefficients are functions of the scores and loadings from the matrix decompositions.

The ability of the PLS approach to predict process behavior for certain choices of the future manipulated input values is impinged by the inherent inability to distinguish between 'input' and 'output' variables, resulting in the need to address this problem by using imputation techniques where the future process output (for a candidate future input) is treated as a 'missing data' problem. The more pressing requirement is that of an alignment variable to handle batches with different batch lengths. While this can be dealt with rather readily in the context of monitoring, developing appropriate models that can deal with this problem in the context of batch process control remains a challenge.

2.2.2 *Classical System Identification*

Classical system identification approaches consist of iterative estimation algorithms to identify a state-space model, for instance, for a linear, deterministic case, a model of the following form is proposed:

$$\mathbf{x}_{k+1} = \mathbf{A}\mathbf{x}_k + \mathbf{B}\mathbf{u}_k, \tag{2.4}$$

$$\mathbf{y}_k = \mathbf{C}\mathbf{x}_k + \mathbf{D}\mathbf{u}_k, \tag{2.5}$$

Given the input and output sequence, the 'unknowns' include the initial state value, and the elements of the matrices A, B, C, and D, along with the choice of the order of the underlying system. One of the most common approaches used in classical system identification consists of prediction error minimization (PEM) methods which include estimation algorithms based on least squares (LS), maximum likelihood estimation (MLE), and Bayesian approaches. The objective in this approach is to compute the unknown values by minimizing prediction errors. Mathematically, this can be stated as follows [8]:

$$\hat{\theta}_l = \arg\min_{\theta \in M} \frac{1}{l} \sum_{L=1}^{l} f(\epsilon_F(L, \theta)) \tag{2.6}$$

where θ denotes the unknowns (including the model parameters), $f(\cdot)$ is a scalar-valued positive function, and ϵ_F denotes the filtered (through a stable linear filter) prediction error sequence:

$$\epsilon(L, \theta_\star) = y(t) - \hat{y}(t|\theta_\star) \tag{2.7}$$

Different formulations under PEM depend on how the prediction error is mathematically quantified (choice of $f(\cdot)$), choice of prediction error sequence filter, and the approach used for realizing the minimization. For a detailed discussion on this approach, the reader is referred to [8].

2.2.3 Subspace Identification

Subspace identification methods can be broadly categorized into deterministic, stochastic, and combined deterministic stochastic methods to identify linear dynamic models from process data. The key difference between these methods is the way in which process noise and disturbances are handled. In the present work, deterministic identification algorithms have been used. The deterministic identification problem can be explained as follows: If s measurements of the input $u_k \in \mathbb{R}^m$ and the output $y_k \in \mathbb{R}^l$ are given, the task is to determine the order n of a linear time invariant system of the following form (and the associated system matrices $\mathbf{A} \in \mathbb{R}^{n\times n}$, $\mathbf{B} \in \mathbb{R}^{n\times m}$, $\mathbf{C} \in \mathbb{R}^{l\times n}$, $\mathbf{D} \in \mathbb{R}^{l\times m}$ up to within a similarity transformation):

$$\mathbf{x}_{k+1} = \mathbf{A}\mathbf{x}_k + \mathbf{B}\mathbf{u}_k, \tag{2.8}$$

$$\mathbf{y}_k = \mathbf{C}\mathbf{x}_k + \mathbf{D}\mathbf{u}_k, \tag{2.9}$$

Unlike classical system identification (see, e.g., [9]), subspace methods are non-iterative and compute the unknowns using matrix algebra (see, e.g., [10, 11]). It is essentially the same problem as classical system identification except without resorting to a direct optimization problem formulation initially. Different algorithms within subspace identification such as canonical variate analysis (CVA) [12], numerical algorithms for subspace state-space system identification (N4SID) [13], and multivariable output error state-space algorithm (MOESP) [14] have been shown to differ only in the weighting of the matrix used at the singular value decomposition step [15]. The deterministic algorithm presented in [16], which is the foundation for the batch subspace identification algorithm proposed latter in this book, is briefly discussed as follows.

Consider the output measurements from a continuous process denoted as $\mathbf{y}[k]$, where k is the sampling instant. The output Hankel matrix is constructed as follows:

$$\mathbf{Y}_{1|i} = \begin{bmatrix} \mathbf{y}[1] & \mathbf{y}[2] & \cdots & \mathbf{y}[j] \\ \vdots & \vdots & & \vdots \\ \mathbf{y}[i] & \mathbf{y}[i+1] & \cdots & \mathbf{y}[i+j-1] \end{bmatrix} \tag{2.10}$$

Similarly, Hankel matrices for input data are constructed. The key step of obtaining state trajectories involves computing SVD of the Hankel matrix as follows. Let SVD of $\mathbf{H} = \begin{bmatrix} H_1 \\ H_2 \end{bmatrix}$ be given by

$$\mathbf{H} = \begin{bmatrix} U_{11} & U_{12} \\ U_{21} & U_{22} \end{bmatrix} \begin{bmatrix} S_{11} & 0 \\ 0 & 0 \end{bmatrix} V^t \tag{2.11}$$

then, the state vector can be computed as $U_q^t U_{12}^t H_1$. Here,

$$\mathbf{H}_1 = \begin{bmatrix} \mathbf{Y}_{1|i} \\ \mathbf{U}_{1|i} \end{bmatrix} \tag{2.12}$$

$$\mathbf{H}_2 = \begin{bmatrix} \mathbf{Y}_{1+i|2i} \\ \mathbf{U}_{1+i|2i} \end{bmatrix} \tag{2.13}$$

and U_q is defined through SVD of $U_{12}^t U_{11} S_{11}$ as

$$U_{12}^t U_{11} S_{11} = \begin{bmatrix} U_q & U_q^{\perp} \end{bmatrix} \begin{bmatrix} S_q & 0 \\ 0 & 0 \end{bmatrix} \begin{bmatrix} V_q^t \\ V_q^{\perp t} \end{bmatrix} \tag{2.14}$$

Mathematically, the identified state trajectories are represented as:

$$\hat{\mathbf{X}}_{i+1} = \begin{bmatrix} \hat{\mathbf{x}}[i+1] \cdots \hat{\mathbf{x}}[i+j] \end{bmatrix} \tag{2.15}$$

Once this state trajectory matrix is obtained, the system matrices can be estimated easily using methods like ordinary least squares as shown below:

$$\mathbf{Y}_{reg} = \begin{bmatrix} \hat{\mathbf{x}}[i+2] & \cdots & \hat{\mathbf{x}}[i+j] \\ \mathbf{y}[i+1] & \cdots & \mathbf{y}[i+j-1] \end{bmatrix} \tag{2.16}$$

$$\mathbf{X}_{reg} = \begin{bmatrix} \hat{\mathbf{x}}[i+1] & \cdots & \hat{\mathbf{x}}[i+j-1] \\ \mathbf{u}[i+1] & \cdots & \mathbf{u}[i+j-1] \end{bmatrix} \tag{2.17}$$

$$\mathbf{Y}_{reg} = \begin{bmatrix} \mathbf{A} & \mathbf{B} \\ \mathbf{C} & \mathbf{D} \end{bmatrix} \mathbf{X}_{reg} \tag{2.18}$$

The existing results, however, are designed primarily for continuous processes (see, e.g., [17]). In these cases, the training data sets are collected from identification experiments done around a desired operating range and can readily be incorporated in the Hankel matrices formation step of the subspace identification algorithm as it is one continuous block of data. However, in case of batch processes, data often consists of different batch runs and cannot simply be concatenated to create one large continuous data set. Therefore, the nature of data collection for batch processes requires specific adaptation of the database and subspace identification algorithm as discussed in Part IV of this book.

2.3 Batch Control

A fundamental objective in a typical batch process is to achieve the final product quality specifications. The measurements related to the terminal quality are usually not available during the batch operation and can only be accessed at the end of batch operation. Thus, it is not feasible to directly measure/control the quality of the product during the operation and track them using classical control techniques.

Traditionally, simple open-loop policy has been used to control batch processes where the idea is to implement the same time-indexed input trajectory from a previous successful batch [18]. The underlying assumption in this approach is that implementing the same, previously successful, input trajectory will lead to similar product quality. Although these approaches are simple to implement and do not require any knowledge of the process model or online measurements, they are incapable of rejecting disturbances and batch-to-batch variations due to absence of feedback. A significant body of literature (see [19] for overview) is also dedicated to develop control strategies to mitigate batch-to-batch variations such as iterative learning control (ILC). The basic idea in these approaches is that it utilizes the information from previous batches to update control policies in the subsequent batches.

A common closed-loop control approach that then emerged is trajectory tracking, where a predefined measured variable is tracked using a PI/PID controller. It is assumed in this approach that if the predefined trajectory, usually from past successful batches, is tightly followed, the desired product quality will be achieved. However, even with perfect tracking, the desired product may not be obtained as the process dynamics may change significantly due to variability in the feedstock properties, process disturbances, etc. Another limitation of this approach is that it does not allow altering the batch duration as they are inherently time-indexed.

The recourse then is to utilize a model that not only predicts the measured variables, but also the quality variables, and use the model within an advanced control design such as model predictive control (MPC).

2.3.1 Model Predictive Control Design

Model predictive control (MPC) is an optimization-based control scheme where the control action is computed by solving an optimization problem that utilizes a process model. In this approach, first the process model is initialized at the current plant measurements (for instance, using a state estimator in case of a state-space model of the process). Then, the input trajectory is computed, using the predictions from the process model that minimizes a predefined objective function along with satisfying any constraints. Lastly, the first input move is implemented on the process and these steps are repeated at every sampling instant. The underlying process model deployed in a MPC design may be first-principles model, a linearized version of first-principles model (for the case of nonlinear models, which is often the case), or a linear identified model of the process.

For the case of batch processes, while the structure of the controller is similar to the standard MPC design, the control objective being different from that of operating around nominal operating condition must be adequately accounted for in the controller implementation. In batch process, usually the objective is to control the properties of the final product which are not directly measurable during the batch operation. Thus, the MPC formulation takes the following form:

$$
\begin{aligned}
\min_{\mathbf{u}} \quad & f(y) \\
s.t. \quad & u_{min} \leq u \leq u_{max}, \\
& \alpha[k] \leq f_2(y[k]) \leq \beta[k], \\
& f_3(y[t_f]) \geq \gamma \\
& \mathbf{x}[k+1] = \mathbf{A}\mathbf{x}[k] + \mathbf{B}\mathbf{u}[k] \\
& \mathbf{y}[k] = \mathbf{C}\mathbf{x}[k] + \mathbf{D}\mathbf{u}[k]
\end{aligned}
\tag{2.19}
$$

where $f(y)$ represents the objective function which, depending on the application, could be 'stage cost' or simply depend on the terminal output $y[t_f]$, u_{min} and u_{max} denote the bounds for the manipulated input. The constraint on $f_2(y[k])$ denotes the output path constraints, while the constraint on $f_3(y[t_f])$ denotes the terminal quality constraint. The optimization problem is solved based on the predictions obtained from the identified (LTI) state-space model, utilizing the current state estimate provided by the state estimator. If the constraints and objective can also be described by linear functions, utilization of the subspace model enables converting the optimization problem to a linear/quadratic program that can be readily implemented online. In contrast to continuous MPC designs, where the prediction horizon is fixed, and the controller implemented in a receding fashion, in batch processes, the prediction horizon invariably shrinks and is implemented accordingly.

Remark 2.1 As with any state-space-based MPC implementation (whether for continuous operation or otherwise), the MPC action is only computed based on the current state estimate, which invariably requires use of an appropriate state estimator/observer. In contrast to continuous operation, where the control action can be implemented continuously after the state estimator has converged, this process needs to be repeated for every batch operation. Thus, for every new batch, the controller is often not implemented online and waits for the state estimation to converge, before bringing the MPC online.

Remark 2.2 The other key difference between continuous and batch control design is around the notion of stability. Given the nonequilibrium nature of batch operation, the question of closed-loop stability needs to be handled with care. The most important point to realize is that the target quality (and the associated point in state space) might simply not be an equilibrium point. Thus, any analysis of stability—which is inherently tied to stability of an equilibrium point—becomes meaningless. The analysis tool that remains most relevant is that of reachable sets, and more importantly finite time reachable sets—the set of states that can be driven to the desired target point in a given time frame—and care must be taken that the control action does not through the process away out of this set, leading to an irrecoverable and wasted batch.

References

1. Flores-Cerrillo, J., MacGregor, J.F.: Control of particle size distributions in emulsion semibatch polymerization using mid-course correction policies. Ind. Eng. Chem. Res. **41**(7), 1805–1814 (2002)
2. Kourti, T.: Multivariate dynamic data modeling for analysis and statistical process control of batch processes, start-ups and grade transitions. J. Chemom. **17**, 93–109 (2003)
3. Kourti, T.: Abnormal situation detection, three-way data and projection methods; robust data archiving and modeling for industrial applications. Annu. Rev. Control **27**(2), 131–139 (2003)
4. Flores-Cerrillo, J., MacGregor, J.F.: Control of batch product quality by trajectory manipulation using latent variable models. J. Process Control **14**, 539–553 (2004)
5. Golshan, M., MacGregor, J.F., Bruwer, M.-J., Mhaskar, P.: Latent variable model predictive control (LV-MPC) for trajectory tracking in batch processes. J. Process Control **20**, 538–550 (2010)
6. Golshan, M., MacGregor, J.F., Mhaskar, P.: Latent variable model predictive control for trajectory tracking in batch processes: alternative modeling approaches. J. Process Control **21**, 1345–1358 (2011)
7. Geladi, P., Kowalski, B.: Partial least-squares regression: a tutorial. Anal. Chim. Acta **185**, 1–17 (1986)
8. Ljung, L.: System Identification: Theory for the User. Pearson Education, Boston (1998)
9. Raghavan, H., Tangirala, A.K., Bhushan Gopaluni, R., Shah, S.L.: Identification of chemical processes with irregular output sampling. Control Eng. Pract. **14**, 467–480 (2006)
10. Tangirala, A.K.: Principles of System Identification: Theory and Practice. Taylor & Francis, London (2014)
11. Qin, S.J.: An overview of subspace identification. Comput. Chem. Eng. **30**, 1502–1513 (2006). 7th International Conference on Chemical Process Control (CPC-7), Lake Louise, Canada, 08–13 January 2006
12. Larimore, W.: Canonical variate analysis in identification, filtering, and adaptive control. In: 29th. IEEE Conference on Decision and Control **2**, 596–604 (1990)
13. Overschee, P.V., Moor, B.D.: Two subspace algorithms for the identification of combined deterministic-stochastic systems. In: 1992 Proceedings of the 31st IEEE Conference on Decision and Control, vol. 311, pp. 75–93 (1992)
14. Verhaegen, M., Dewilde, P.: Subspace model identification Part 1. The output-error state-space model identification class of algorithms. Int. J. Control **56**, 1187–1210 (1992)
15. VanOverschee, P., DeMoor, B.: A unifying theorem for three subspace system identification algorithms. Automatica **31**, 1853–1864 (1995). 10th IFAC Symposium on System Identification, Copenhagen, Denmark, 04–06 July 1994
16. Moonen, M., Moor, B.D., Vandenberghe, L., Vandewalle, J.: On-and off-line identification of linear state-space models. Int. J. Control **49**, 219–232 (1989)
17. Wang, J., Qin, S.J.: A new subspace identification approach based on principal component analysis. J. Process Control **12**, 841–855 (2002)

18. Bonvin, D., Srinivasan, B., Hunkeler, D.: Control and optimization of batch processes. IEEE Control Syst. Mag. **26**, 34–45 (2006)
19. Lee, J.H., Lee, K.S.: Iterative learning control applied to batch processes: an overview. Control Eng. Pract. **15**, 1306–1318 (2007). IFAC International Symposium on Advanced Control of Chemical Processes, Gramado, Brazil, 02–05 April 2006

Part II
First-Principles Model Based Control

Chapter 3
Safe-Steering of Batch Processes

3.1 Introduction

Faults in processing or auxiliary equipment (sensors, actuators, etc.) are ubiquitous in the chemical process industry, can have a serious impact on product quality, lead to undesirable pollutant emissions, and negatively impact the overall plant productivity and economy. Batch and fed-batch process productivity are particularly susceptible to faults as there is an emphasis on final product quality, and a fault during a batch may ruin the entire batch product. Specialized sectors of the chemical industry operate exclusively in batch or fed-batch modes. Key examples include (1) the production of biochemicals (e.g., ethanol), polymers, and pharmaceutical products; (2) the separation and transformation of materials by batch distillation; and (3) the processing of materials by injection molding. Additionally, for continuous processes with an initial batch mode of operation, namely bio-processes, conditions at batch termination can ultimately dictate process performance upon the transition to continuous mode [1]. In addition to faults, other operational issues such as constraints and disturbances must be accounted for with suitable batch operating policies, appropriate monitoring schemes, and well-designed batch process control structures.

The primary control objective in batch processes is to reach a desired product quality by batch termination. As a result, batch process control structures are designed using end-point-based control approaches. Early approaches to batch process operation consisted of open-loop policies in which optimal state trajectories were determined off-line by solving a dynamic optimization (DO) problem (see, e.g., [2]) or through past experience. These state trajectories, which incorporated desired end-point properties, were subsequently tracked using local PID or predictive controllers (see, e.g., [3–5]).

In many instances, however, process noise and disturbances can render the optimal state trajectories suboptimal or the tracking problem infeasible. In response to this, one modification to a completely open-loop control policy has been to correct optimal state trajectories midcourse at specified decision points in the batch based on the projected end-point properties [6]. With increased computational power together with

P. Mhaskar et al., *Modeling and Control of Batch Processes*,
Advances in Industrial Control, https://doi.org/10.1007/978-3-030-04140-3_3

more efficient optimization algorithms, shrinking horizon model predictive control (MPC) of batch processes is becoming commonplace. In this approach, the optimization problem is solved at each sampling instance until batch termination, using plant measurements (or state variable estimates) to update model predictions at each sampling time. Rather than relying on tracking controllers, the resulting optimal input changes are then implemented directly on the process. Differentiating features among these MPC approaches include the complexity of the model used for predictions and the nature of the input parametrization in the dynamic optimization problem. With respect to the former, the predictive model may be a linearized version of the full nonlinear model or the rigorous, first-principles nonlinear model. Due to strong nonlinearities present in most batch processes, performance using linear models is severely limited. Successive linearization techniques or scheduling of multiple linear models represents a work-around to nonlinear MPC in these cases, but with the significant reduction of computational times of real-time optimization algorithms MPC based on full nonlinear models is becoming increasingly tractable [7–10]. Another factor affecting the computational effort in an MPC approach is input parametrization. Inherent to the typical MPC algorithm, the optimal solution consists of the entire input trajectory from the time at which the problem is solved to batch termination, implying significant computational effort, especially during the start of the batch. In fact, a majority of the developments in real-time optimization algorithms are based on improved input parametrization techniques, which can be derived from an off-line characterization of an optimal solution and strive to reduce the number of decision variables in the optimization problem (e.g., see [11–14]). However, with modeling errors and process noise, actual process trajectories may deviate considerably from the nominal one, rendering a certain input parametrization suboptimal or infeasible in the worst case. One contribution of the present chapter is a computationally efficient model predictive control scheme which only requires the online computation of the current control action, while guaranteeing end-point reachability, if achievable.

For batch (as well as continuous) processes, the occurrence of a fault can invalidate the desirable properties of a control design. Compared to batch systems, there has been extensive research on fault-tolerant control structures (FTCSs) for continuous processes. Most of the existing methods for FTC rely on the assumption of the availability of sufficient control effort or redundant control configurations to maintain operation at the nominal equilibrium point in the presence of faults. These methods can be categorized within robust/reliable control approaches (also called passive FTC; see, e.g., [15]) and reconfiguration-based fault-tolerant control approaches (also called active FTC, see, e.g., [16–22]). More recently, the control of nonlinear (continuous) processes subject to input constraints and faults that preclude the possibility of operation at the nominal equilibrium point during a fault has been studied. This led to the development of a safe-parking framework in [23]. The safe-parking fault-tolerant control framework specifically considers the class of equipment failure that does not allow continued operation at the nominal operating point due to input constraints. The framework answers the problem of choosing what steady state to operate the plant during fault rectification such that a smooth transition back to the nominal (i.e., fault-free) equilibrium point is feasible and optimal with respect to some measure of plant economics. The safe-parking framework was recently generalized to address availability of limited measurements and uncertainty [24].

The extensive results for handling faults for continuous processes, however, do not carry over to batch processes. Specifically, the absence of equilibrium points in batch processes and fundamental differences in the control objectives between batch and continuous processes [4] prevent the direct applicability of research results from continuous processes. As with the robust and reconfiguration-based approaches for continuous processes, although several robust nonlinear MPC structures for batch processes are available [25, 26] (as well as dedicated fault monitoring designs for batch systems; see, e.g., [27, 28]), the fault-tolerant characteristic in these formulations stems from the underlying assumption of availability of sufficient control effort such that the desired objective of the batch, or the tracking of desired state trajectories, remains achievable following fault repair. The recently developed safe-parking framework [23, 24] does not remain applicable due to the absence of nominal equilibrium point or safe-parking points. In particular, the requirement to safe-park in [23, 24] is that of the existence of operating points where the process can be stabilized in the event of a fault, and this requirement does not hold in the event of a fault in a batch process. With the basic premise of the safe-parking framework not holding for batch processes, the subsequent method of handling faults in [23, 24] does not apply to batch processes.

In the absence of a framework for handling faults in batch processes, the continuation of the implementation of controllers to drive the process to the desired end-point may not be the best option. For instance, if one of the inputs in a batch process fails (i.e., an actuator is 'stuck' at a fail-safe value), it is likely that the optimization problem in a conventional end-point-based MPC may become infeasible during the faulty period because the desired end-point properties can no longer be reached with limited available input for the rest of the batch duration. If the fault is repaired sufficiently fast, on the other hand, it may still be possible to reach the desired end-point. However, without the knowledge of the fault repair time, traditional MPC approaches (during fault rectification) would dictate computing the manipulated input trajectory using the reduced control action till batch termination (therefore yielding an infeasible solution). By repeatedly applying a saturated version of an infeasible input trajectory, the process can be driven to a point where it is no longer possible to meet desired end-point properties even if the fault is repaired in due time. Therefore, the batch process control problem may continue to remain infeasible even after fault rectification and the desired end-point properties will not be reached. This could result in the loss of the batch product, as well as significant wastage of time and money for reactor cleanup, if required.

Motivated by the above considerations, this chapter considers the problem of control of batch processes to achieve a desired final product quality subject to input constraints and faults in the control actuators. Specifically, faults are considered that cannot be handled via robust control approaches. A safe-steering framework is developed to address the problem of determining how to utilize the functioning inputs during fault rectification in order to achieve the desired product properties at batch termination. The rest of this manuscript is organized as follows: The class of processes considered is first presented followed by a review of a representative end-point-based nonlinear MPC formulation for batch process control. Next, we design a reachability region-based predictive controller that requires the online computation of only the immediate control move. The safe-steering problem is then formulated, and we use the reachability region-based controller design to come up with the safe-steering framework. Then, a fed-batch bioreactor example is used to illustrate the details of the safe-steering framework. Finally, we summarize our results.

3.2 Preliminaries

In this section, the class of batch processes considered is presented followed by a representative formulation of existing nonlinear, shrinking horizon predictive controller used for batch process control.

3.2.1 Process Description

We consider batch process systems subject to input constraints and failures described by:

$$\begin{aligned} \dot{x} &= f(x(t), u_\sigma(t)) \\ t &\in [t_0, t_f], u_\sigma(\cdot) \in \mathbf{U}_\sigma, x(t_0) = x_0 \end{aligned} \tag{3.1}$$

where $x \in \mathbb{R}^n$ denotes the vector of state variables and $u_\sigma(t) \in \mathbb{R}^m$ denotes the vector of constrained manipulated inputs, taking values in a nonempty convex subset $\mathbf{U}_\sigma$ of $\mathbb{R}^m$, where $\mathbf{U}_\sigma = \{u \in \mathbb{R}^m : u_{min,\sigma} \leq u \leq u_{max,\sigma}\}$, where $u_{min,\sigma}, u_{max,\sigma} \in \mathbb{R}^m$ denote the constraints on the manipulated inputs. The times t_0 and t_f denote the initial time and batch termination times, respectively. The variable, $\sigma \in \{1, 2\}$, is a discrete variable that indexes fault-free and faulty operation; specifically, $\sigma = 1$ denotes fault-free operation and $\sigma = 2$ denotes a faulty operation. The faulty operation specifically considered in this chapter involves actuator failure for a finite duration of time defined by the time of fault occurrence, t^{fault}, and time of fault recovery, $t^{recovery}$. An example of such a failure condition is a flow valve feeding a fed-batch reactor becoming 'stuck' at its fail-safe value between t^{fault} and $t^{recovery}$ while remaining operational at all other times. According to the definition of σ, $\sigma = 1$ for $t \in [t_0\ t^{fault})$, $\sigma = 2$ for $t \in [t^{fault}\ t^{recovery})$, and $\sigma = 1$ for $t \in [t^{recovery}\ t_f]$. The vector function $f : \mathbb{R}^n \times \mathbf{U}_\sigma \rightarrow \mathbb{R}^n$ is assumed to be sufficiently smooth on its domain of definition. The notation, $\|\cdot\|_Q$, refers to the weighted norm, defined by $\|x\|_Q = x^T Q x$ for all $x \in \mathbb{R}^n$, where Q is a positive-definite symmetric matrix and x^T denotes the transpose of x. Throughout the manuscript, we assume for any

$u \in \mathbf{U}_\sigma$, the solution of the batch system of Eq. (3.1) exists and is continuous for all $t \in [t_0, t_f]$. For this chapter, we focus on the state feedback problem where $x(t)$ is assumed to be available for all $t \in [t_0, t_f]$.

3.2.2 End-Point-Based MPC

In this section, a representative formulation of a shrinking horizon, nonlinear predictive controller is presented. Note that the representative formulation is not meant to generalize all variations of published MPC formulations for batch processes, but only meant to capture the key idea in most existing formulations, which is the computation of the entire manipulated trajectory from the current time to the end of the batch. To this end, consider the batch process described by Eq. (3.1) for $\sigma(t) = 1$ (i.e., a fault-free environment where all manipulated inputs can be changed via a feedback law), for which the control action at each sampling instance is computed by solving a dynamic optimization problem of the form:

$$u_{MPC}(x) := \arg min\{J_{EP}(x, t, u(\cdot))|u(\cdot) \in S\} \tag{3.2}$$

subject to:

$$\dot{\tilde{x}} = f(\tilde{x}(\tau), u(\tau)) \tag{3.3}$$

$$\tilde{x}(0) = x(t) \tag{3.4}$$

$$\tilde{x}(t_f - t) = x_d \tag{3.5}$$

where $S = S(t, T)$ is the family of piecewise continuous functions (functions continuous from the right), with period Δ, mapping $[t, t + T]$ into U, x_0 represents the initial condition of the batch, and x_d denotes specified desired states at t_f. A control $u(\cdot)$ in S is characterized by the sequence $\{u[j]\}$ where $u[j] := u(j\Delta)$ and satisfies $u(t) = u[j]$ for all $t \in [j\Delta, (j + 1)\Delta)$. The minimizing control $u_{MPC}(\cdot) \in S$ is then applied to the plant over the interval $[t, t + \Delta)$, and the procedure is repeated until batch termination. The performance objective, $J_{EP}(x, t, u(\cdot))$, in the optimization problem can have the following (general) form:

$$J_{EP}(x, t, u(\cdot)) = M(t_0, \tilde{x}(t_0), t_f, \tilde{x}(t_f), x_d) + \int_t^{t_f} L(\tilde{x}(\tau), u(\tau))\, d\tau \tag{3.6}$$

where $M(\cdot)$ and $L(\cdot)$ represent the Mayer and Lagrangian terms, respectively. For end-point optimization or end-point constrained problems, Eq. (3.6) typically simplifies to the Mayer form (i.e., $L = 0$), a functional term that explicitly involves initial and final conditions. The Lagrangian term, L, is frequently used to implement soft constraints on the control rate or minimize deviations from nominal optimal state and input trajectories.

Remark 3.1 Note that evaluating the end-point constraint (Eq. (3.5)) and the objective function in the above formulation (Eq. (3.6)) requires state evaluation at batch termination. This, in turn, implies the nonlinear model (Eq. (3.3)) must be integrated at each sampling time to t_f and the input trajectory up to t_f must be computed by the optimizer at each time step. Consequently, both integration and optimization become computationally expensive (whether using a sequential or simultaneous approach to solving the optimization problem). Note that while the solution to the optimization problem at a certain time step is a good initial guess for the next time step (exactly in the absence of uncertainties), modeling errors and disturbances may require significant computational effort in computing the manipulated input trajectory at every time step. We present in the next section an inherently different predictive controller formulation that does not require the online integration of the process model nor the online computation of the input trajectory up to the final time at each sampling instance and thus significantly reduces online computational demands (as expected, much of the computational expense is transferred off-line; see the next section).

3.3 Reverse-Time Reachability Region-Based MPC

In this section, we present a nonlinear model predictive controller for batch processes. The key idea behind this design is to require the computation of only the immediate value of the manipulated variable while ensuring the desired end-point properties remain attainable throughout the batch. Preparatory to the controller design, we first review the notion of reverse-time reachability regions, which are essential in the control design and analysis.

3.3.1 Reverse-Time Reachability Regions

As previously discussed, the objective in batch processes is to reach a desired end-point, and of interest are the set of states from where the desired end-point can be reached. This set can be expressed in the form of reverse-time reachability regions, formally defined below:

Definition 3.1 For the process described by Eq. (3.1), the reverse-time reachability region at time, t, $R^{-1}(t)$ is the set:

$$R^{-1}(t) = \{x_0 \mid x(t) = x_0,\ u(\tau) \in \mathbf{U}\ \forall\ \tau \in [t, t_f],\ \dot{x} = f(x, u),\ x(t_f) = x_d\} \tag{3.7}$$

The reverse-time reachability region at time t, $R^{-1}(t)$, therefore, consists of all process states from which the batch process can be steered to x_d, the desired end-point, by the end of the batch (i.e., in a time $t_f - t$) while satisfying the input constraints. The reason behind naming this set the 'reverse-time reachability region' is as follows: Note that a reachability region, $R(t)$, for both batch and continuous processes is defined as the set of states that can be reached from a given initial condition in a time, t, subject to constraints on the manipulated inputs. If the 'reverse-time' version of the process is considered (i.e., $\dot{x} = -f(x, u)$), and the reachability region for this reverse-time process is computed (setting the initial condition as the desired end-point of the original process), this in turn yields the set of states from where the desired end-point can be reached for the original process (and hence the name reverse-time reachability region). An algorithm to generate reverse-time reachability region and the associated computational issues are discussed in the following remarks.

While the set is defined allowing $u(t)$ to take values in U, computation of the reachability region can only be carried out by discretizing (in time) the control action (i.e., subject to a control action held constant for a predefined period of time). Below, we define the discrete time version (where the control action is held for a time Δ till batch termination)

Definition 3.2 The reverse-time reachability region at a time $t = t_f - k\Delta$ (where $k \leq \dfrac{t_f - t_0}{\Delta} \in Z$), indexed by k is defined as:

$$R_k^{-1} = \{x_0 \mid x(t) = x_0,\ u_i \in \mathbf{U}\ \forall i = 0, \ldots, k,\ \dot{x} = f(x, u),\ x(t_f) = x_d\} \quad (3.8)$$

One way to compute R_k^{-1} is to scan the state space and test the feasibility of an optimization problem that requires the end-point constraint to be met subject to input constraints and to include every state for which the optimization has a feasible solution. However, the understanding of these sets as being the reverse-time reachability regions allows the sequential determination (of an estimate) of these sets without having to solve optimization problems. In particular, for a given x_d, the reverse-time process model can be integrated backward in time for the duration of Δ, holding the value of the manipulated input constant. Performing this integration for all possible (appropriately discretized) values of the manipulated inputs in turn yields an (under) estimate of R_1^{-1} (the fact that the computation yields an underestimate, however, does not negatively impact its use within the controller design; see Remark 3.8 for a discussion). A finer discretization in terms of the manipulated input naturally yields a better estimate of the reverse-time reachability region. R_2^{-1} can in turn be determined by repeating the process for all elements in R_1^{-1} and the process repeated to yield the reverse-time reachability region for the initial time.

A pseudo-code on the construct of RTRRs is presented below to clarify the described algorithm above. Let the number of sampling instances during the batch be indexed by $k = 1, \ldots, K$ with $k = 1$ corresponding to t_f and $k = K$ corresponding to t_0. Additionally, let u_{ALL} be a $m \times N$ matrix where m denotes the number of inputs as before and N is the total number of possible input combinations following discretization of the m inputs. Finally, let x_d be a vector with dimensions $n \times 1$.

Set $IC_1 = x_d$
for i from 1 to $I - 1$:
    Set L as the current number of columns of IC_i
    for l from 1 to L
        for j from 1 to N
            Integrate reverse time model using jth column of u_{ALL} with lth column of L as initial conditions
            Store final conditions from integration as column in IC_{i+1}
        end
    end
end

Using the algorithm just described, the computational demands of generating reachability regions increase when computing R_2^{-1} compared to R_1^{-1} (since R_2^{-1} is the set of initial conditions from where a state in R_1^{-1} can be reached, compared to R_1^{-1} which is the set of initial conditions from where a point x_d can be reached). In general, the increase in computational demand is related to the increase in the size of the reverse-time reachability regions, as we go back in time. However, it is worth noting that the size of the reachability region does not necessarily grow as fast when going back in time for systems where the desired end-point is an equilibrium point (i.e., continuous processes). To understand this, consider reverse-time reachability region for a desired end-point which is an equilibrium of the process (denoted by x_e). Every state in R_i^{-1} would be included in R_{i+1}^{-1} because it is possible for the process to stay at x_e for another Δ, having reached x_e in i time steps (by virtue of x_e being an equilibrium point). This is not true when the desired end-point is not an equilibrium point and therefore the size of the reverse-time reachability regions, and therefore the associated computational cost, does not prohibitively increase as we proceed further toward the initial time.

In addition to being dependent on the size of the previously generated RTRR, the computational demands of the generation algorithm are also generally dependent on the number of states and inputs in the system. For systems with a high number of state variables, the required computational effort is influenced by the efficiency of the integrator as well as its ability to handle large-scale systems. In these cases, one of the hosts of efficient large-scale integrators available in the public domain (i.e., DASSL, CVODE, etc.) can be utilized. While increasing the number of states makes the RTRR generation more complex through internal computations (i.e., the Jacobian) performed by the integrator, the number of inputs affects the number of necessary integrations. The number of integrations, in fact, grows exponentially with the number of manipulated inputs. For instance, consider a system with two bounded inputs, which are both discretized at ten different points that spans the bounds. Taking R_1^{-1} to be one point, the number of required integrations to generate R_2^{-1} is 100 as the reverse-time model of the system must be integrated for every discretization point. Note, however, that the generation algorithm has an important feature in that integrations of the model equations may be done independently for different values of the input and initial conditions. Accordingly, to alleviate potential computational issues associated with having a high number of inputs, starting from a given reverse-time reachability region, integrations of the reverse-time model may be done independently, which in turn implies parallel computing schemes can be readily employed to generate reverse-time reachability regions to significantly reduce the computation times.

Remark 3.2 The RTRR generation algorithm outlined above describes how RTRRs can be constructed via only integrations of the reverse-time model of the system, $\dot{x} = -f(x, u)$. With the assumptions that $-f(x, u)$ is continuous on (x, u) and is locally Lipschitz in x on $D \times \mathbf{U}$, where $D \subset \mathbb{R}^n$, the continuity of solutions of $\dot{x} = -f(x, u)$ in terms of the initial condition and input is guaranteed. As a result, these assumptions ensure that RTRRs generated at each time step will be compact sets. While these continuity assumptions ensure against disjointed sets, no such general assumptions can be made to guarantee the convexity of RTRRs generated at each time step. Note that we do not require the sets to be convex for the implementation of our results. Also note that for batch processes which can be approximated well using convex differential equations in (x, u) (i.e., linear or linearized models), RTRRs generated via the algorithm in this section will be convex and this can lead to a significant reduction in the computational requirements for generating RTRRs. For the current chapter, however, we focus on nonlinear processes and only the continuity of the model equations is assumed.

Remark 3.3 While existing Lyapunov-based control designs can be very well used in the context of tracking state trajectories in batch processes, the fact that the desired end-point in a batch is typically not an equilibrium point precludes the use of Lyapunov-based techniques to determine the set of initial conditions from where a desired end-point can be reached in finite time. To begin with, the basic assumption in Lyapunov-based control designs that of $f(x_d, 0) = 0$ is not satisfied in the case of batch processes. Note that this cannot be achieved by a coordinate transformation either, because x_d is simply not an equilibrium point of the process. Of course, a positive-definite function $V_{candidate}$ can be defined such that $V_{candidate}(x_d) = 0$. The set of states for which $\dot{V}_{candidate}$ can be made negative, however, does not form a neighborhood around x_d, which in turn precludes the construction of 'invariant' sets around x_d. In summary, in contrast to continuous processes where the desired operating point is an equilibrium point, Lyapunov-based techniques do not allow for computing the set of states from where the process can be guaranteed to be steered toward the desired end-point.

Remark 3.4 While seemingly conceptually similar, the reverse-time reachability regions in the context of continuous systems are inherently different from those in the context of batch processes. In particular, when considering stabilization to an equilibrium point, the reverse-time reachability regions, with the time tending to infinity, yield the so-called null-controllable regions (the set of initial conditions from where a process can be stabilized at an equilibrium point). For stabilization at an equilibrium point, $R^{-1}(t_2) \subset R^{-1}(t_1)$ when $t_2 < t_1$. To understand this, consider the set of states which constitute $R^{-1}(t_1)$; there naturally exists a subset of states within $R^{-1}(t_1)$ that can be steered to x_d in a time $t_f - t_1$ (since $t_f - t_1$ is less than $t_f - t_2$) and simply kept there until t_f as x_d is an equilibrium point. In contrast, in the context of a desired end-point that is not an equilibrium point, just because a point is in $R^{-1}(t_2)$ does not ensure that it is in $R^{-1}(t_1)$ since the process cannot be 'maintained' at x_d.

Remark 3.5 The presence of constraints on the manipulated input has significant implications on the ability to control continuous (e.g., see [29]) as well as batch systems. Despite their differences, one common property among null-controllable regions for continuous systems and reverse-time reachability regions in the context of batch systems is that they are both dependent only on system dynamics and input constraints and do not depend on a specific control law.

3.3.2 MPC Formulation

An MPC formulation that utilizes the reverse-time reachability region characterization is presented in this section. To this end, consider the process described by Eq. (3.1), for which the reverse-time reachability regions have been characterized for a given hold and implement time Δ. The control action at a time t (with $k = (t_f - t)/\Delta$) is computed by solving the optimization problem below:

$$u_{MPC}(x) := \arg min\{J(x, t, u(\cdot))|u \in U\} \tag{3.9}$$

subject to:

$$\dot{\tilde{x}} = f(\tilde{x}(\tau), u) \tag{3.10}$$

$$\tilde{x}(0) = x(t) \tag{3.11}$$

$$\tilde{x}(\Delta) \in R_{k-1}^{-1} \tag{3.12}$$

The objective function can be appropriately chosen to meet desired performance objectives. For instance, if the objective is to minimize discrepancies between actual state trajectories and some nominal optimal state trajectories (x_{nom}^*), and the variation in the control moves, the performance objective could be formulated as:

$$J = \int_t^{t+\Delta} \|\tilde{x}(\tau) - x_{nom}^*\|_Q + \|\Delta u\|_R \mathrm{d}t \tag{3.13}$$

where Q and R are weighting matrices.

As evidenced by the constraint shown in Eq. (3.12), implementation of the reverse-time reachability region-based controller necessitates an explicit characterization of reachability regions. In this manuscript, the constraint in Eq. (3.12) is chosen to be represented as an inequality constraint (i.e., $h(x(t), u(t)) \leq 0$; see Remark 3.8).

Note that by definition, reverse-time reachability regions take into account desired end properties of a batch. Consequently, any existing end-point constraints in the end-point-based MPC scheme can be replaced with a constraint that requires, at each sampling instance, the process states to remain in the reverse-time reachability region at the next time step (Eq. (3.12)). Implications on the guarantees of reachability to the desired end-point through this replacement are formalized in Theorem 3.3 below.

Theorem 3.3 *Consider the constrained system of Eq. (3.1) with* $\sigma = 1$ *under the MPC law of Eqs. (3.9)–(3.12). If and only if* $x(t_0) \in R^{-1}(t_0)$*, the optimization problem defining the MPC law in Eqs. (3.9)–(3.12) remains feasible for all* $t \in [t_0, t_f]$ *and* $x(t_f) = x_d$.

Proof of Theorem 3.3: We first show the only if part of the theorem. To this end, consider an initial condition $x_0 \notin R^{-1}(t_0)$. If the constraint of Eq. (3.12) is feasible and it is implemented in the closed loop, then there exists a sequence of control moves that take the state to x_d by t_f, in turn implying that $x_0 \in R^{-1}(t_0)$. This argument can be repeated at every time step, eventually leading to the MPC laws in Eqs. (3.9)–(3.12) remaining feasible for all $t \in [t_0, t_f]$ and $x(t_f) = x_d$ only if $x_0 \in R^{-1}(t_0)$. We now show the sufficiency of the condition. To this end, consider the case when $x(t_0) \in R^{-1}(t_0)$. By definition, there exists a set of control moves that takes the state to x_d by t_f. For such a set of control moves (and the associated state trajectory), this must imply that the state trajectory at $t_0 + \Delta$ resides in $R^{-1}(t_0 + \Delta)$ (invoking the necessity of the condition proved earlier for $x(t_0 + \Delta)$). In essence, this implies that there exists a feasible solution to the constraint of Eq. (3.12). This completes the proof of Theorem 3.3. ■

Remark 3.6 The statement of Theorem 3.3 essentially formalizes that the existence of the state in the reverse-time reachability region is a necessary and sufficient condition for the states to be steered to the desired end-point. The necessity of the condition has an important implication in that if at any time during the batch the states are driven outside the reverse-time reachability region, the desired end-point simply cannot be reached. In other words, the condition of continued existence in successive reverse-time reachability region cannot be relaxed because if the state goes outside the reverse-time reachability region, it is simply not possible (whether using the proposed reverse-time reachability region-based predictive controller or any other control law) to steer the states back into the reachability region and then to the desired end-point by the batch termination time.

Remark 3.7 In this chapter, we assume that the batch process design includes a specification for the desired end-point properties, x_d. This allows for the computation of reverse-time reachability regions using the algorithm presented earlier in this section. In some cases, not all the desired state values at batch termination are explicitly constrained at the process design phase. In these cases, the objective may be to maximize or minimize a certain product concentration. Accordingly, for these cases, a nominal optimization problem can be solved off-line with a given performance objective, and x_d can be taken to be the state variable vector at t_f from the optimal state trajectories. Additionally, rather than using a single process state value as the desired end-point, a suitable neighborhood around x_d, B_{x_d}, can be the desired neighborhood to be reached. The proposed MPC naturally allows for incorporating such a scenario by computing $R^{-1}(t_f - \Delta)$ as the set of states that can be steered to the neighborhood B_{x_d} and successively continuing backward.

Remark 3.8 When using the algorithm outlined for generating the reverse-time reachability regions, the true reverse-time reachability regions are estimated as point sets. Depending on the specific system under investigation, the shape and orientation of the point set may permit different strategies for explicitly characterizing the point set. In any case, the explicit characterization must be either an exact characterization, which is unlikely, or an underestimate of the true region. With characterizations that represent an overestimate, process states may be incorrectly identified as belonging to the true reachability region. Hence, the constraint in Eq. (3.12) may be satisfied initially even when the states are not contained in the true reverse-time reachability region, invalidating guarantees of successive feasibility. In contrast, if the explicit characterization is an underestimate (generated appropriately), successive feasibility of the optimization problem can be still guaranteed.

Remark 3.9 One approach to generate the explicit characterization of reverse-time reachability regions is considering 2-D projections of the reverse-time reachability regions, posing it as a geometric optimization problem, and using polynomial discrimination as the explicit characterization method. Briefly, polynomial discrimination finds the t-level set of a polynomial with degree less than or equal to d; i.e., consider $\{x \mid \sum_{i_1+\cdots+i_n \leq d} a_{i_1 \ldots i_d} x_1^{i_1} x_n^{i_n} = t\}$ that discriminates two sets of points [30]. An underestimate of the true reverse-time reachability regions can be obtained through specifying an interior subset of the point set as one class of points and the complements as the other class. Such subsets can be identified using, e.g., the hidden point removal (HPR) algorithm presented in [31]. In the HPR algorithm for 2-D point sets, points visible from a certain perspective location can be identified. Thus, by appropriately selecting the perspectives from which to view the point set, boundary points of the set can be identified and removed. It should be noted that the distance between the viewpoint and point set should also be chosen carefully, and the results in [31] include an optimal way to choose such a distance such that the number of false negatives is minimized. By successively applying the HPR algorithm to a 2-D point set and removing boundary points at each iteration, a strictly interior subset of the point set can be identified. As the results from the HPR algorithm are not exact, visual verification can be used to ensure that strictly interior points are being identified with the HPR algorithm. Geometrically, polynomial discrimination seeks an algebraic surface (as defined by the degree and coefficients of the polynomial) that can separate the two classes. If it is observed that the two classes cannot be separated by a polynomial of a given degree, one natural step to take is to increase the degree of the polynomial and retry the fit. The details regarding polynomial discrimination and approximate discrimination approaches are available in [30].

Remark 3.10 With the attainment of the desired end-point achieved via the constraint, the objective function in the MPC formulation can be utilized to satisfy performance (as shown in Eq. (3.13)) or even robustness objectives. Specifically, to enhance disturbance rejection and robustness of the closed-loop batch system, the objective function can be used to penalize the Euclidean distance between a process state and the analytical center of a reverse-time reachability region. This would tend to drive the process state toward the center of the reverse-time reachability region, thereby reducing the chances of disturbances driving the process to a point from where the end-point is no longer reachable. Also, worth noting is that it is possible to employ a multi-objective function which consists of a weighted sum of performance and robustness terms.

3.4 Safe-Steering of Batch Processes

In the previous section, we presented a reverse-time reachability region-based predictive controller that was designed with a fault-free assumption. As discussed earlier, in contrast to continuous processes, the problem of fault tolerance is fundamentally different in batch processes. Specifically, a fault-tolerant control framework for batch processes must be designed with the desired end-point in mind, which is rarely an equilibrium point. In this section, we utilize the reverse-time reachability region-based MPC scheme to develop what we call the safe-steering framework. To this end, the safe-steering problem is initially formulated and then the safe-steering framework is presented.

3.4.1 Problem Definition

We consider the class of faults where a control actuator fails and reverts to its fail-safe value. Fail-safe positions are intended to minimize the possibilities of excursions to dangerous conditions such as high temperatures and pressures. Examples of fail-safe positions are fully open for a coolant valve and fully closed for a steam flow valve. More specifically, w.l.o.g., we characterize the fault occurring in the first control actuator at a time t^{fault} which is subsequently rectified at a time, t^{repair} (i.e., for $t \leq t^{fault}$ and $t \geq t^{repair}$, $\sigma(t) = 1$ while for $t^{fault} \leq t \leq t^{repair}$, $\sigma(t) = 2$) as $u_2^1(t) = u_{failed}^1 \forall t \in [t^{fault}, t^{repair}]$, with $u_{min}^1 \leq u_{failed}^1 \leq u_{max}^1$, where u^i denotes the ith component of the input vector u, and the only inputs available for control between t^{fault} and t^{repair} are $u_2^i, i = 2, \ldots, m$. Note that if t^{fault} is at a time which is not an integer multiple of the batch sampling time, Δ, t^{fault} can be taken to be at the upcoming integer multiple of Δ, and the safe-steering framework can be implemented as presented in this section. With one of the inputs fixed at a fail-safe position, while it may still be possible to reach the desired end-point if the fault is rectified sufficiently fast, continued operation of the representative end-point-based

MPC scheme may drive the states to a point from where it is not possible to reach the desired end-point even upon fault repair. We define the safe-steering problem as one of the identifying trajectories of the functioning inputs between the time of fault occurrence and repair (without requiring the value of t^{repair}, or an estimate thereof, to be known a priori) that will preserve reachability to the desired end-point properties.

3.4.2 Safe-Steering to Desired End-Point Properties

The main requirement for safely steering a batch during fault rectification is to repeatedly maintain states in regions from which the desired end-point remains reachable. To this end, the reverse-time reachability region-based predictive controller can be utilized. Consider the batch system described by Eq. (3.1) for which the first control actuator faults at t^{fault} and are repaired at t^{repair}. Also, assume the reverse-time reachability regions for fault-free operation have been characterized for all sampling instances between t^{fault} and t^{repair}. The following theorem addresses the safe-steering problem.

Theorem 3.4 *Consider the constrained system of Eq. (3.1) under the reverse-time reachability region-based model predictive controller given by Eqs. (3.9)–(3.12). In the event of a fault at* t^{fault}*, if and only if the optimization problem remains feasible* $\forall t \in [t^{fault}, t^{repair}]$*, then the optimization problem continues to be feasible* $\forall t \in [t^{repair}, t_f]$ *and* $x(t_f) = x_d$.

Sketch of Proof of Theorem 3.4: The proof of this theorem follows from Theorem 3.3. Essentially equating the time t^{repair} to t_0, results in satisfaction of the requirements of Theorem 3.3 and the optimization problem continue to be feasible $\forall t \in [t^{repair}, t_f]$ and $x(t_f) = x_d$ follows. This completes the proof of Theorem 3.4. ■

Remark 3.11 The key idea in the safe-steering framework, as formally expressed in Theorem 3.4, is to continuously maintain the process (if possible) in the reverse-time reachability regions after t^{fault}. If the reverse-time reachability region-based controller continues to be feasible after t^{fault} up until the fault gets rectified at t^{repair}, this implies that at t^{repair}, the states will be contained within $R^{-1}(t^{repair})$, and according to the definition of reverse-time reachability regions, the desired end-point will be reachable. Note that during the fault rectification period, there is no guarantee that the states can be maintained within the (fault-free) reachability regions. The result of Theorem 2, therefore, relies on the assumption of the continued feasibility of the optimization problem after t^{fault}. This assumption, however, is not restrictive for the reason that it simply cannot be relaxed. If any time during fault repair, the state moves outside the reverse-time reachability region, the end-point state properties cannot be reached even if the fault is immediately repaired.

Remark 3.12 Note also that the key reasoning behind the statement of Theorem 3.4 is as follows: Once a fault takes place, since the desired end-point cannot be reached with the limited available control action, if one continues to try and implement the end-point-based MPC it would yield an infeasible solution. (Note that if the end-point-based MPC continues to be feasible following a fault, then it could be continued to be used for the duration of the batch. This chapter looks at the specific class of input failures for which an end-point-based MPC will become infeasible due to the limited available control action.) With the limited control action available, while it may not be possible to drive the state to the desired end-point, it may still be possible to maintain the state for some time in the successive reverse-time reachability regions. Implementation of a truncated version of the infeasible solution from an end-point-based MPC may not retain the states in the reverse-time reachability region; however, the implementation of the reverse-time reachability region-based MPC (if feasible) allows for finding such a control action that preserves the states within the reverse-time reachability regions and subsequent end-point being reached upon fault recovery. Note that if the fault repair time was known ahead of time (which of course, is not), then in principle the end-point-based MPC would also have been able to find the manipulated input trajectory utilizing the reduced available control action between t^{fault} and t^{repair} and full control action thereafter to achieve the desired end-point. The utilization of the reverse-time reachability region-based MPC allows to achieve this without the need for knowing t^{repair} ahead of time.

Remark 3.13 The safe-parking framework and safe-steering framework for continuous and batch systems, respectively, both address the problem of how to operate a plant during fault recovery and both frameworks address the kind of faults that prevent the desired plant operation under nominal control laws. Safe-parking for continuous processes handles faults that preclude operation at a nominal equilibrium point, while safe-steering for batch processes addresses the kind of faults that preclude driving the states to the desired end-point. The safe-steering and safe-parking framework both answer the question of how to compute the input trajectories between fault occurrence and recovery such that the desired nominal operation can be preserved or resumed. This is where the similarity between the approaches ends. As discussed previously in the introduction, in the presence of faults that prevent operation at a desired equilibrium point, safe-parking [23, 24] involves a transition to a new safe-park (equilibrium) point which allows the transition back to the nominal equilibrium in some optimal way following fault recovery. Prior to considering any optimality criteria regarding the transitions, the safe-parking framework must locate the feasible operating points which allow such transitions, and the main criteria used in locating these 'safe-park points' are the existence of such equilibrium points and then preserving process stability by utilizing stability region characterization [32, 33] for the nominal and safe-park points. In particular, the nominal equilibrium point must be contained within the stability region of the safe-park point and vice versa. Neither the safe-park points, nor the controller designs for continuous systems or the associated stability regions remain applicable in the context of batch processes. In particular, the primary concern in the safe-steering framework is reachability. In the absence of an equilibrium point, the process cannot be 'parked,' but possibly 'steered' in a way that allows for end-point reachability if the fault is rectified sufficiently fast. This is achieved in the safe-steering framework via using the proposed reverse-time reachability region-based MPC.

Remark 3.14 While in this chapter we assume full state availability to enable clear presentation of the key ideas, the framework can readily incorporate the use of state observers that utilize the measurements to generate estimates of the state which can in turn be used for implementing the proposed safe-steering mechanisms. The effect of possible errors in the state estimates can be mitigated by requiring the states (predicted using the estimates) to go to subsets of the reverse-time reachability regions. A detailed analysis of the output feedback problem, however, remains outside the scope of the present chapter.

Remark 3.15 Note that the RTRR regions are estimated by integrating the process model equations. Consequently, the shapes and sizes of the estimated regions are sensitive to modeling errors. Due to the possibility of discrepancies between the estimated and true RTRR sets in the presence of modeling errors, there is the potential of misclassifying batch conditions as being contained within a true RTRR of the process. In such cases, when process states have escaped the true RTRR (even though the states may be classified as lying within the estimated RTRR), reaching the desired end-point becomes impossible. One potential solution to this problem is to utilize more conservative characterizations of RTRRs to eliminate the possibility of misclassification (based on bounds on the modeling errors, if known). A detailed exploration of the robustness of the proposed framework and the sensitivity of RTRR regions to modeling errors is outside the scope of the current chapter.

Remark 3.16 Note also that the framework can be utilized to exploit scenarios where the batch termination time is allowed to be changed or, at the very least, allowed to be changed over a reasonable range. This can be achieved as follows: At the time of fault repair, if the state is not present in the reverse-time reachability region to go to the desired end-point by batch termination (due to it taking exceedingly long for fault rectification), the explicit characterization of the reverse-time reachability region can be utilized to compute the value of the (least) batch termination time for which the state would reside in the reverse-time reachability region. If such a time exists, the batch can possibly be run for a longer duration (if economic and operation constraints permit) to yield the desired product. If no such time exists, this points to the fact that the batch may have gone beyond redemption and again helps in taking the immediate (and necessary) action of discarding the batch, instead of unsuccessfully trying to reach the desired end-point and wasting valuable time.

3.5 Simulation Example

We illustrate in this section the proposed reverse-time reachability region-based MPC and the safe-steering framework via a fed-batch process. To this end, consider an isothermal fed-batch fermentation process for which the mathematical model takes the following form:

$$\dot{x_1} = \frac{\mu_{max} x_2 x_1}{K_1 + x_2 + K_2 x_2^2} - \frac{x_1}{x_3}(F_1 + F_2) \tag{3.14}$$

$$\dot{x_2} = -\frac{\mu_{max} x_2 x_1}{Y(K_1 + x_2 + K_2 x_2^2)} - \frac{S_f - x_2}{x_3}(F_1 + F_2) \tag{3.15}$$

$$\dot{x_3} = F_1 + F_2 \tag{3.16}$$

where x_1, x_2, and x_3 denote the biomass concentration, substrate concentration, and volume (respectively) and constitute the state vector, x. The physical meaning of the parameters in the model equations and their nominal values can be found in Table 3.1. The primary control objective for this fed-batch process is to achieve the arbitrarily chosen desired end-point of $x_d = [8\ 7.5\ 30]^T$. The fermentation time is taken to be one hour with a controller execution period of 0.05 h. The manipulated variables in the fermentation process are the two flow rates, F_1 and F_2, with constraints: $0 \leq F_1 \leq 10$ L/h and $0 \leq F_2 \leq 10$ L/h.

3.5.1 *Calculation of Reverse-Time Reachability Region*

The reverse-time reachability regions, $R^{-1}(t)$, for this system are first generated as three-dimensional point sets using the procedure described earlier in the reverse-time reachability region section. To aid in the visualization and the eventual explicit characterization, the point set corresponding to a given reverse-time reachability region is first projected onto three two-dimensional spaces, namely the $x_1 - x_2$, $x_1 - x_3$, and $x_2 - x_3$ spaces. Thus, $R^{-1}(t)$ would be characterized through $R^{-1}_{x_1-x_2}(t)$, $R^{-1}_{x_1-x_3}(t)$, and $R^{-1}_{x_2-x_3}(t)$, where the subscripts denote the states considered in each projection.

Table 3.1 Process parameters for the fed-batch bioreactor of Eqs. (3.14)–(3.16)

Parameter	Description	Value	Unit
μ_{max}	Maximum growth rate	25	h^{-1}
K_1	Monod constant	0.03	g/L
K_2	Kinetic parameter	0.25	L/g
Y	Yield coefficient	0.25	
S_f	Influent substrate concentration	10	g/L

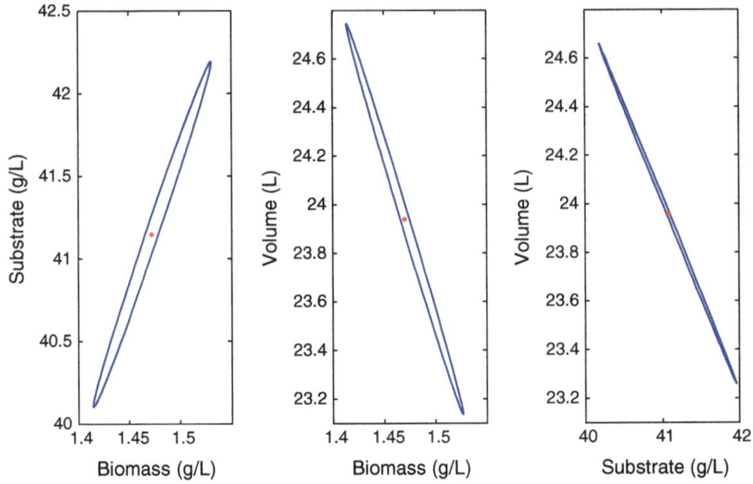

Fig. 3.1 Reverse-time reachability region projections at $t = 0.4\,\text{h}$

At a given point in time, $x(t)$ is considered to be within $R(t)$ if the three projected reverse-time reachability regions contained the corresponding projections of $x(t)$ simultaneously. Ellipses were utilized for the explicit characterization of the two-dimensional (2-D) point sets in the three projections. Figures 3.1 and 3.4 illustrate the results of using ellipses for all three projections at two distinct times in the batch, $t = 0.40\,\text{h}$ and $t = 0.85\,\text{h}$.

3.5.2 Implementing the Reverse-Time Reachability Region-Based MPC

The effectiveness of the representative end-point controller and the reverse-time reachability region-based controller is compared first during fault-free operation. The objective functions used in the end-point and reverse-time reachability region-based formulations, J_{EP} and J_R, were:

$$J_{EP} = \int_t^{t_f} \Delta u \mathrm{d}t \tag{3.17}$$

$$J_R = \|x_{1,2}(t+\Delta) - x_{1,2}^c(t+\Delta)\| + \|x_{1,3}(t+\Delta) - x_{1,3}^c(t+\Delta)\| + \|x_{2,3}(t+\Delta) - x_{2,3}^c(t+\Delta)\| \tag{3.18}$$

where $x_{i,j}(t)$ represents the projection of the states at time t onto the $x_i - x_j$ space and $x_{i,j}^c(t)$ represents the center point of the reverse-time reachability region at time t in the $x_i - x_j$ space. The initial condition for both controllers was specified to

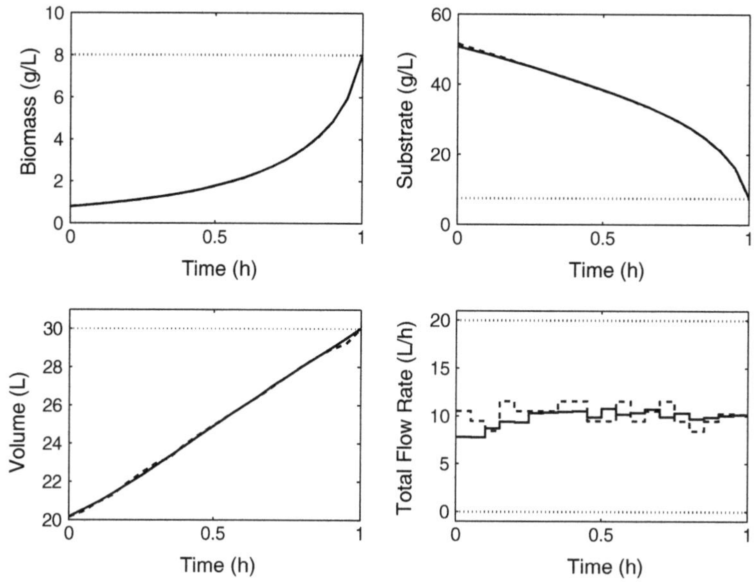

Fig. 3.2 State and input profiles of the fed-batch bioreactor system under the end-point-based MPC (solid) and reverse-time reachability region-based MPC (dashed) schemes during a fault-free batch. Both controllers successfully drive the process to the desired end-point, but the reverse-time reachability region-based MPC takes significantly less online computational effort

be $x(0) = [0.7853\ 50.7515\ 20.1629]^T$. This initial condition was verified to be in $R^{-1}(0)$, guaranteeing the desired end-point is reachable in a fault-free batch.

The state and input profiles resulting from the implementation of both controllers are presented together in Fig. 3.2. In both cases, the desired end-point properties of the batch are reached at batch termination, meeting the primary control objective. Note that the input variation observed for the reachability region-based controller could have been reduced further via a penalty term for Δu. Using the MATLAB functions, `tic` and `toc`, the required run times on an Intel P4 3.0 Ghz machine for the end-point-based predictive controller (with a sequential solution strategy) and the reverse-time reachability region-based controller were 30.47 min and 26.91 s, respectively. The significant reduction in the computational effort from the reverse-time reachability region-based MPC scheme is evident from these results. For batch processes with additional inputs and states and/or a longer duration, the integration and optimization routines in an end-point-based MPC strategy will demand additional computational effort; therefore, the reduction in the run times will be even more substantial with reverse-time reachability region-based MPC.

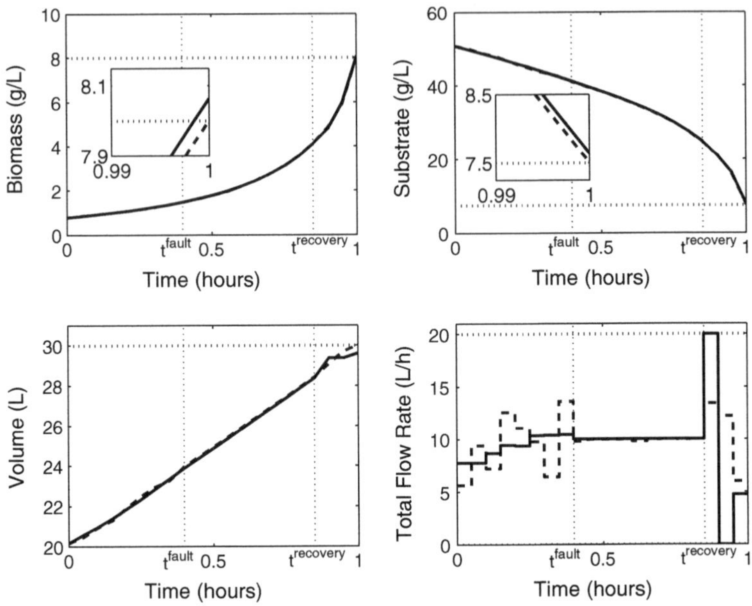

Fig. 3.3 State and input profiles of the fed-batch bioreactor system under the end-point-based MPC (solid) and reverse-time reachability region-based MPC (dashed) schemes with input failure of F_1 between 0.4 and 0.85 h. The insets show the desired end-point being reached using the proposed safe-steering framework

3.5.3 *Safe-Steering of the Fed-Batch Process*

Having demonstrated the computational efficiency of the reverse-time reachability region-based MPC, we now consider the application of the two MPC schemes in the presence of faults. To this end, consider the case where at $t = 0.4$ h, the valve regulating F_1 fails and reverts to its fail-safe position (completely shut), resulting in $F_1 = 0$. Thus, the inputs to the system become restricted to $F_1 = 0$ and $0 \leq F_2 \leq 10$. At $t = 0.85$ h, the fault is repaired. The state and input profiles resulting from the use of an end-point-based predictive controller and the reverse-time reachability region-based controller are presented in Fig. 3.3.

We observe that even after fault recovery, the end-point-based MPC scheme is not able to drive the process to the desired end-point states. Instead, due to repeated infeasibleness during fault repair ($t^{fault} < t < t^{recover}$) as a consequence of limited available control effort (with the controller requiring the states to go to the desired end-point by batch termination using the reduced control effort for the entire remaining batch duration), the controller steers the system to the end-point of $x(t_f) = [8.0634\ 7.6367\ 29.6084]^T$. The states at the time of fault recovery for the end-point-based MPC are $x(t = 0.85) = [4.1155\ 24.7323\ 28.3706]^T$. Examining the reverse-time reachability regions at $t = 0.85$ h, this process state is not

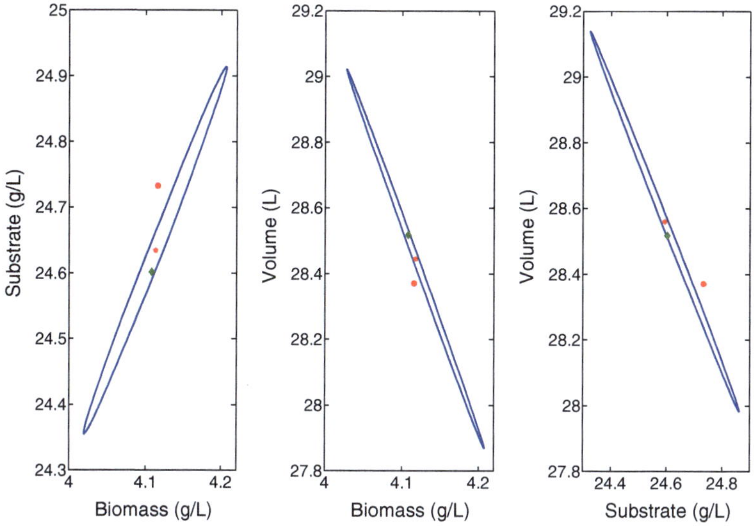

Fig. 3.4 Reverse-time reachability region projections at $t = 0.85\,\mathrm{h}$. The $\circ$ denotes the state values at t^{repair} using the end-point-based MPC, while the $\diamond$ shows the state values using the proposed reverse-time reachability region-based MPC

included in any of the three projections of the reverse-time reachability region (see Fig. 3.4). Hence, upon fault recovery, the end-point-based MPC scheme prescribes extreme control action in an (unsuccessful) attempt to arrive at the desired end-point by the end of the batch. In contrast, in the reverse-time reachability region-based MPC scheme, there is no need to resort to extreme values of the input to recover reachability to the desired end-point because the input moves during fault duration are computed to preserve reachability. While the end-point-based MPC scheme ends up implementing control action that drives the states to a point by $t = 0.85\,\mathrm{h}$ from where it is no longer possible to reach the desired end-point, the reverse-time reachability region-based controller, by only considering one time step ahead, remains feasible during the failure period, ensuring the states evolve within reverse reachability regions. Consequently, the desired end-point is reachable upon fault rectification. Note that if the fault was not rectified by $t = 0.85\,\mathrm{h}$, neither the end-point-based MPC nor the reverse-time reachability region-based MPC would have been able to reach the desired end-point. However, the key point is that in this case (with the fault being rectified at $t = 0.85\,\mathrm{h}$), there exists a set of control moves utilizing reduced control action during faulty operation and full control action after fault repair that allows reaching the desired end-point, and the reverse-time reachability region-based MPC is able to find such a manipulated input trajectory (without a priori knowing t^{repair}).

3.6 Conclusions

This chapter considered the problem of control of nonlinear batch processes to achieve a desired final product quality subject to input constraints and faults in the control actuators. Specifically, faults were considered that cannot be handled via robust control approaches, and preclude the ability to reach the desired end-point, necessitating fault rectification. A safe-steering framework was developed to address the problem of determining how to utilize the functioning inputs during fault rectification in order to ensure that after fault rectification, the desired product properties can be reached upon batch termination. To this end, first a novel reverse-time reachability region-based MPC was formulated that reduces online computations, as well as provides a useful tool for handling faults. Next, a safe-steering framework was developed that utilizes the reachability region-based MPC in steering the state trajectory during fault rectification to enable (upon fault recovery) the achieving of the desired end-point properties by batch termination. The online efficiency of the proposed controller and its utility in the context of safe-steering framework were illustrated using a fed-batch process example.

References

1. Mhaskar, P., Aumi, S.: Transition from batch to continuous operation in bio-reactors: a model predictive control approach and application. Can. J. Chem. Eng. **45**, 416–423 (2007)
2. Cruickshank, S.M., Daugulis, A.J., McLellan, P.J.: Dynamic modeling and optimal fed-batch feeding strategies for a two-phase partitioning bioreactor. Biotech. Bioeng. **67**, 224–233 (2000)
3. Zhang, G.P., Rohani, S.: On-line optimal control of a seeded batch cooling crystallizer. Chem. Eng. Sci. **58**, 1887–1896 (2003)
4. Soroush, M., Kravaris, C.: Optimal-design and operation of batch reactors. 1. Theoretical framework. Ind. Eng. Chem. Res. **32**, 866–881 (1993)
5. Soroush, M., Kravaris, C.: Optimal-design and operation of batch reactors. 2. A case-study. Ind. Eng. Chem. Res. **32**, 882–893 (1993)
6. Yabuki, Y., Nagasawa, T., MacGregor, J.F.: An industrial experience with product quality control in semi-batch processes. Comput. Chem. Eng. **24**, 585–590 (2000)
7. Hua, X.M., Rohani, S., Jutan, A.: Cascade closed-loop optimization and control of batch reactors. Chem. Eng. Sci. **59**, 5695–5708 (2004)
8. Shi, D., Mhaskar, P., El-Farra, N.H., Christofides, P.D.: Predictive control of crystal size distribution in protein crystallization. Nanotechnology **16**, S562–S574 (2005)
9. Shi, D., El-Farra, N.H., Li, M., Mhaskar, P., Christofides, P.D.: Predictive control of particle size distribution in particulate processes. Chem. Eng. Sci. **61**, 268–281 (2005)
10. Sheikhzadeh, M., Trifkovic, M., Rohani, S.: Real-time optimal control of an anti-solvent isothermal semi-batch crystallization process. Chem. Eng. Sci. **63**, 829–839 (2008)
11. Palanki, S., Vemuri, J.: Optimal operation of semi-batch processes with a single reaction. Int. J. Chem. React. Eng. **3**, 1219 (2005)
12. Pistikopoulos, E.N., Dua, V., Bozinis, N.A., Bemporad, A., Morari, M.: On-line optimization via off-line parametric optimization tools. Comput. Chem. Eng. **26**, 175–185 (2002)
13. Huynh, N., Kazantzis, N.: Parametric optimization of digitally controlled nonlinear reactor dynamics using Zubov-like functional equations. J. Math. Chem. **38**, 499–519 (2005)
14. Mhaskar, P.: Robust model predictive control design for fault-tolerant control of process systems. Ind. Eng. Chem. Res. **45**, 8565–8574 (2006)

15. Wang, Z.D., Huang, B., Unbehauen, H.: Robust reliable control for a class of uncertain nonlinear state-delayed systems. Automatica **35**, 955–963 (1999)
16. Cheng, L.L., Kwok, K.E., Huang, B.: Closed-loop fault detection using the local approach. Can. J. Chem. Eng. **81**, 1101–1108 (2003)
17. Mehranbod, N., Soroush, M., Panjapornpon, C.: A method of sensor fault detection and identification. J. Process Control **15**, 321–339 (2005)
18. Mhaskar, P., Gani, A., El-Farra, N.H., McFall, C., Christofides, P.D., Davis, J.F.: Integrated fault-detection and fault-tolerant control for process systems. AIChE J. **52**, 2129–2148 (2006)
19. El-Farra, N.H.: Integrated fault detection and fault-tolerant control architectures for distributed processes. Ind. Eng. Chem. Res. **45**, 8338–8351 (2006)
20. Mhaskar, P., McFall, C., Gani, A., Christofides, P.D., Davis, J.F.: Isolation and handling of actuator faults in nonlinear systems. Automatica **44**, 53–62 (2008)
21. El-Farra, N.H., Giridhar, A.: Detection and management of actuator faults in controlled particulate processes using population balance models. Chem. Eng. Sci. **63**, 1185–1204 (2008)
22. Armaou, A., Demetriou, M.A.: Robust detection and accommodation of incipient component and actuator faults in nonlinear distributed processes. AIChE J. **54**, 2651–2662 (2008)
23. Gandhi, R., Mhaskar, P.: Safe-parking of nonlinear process systems. Comput. Chem. Eng. **32**, 2113–2122 (2008)
24. Mahmood, M., Gandhi, R., Mhaskar, P.: Safe-parking of nonlinear process systems: handling uncertainty and unavailability of measurements. Chem. Eng. Sci. **63**, 5434–5446 (2008)
25. Alamir, M., Balloul, I.: Robust constrained control algorithm for general batch processes. Int. J. Control **72**, 1271–1287 (1999)
26. Terwiesch, P., Agarwal, M., Rippin, D.W.T.: Batch unit optimization with imperfect modelling: a survey. J. Process Control **4**, 238–258 (1994)
27. Undey, C., Ertunc, S., Cinar, A.: Online batch/fed-batch process performance monitoring, quality prediction, and variable-contribution analysis for diagnosis. Ind. Eng. Chem. Res. **42**, 4645–4658 (2003)
28. Undey, C., Tatara, E., Cinar, A.: Intelligent real-time performance monitoring and quality prediction for batch/fed-batch cultivations. J. Biotech. **108**, 61–77 (2004)
29. Kapoor, N., Daoutidis, P.: Stabilization of nonlinear processes with input constraints. Comput. Chem. Eng. **24**, 9–21 (2000)
30. Boyd, S., Vandenberghe, L.: Convex Optimization. Cambridge University Press, Cambridge (2004)
31. Katz, S., Tal, A., Basri, R.: Direct visibility of point sets. ACM Trans. Graph. **26**, 1–11 (2007)
32. Mhaskar, P., El-Farra, N.H., Christofides, P.D.: Predictive control of switched nonlinear systems with scheduled mode transitions. IEEE Trans. Autom. Control **50**, 1670–1680 (2005)
33. Mhaskar, P., El-Farra, N.H., Christofides, P.D.: Stabilization of nonlinear systems with state and control constraints using Lyapunov-based predictive control. Syst. Control Lett. **55**, 650–659 (2006)

Chapter 4
Robust Model Predictive Control and Fault-Handling of Batch Processes

4.1 Introduction

Batch and fed-batch processes have widespread applications within a variety of sectors especially those that are used for the manufacture of high-quality low-volume products such as bio-chemicals and polymers. The primary control objective in batch processes is to reach a specified product quality by batch termination. Consequently, batch control strategies are typically designed to reach a desired end-point and can be categorized into trajectory tracking approaches or end-point-based approaches. In trajectory tracking approaches, optimal state trajectories that terminate at the desired end-point are generated off-line by solving a dynamic optimization problem (see, e.g., [1]) or using iterative dynamic programming (see, e.g., [2]), and subsequently tracked online using local PID or predictive controllers (see, e.g., [3–6]). A drawback in these approaches is that process noise and disturbances can render optimal state trajectories suboptimal or in the worst case, the tracking problem infeasible.

With increased quality of computational resources and more efficient optimization algorithms, shrinking horizon end-point-based model predictive control (MPC) is becoming a possibility for batch process control. In computing the control action in this approach, a dynamic optimization problem that incorporates the desired end-point (in the objective function and/or constraints) is solved at each sampling instance until batch termination, using plant measurements (or state variable estimates) to update model predictions at each sampling time. A truncated version of the resulting optimal input changes is then implemented directly on the process, the prediction horizon is shifted one sampling instance, and the process is repeated at the next sampling instance. One drawback of end-point-based MPC approaches is the significant computational demand associated with solving the dynamic optimization problem at each sampling instance. More specifically, the solution of the dynamic optimization problem consists of the entire input trajectory from the time at which the problem is solved to batch termination, implying significant computational demands especially during the early stages of the batch. Additionally, with model uncertainty, significant discrepancies can result between the predicted and actual behavior of a process. For

P. Mhaskar et al., *Modeling and Control of Batch Processes*,
Advances in Industrial Control, https://doi.org/10.1007/978-3-030-04140-3_4

instance, a predictive model integrated with nominal values of the uncertainty can indicate the process will be driven to the desired end-point for a specific control move, but applying the identical control move on the actual process can lead to a violation of product end-use properties and/or safety constraints due to inaccurate parameter values. Therefore, incorporating model uncertainty in the control scheme either by modifying the conventional dynamic optimization problem (e.g., see [7–9]) or by reducing biases in state estimates arising from model uncertainty is essential (e.g., see [10–13]) for obtaining acceptable controller performance in the presence of uncertainty.

While several end-point-based MPC formulations that explicitly account for model uncertainty using a min-max optimization framework are available (e.g., see [8, 9]), these approaches are often more computationally prohibitive. This is because the control moves are computed by taking into account the worst-case values of the uncertainty which are computed using an embedded optimization problem within the dynamic optimization. Techniques to reduce the real-time optimization computational demands include those that attempt to reduce the complexity of the model used for predictions and those in which a parametrization of an open-loop optimal input profile is derived off-line. With respect to the former, some of these approaches include successive linearization of the non-linear model equations at each sampling instance and the scheduling of multiple linear models (see [9] for a review). However, with significant improvements in real-time optimization algorithms and performance limitations associated with using linear MPC for highly non-linear batch processes, MPC using the full non-linear model is becoming increasingly common (e.g., see [14–17]). Input parametrization techniques (e.g., see [18–20]), on the other hand, strive to reduce the number of decision variables in the dynamic optimization problem. However, with modeling errors and process noise, plant process trajectories can deviate considerably from the nominal optimal trajectories on which the input parametrization is based, rendering a certain input parametrization suboptimal or infeasible.

The variability in the raw material availability adds another layer of complexity to batch process control and motivates designing methods for determining the suitability of running a batch with the given raw material. In particular, for a given control law, it is important to ascertain the initial conditions (without running the batch in its entirety) for which the desired control objectives are obtainable to minimize resource and time wastage. While there exist MPC designs for continuous systems that allow the explicit characterization of the set of initial conditions from where stability is achievable [21–23], these results are not applicable for batch systems because the desired end-point is usually not an equilibrium point. Currently, there exist no end-point-based MPC designs for batch systems that provide an explicit characterization of a feasibility region from where it can be guaranteed that the desired control objectives can be met. Recently, in [24], a computationally efficient, non-linear MPC design based on the concept of 'reverse-time reachability regions (RTRRs)' was presented. RTRRs were defined as the set of states from where the desired end-point can be reached by batch termination. In contrast to existing results, the MPC formulation in [24] provides an explicit characterization of the set of initial conditions from where a desired end-point is achievable. The reachability guarantees provided by the RTRR-based controller however do not hold in the presence of model uncertainty owing to the significant impact of the presence of uncertainty on RTRRs. In particular, a direct application of RTRR-based MPC of [24] could very likely result in off-spec product and a wasted batch. One contribution of the current chapter is the redesign of the RTRR-based predictive controller to account for model uncertainty by explicitly incorporating uncertainty bounds in the generation of RTRRs and their subsequent use within a robust MPC formulation.

Faults in processing or auxiliary equipment are ubiquitous in the chemical process industry and are an aspect of plant operation that requires special attention in the control design. The emphasis on the final product quality in batch and fed-batch processes makes their productivity particularly susceptible to faults as a fault can invalidate the desirable properties of a control design. While there has been significant work on fault detection and isolation for batch processes (see, e.g., [25–29]), fault tolerant control structures (FTCS) specific to batch systems have received limited research attention. The majority of the extensive research on FTCS for continuous processes (see, e.g., [30]) including recent results that address how to operate a plant during temporary faults (e.g., [31]) cannot be applied to batch processes due to the absence of equilibrium points and fundamental differences in the control objectives between batch and continuous processes [4].

As is the case for continuous processes, the majority of the FTCS for batch processes (e.g, see [8, 32]) essentially relies on the robustness of the control design to handle faults as disturbances during the failure period. The fault-tolerant characteristic in these formulations stems from the underlying assumption of availability of sufficient control effort such that the primary control objective remains achievable even in the presence of the fault. However, processes often encounter faults where the nominal control objective cannot be achieved if the fault persists, and furthermore, in the absence of a framework for explicitly handling such faults in batch processes, continuation of the implementation of controllers with limited fault-tolerant properties could lead to a missed opportunity to implement control action that could enable achieving the primary control objective after fault repair. A desirable property in a framework for handling faults in the context of batch systems, therefore, would be one that can identify an input trajectory (if it exists) without requiring any prior knowledge of the fault repair time to ensure end-point reachability upon fault repair. A safe-steering framework was recently developed in [24] that addressed the problem of determining how to utilize functioning inputs during fault rectification to enable desired product properties reachability following fault repair. The safe-steering framework in [24], however, does not account for the presence of uncertainty, and neither the controllability nor the safe-steering guarantees of [24] remain applicable in the presence of uncertainty.

Motivated by the above considerations, this chapter considers the problem of designing computationally efficient non-linear MPC for batch processes subject to input constraints, faults in the control actuators, and model uncertainty. Specifically, faults are considered that cannot be handled via robust control approaches and (if not rectified) preclude the reachability to the desired end-point with limited control effort. The rest of this manuscript is organized as follows: First, the class of processes considered is presented. Next, we generalize the RTRR-based predictive controller to explicitly account for model uncertainty by first developing a methodology and algorithm to generate robust RTRRs for their subsequent use within a robust MPC design for batch processes. Then, after formulating the safe-steering problem, a safe-steering framework is developed that utilizes the robust RTRR-based MPC to ensure states can be driven inside a desired end-point neighborhood if the fault is repaired sufficiently fast. It is shown that the proposed safe-steering framework is guaranteed to find a viable input trajectory if it exists (i.e, if the fault is repaired sufficiently fast) that can drive the system to a desired end-point neighborhood following fault repair. Then, closed-loop simulation results of a fed-batch reactor subject to actuator failure, model uncertainty, limited availability of measurements and sensor noise are presented to illustrate the efficacy of the proposed robust MPC formulation and the details of the safe-steering framework. Finally, we summarize our results.

4.2 Preliminaries

In this section, we describe the class of batch processes considered and review a representative end-point-based non-linear, shrinking horizon predictive controller, the RTRR-based predictive controller of [24], and the safe-steering framework developed in [24] without accounting for uncertainty.

4.2.1 Process Description

We consider batch process systems with uncertainty subject to input constraints and failures described by:

$$\begin{aligned}&\dot{\mathbf{x}}(t) = \mathbf{f}(\mathbf{x}(t), \mathbf{u}_\sigma(t), \boldsymbol{\theta}(t))\\&t \in [t_0, t_f], \mathbf{u}_\sigma(\cdot) \in \mathbf{U}_\sigma, \boldsymbol{\theta}(\cdot) \in \boldsymbol{\Theta}, \mathbf{x}(t_0) = \mathbf{x}_0\end{aligned} \tag{4.1}$$

where $\mathbf{x}(t) \in \mathbb{R}^n$ denotes the vector of state variables, $\mathbf{u}_\sigma(t) \in \mathbb{R}^m$ denotes the vector of constrained manipulated inputs, taking values in a nonempty convex subset $\mathbf{U}_\sigma$ of $\mathbb{R}^m$, where $\mathbf{U}_\sigma = \{\mathbf{u} \in \mathbb{R}^m \mid \mathbf{u}_{\min,\sigma} \leq \mathbf{u} \leq \mathbf{u}_{\max,\sigma}\}$, where $\mathbf{u}_{\min,\sigma}, \mathbf{u}_{\max,\sigma} \in \mathbb{R}^m$ denote the constraints on the manipulated inputs, and $\boldsymbol{\theta}(t) = [\theta^1(t) \;\cdots\; \theta^q(t)]^T \in \boldsymbol{\Theta} \subset \mathbb{R}^q$ where $\boldsymbol{\Theta} = \{\boldsymbol{\theta} \in \mathbb{R}^q \mid \boldsymbol{\theta}_{\min} \leq \boldsymbol{\theta} \leq \boldsymbol{\theta}_{\max}\}$, where $\boldsymbol{\theta}_{\min}, \boldsymbol{\theta}_{\max} \in \mathbb{R}^q$ denote the bounds on the vector of uncertain (possibly time-varying) but bounded variables taking values in a nonempty compact convex subset of $\mathbb{R}^q$. The times t_0 and t_f denote the initial time and batch termination times, respectively. The variable, $\sigma \in \{1, 2\}$, is a discrete variable that indexes fault-free and faulty operation with $\sigma = 1$ signifying fault-free operation and $\sigma = 2$ signifying faulty operation. The fault scenarios considered in this chapter involve actuator failure for a finite duration of time defined by the time of fault occurrence, t^{fault}, and time of fault repair, t^{repair}. An example of such a failure scenario is a flow valve feeding a fed-batch reactor becoming 'stuck' at its fail-safe value between t^{fault} and t^{repair} while remaining operational at all other times. According to the definition of σ, $\sigma = 1$ for $t \in [t_0, t^{\text{fault}})$, $\sigma = 2$ for $t \in [t^{\text{fault}}, t^{\text{repair}})$, and $\sigma = 1$ for $t \in [t^{\text{repair}}, t_f]$.

The vector function $\mathbf{f} : \mathbb{R}^n \times \mathbf{U}_\sigma \times \boldsymbol{\Theta} \rightarrow \mathbb{R}^n$ is assumed to be continuous on $(\mathbf{x}, \mathbf{u}, \boldsymbol{\theta})$ and locally Lipschitz in $\mathbf{x}$ on $\mathcal{D} \times \mathbf{U} \times \boldsymbol{\Theta}$, where $\mathcal{D} \subset \mathbb{R}^n$. The notation, $\| \cdot \|_{\mathbf{Q}}$, refers to the weighted norm, defined by $\|\mathbf{x}\|_{\mathbf{Q}} = \mathbf{x}^T\mathbf{Q}\mathbf{x}$ for all $\mathbf{x} \in \mathbb{R}^n$, where $\mathbf{Q}$ is a positive-definite symmetric matrix and $\mathbf{x}^T$ denotes the transpose of $\mathbf{x}$. Throughout the manuscript, we assume that for any $\mathbf{u}(\cdot) \in \mathbf{U}_\sigma$ and $\boldsymbol{\theta}(\cdot) \in \boldsymbol{\Theta}$, the solution of the batch system of Eq. (4.1) exists and is continuous for all $t \in [t_0, t_f]$ and we focus on the state feedback problem, i.e., $\mathbf{x}(t)$ is assumed to be available for all $t \in [t_0, t_f]$.

4.3 Robust Reverse-Time Reachability Region-Based MPC

In this section, we present a computationally efficient robust non-linear predictive controller for batch processes. Preparatory to the robust controller formulation, we introduce the notion of robust RTRRs, which are essential in the control design and analysis, and then present an algorithm to generate these regions. This is followed by the formulation of a robust RTRR-based MPC design.

4.3.1 Robust Reverse-Time Reachability Regions

In order to define robust RTRRs, we first note that in batch process control, the implication of the presence of model uncertainty is that in general, exact end-point reachability guarantees cannot be made regardless of the control law. Instead, only reachability to a certain neighborhood of the end-point can be guaranteed. To understand this, consider the batch system described by Eq. (4.1) subject to a predictive control law. At time $t = t_f - \delta$, a dynamic optimization problem is solved to compute the set of input moves that drives the system to the desired end-point in time δ for nominal values of the uncertainty, $\boldsymbol{\theta}_{\text{nom}}$. However, if this same control move is implemented on the plant, there is no guarantee that the process will be driven to $\mathbf{x}_d$ because it is unknown if $\boldsymbol{\theta}_{\text{nom}}$ represents the true values of the plant parameters. Based on this argument, one desirable property of a robust MPC formulation for batch process is to guarantee the existence of inputs to drive the process states inside a desired neighborhood around the end-point, which we denote by $\mathcal{B}(\mathbf{x}_d)$, that can be chosen based on the acceptable level of variance in the final product quality. Accordingly, of interest is the set of process states from where a desired end-point neighborhood can be reached in the presence of model uncertainty while satisfying all input constraints. These sets are denoted as robust RTRRs and are formally defined below:

Definition 4.1 For the process described by Eq. (4.1), the robust RTRR at time, t, is the set:

$$\tilde{\mathcal{R}}(t) = \{\mathbf{x}_0 \mid \mathbf{x}(t) = \mathbf{x}_0,\ \dot{\mathbf{x}}(t) = \mathbf{f}(\mathbf{x}(t), \mathbf{u}(t), \boldsymbol{\theta}(t))\ \exists\, \mathbf{u}(\tau) \in \mathbf{U}\ \forall \tau \in [t, t_f],\ \text{such that}\ \forall \boldsymbol{\theta}(\tau) \in \boldsymbol{\Theta}\ \forall \tau \in [t, t_f], \mathbf{x}(t_f) \in \mathcal{B}(\mathbf{x}_d)\} \tag{4.2}$$

To account for the discrete nature of control implementation, we define the discrete time version of robust RTRRs below:

Definition 4.2 For the process described by Eq. (4.1), the robust RTRR at a time $t = t_f - k\delta$, indexed by k is the set:

$$\tilde{\mathcal{R}}_k = \{\mathbf{x}_0 \mid \mathbf{x}(t) = \mathbf{x}_0,\ \dot{\mathbf{x}}(t) = \mathbf{f}(\mathbf{x}(t), \mathbf{u}(t), \boldsymbol{\theta}(t))\ \exists\, \mathbf{u}(t) = \{\mathbf{u}[i]\} \in \mathbf{U}\ \forall i = 1, \ldots, k\ \text{where}\ \mathbf{u}[i] = \mathbf{u}(i\delta)\ \text{and satisfies}\ \mathbf{u}(t) = \mathbf{u}[i]\ \forall t \in [i\delta, (i+1)\delta)\ \text{such that}\ \forall \boldsymbol{\theta}(\tau) \in \boldsymbol{\Theta}\ \forall \tau \in [t, t_f],\ \mathbf{x}(t_f) \in \mathcal{B}(\mathbf{x}_d)\} \tag{4.3}$$

Note that for the special case of $k = 0$, $\tilde{\mathcal{R}}_k$ is defined to be the desired end-point neighborhood or $\mathcal{B}(\mathbf{x}_d)$. Prior to presenting an algorithm to generate robust RTRRs, the existence of these regions must first be established. This is formalized below in Theorem 4.3. The proof is available in the Appendix.

Theorem 4.3 *For the batch system described by Eq. (4.1), given* $\boldsymbol{\Theta}$ *and a nonempty* $\tilde{\mathcal{R}}_{k-1}$, *there exists a* δ^* *such that for any* $\delta \leq \delta^*$, $\tilde{\mathcal{R}}_k \neq \emptyset$ *(i.e,* $\tilde{\mathcal{R}}_k$ *is nonempty).*

Remark 4.1 In the absence of model uncertainty, the existence of nonempty RTRRs at each sampling instance of the batch is guaranteed simply from the existence of a solution for the system over a finite time. When considering uncertainty, however, the size of robust RTRRs depends on the size of the desired end-point neighborhood and also the sampling period in the batch. For example, if we consider a fixed $\mathcal{B}(\mathbf{x}_d)$, robust RTRRs may cease to exist as we proceed toward the initial time if the given sampling time is too large. Theorem 4.3 is therefore important in establishing the trade-off between $\mathcal{B}(\mathbf{x}_d)$ and the sampling period. From a practical perspective, for a specified desired end-point neighborhood, the result of Theorem 4.3 implies that the sampling time can be used to mitigate the reduction in the size of robust RTRRs as we proceed toward the initial time.

Remark 4.2 While a 'nominal' MPC formulation was presented in Sect. 3.2.2 to emphasize the fact that the control calculation at every instance requires the solution of the entire input trajectory, a min-max dynamic optimization problem can in principle be employed to handle the problem of uncertainty. The solution to such a min-max problem would be a manipulated input trajectory from the current time to the end of the batch that minimizes (using the inputs as decision variables) the maximum (over all realizations of the uncertainty) value of the objective function. This added layer of optimization renders min-max-based MPC approaches for batch systems even more computationally expensive than the nominal end-point-based MPC formulation and motivates the development of a computationally efficient robust non-linear MPC formulation for batch processes.

Remark 4.3 An algorithm to generate estimates of RTRRs as point sets at each sampling instance for the batch was proposed in [24] that consisted entirely of successive off-line integrations of the reverse-time model of the system (i.e., $\dot{\mathbf{x}}(t) = -\mathbf{f}(\mathbf{x}(t), \mathbf{u}(t))$). Specifically, for a given $\mathbf{x}_d$, the reverse-time process model can be integrated for the duration of δ, holding the value of the manipulated inputs constant over δ. Performing this integration for all possible appropriately discretized values of the manipulated inputs in turn yields an (under) estimate of $\mathcal{R}_1$ as a point set. A point-set estimate of $\mathcal{R}_2$ can then be determined by repeating the process for all elements in $\mathcal{R}_1$, and the process repeated to yield RTRRs up to the initial time. The details regarding alleviating the associated computational costs with this algorithm, assumptions to ensure the generation of compact sets, and characterization strategies for the point-set estimates are addressed in [24]. Note also that the presence of uncertainty cannot be accounted for by simply repeating the above set of calculations for all allowable values of the uncertainty and requires redesigning the RTRR generation algorithm.

Generating Robust Reverse-Time Reachability Regions

In this section, given a desired end-point neighborhood, $\mathcal{B}(\mathbf{x}_d)$, we develop a methodology to sequentially generate robust RTRR estimates off-line. More specifically, starting at $k = 1$, for a given $\mathcal{B}(\mathbf{x}_d)$, we identify an explicitly characterized estimate of $\tilde{\mathcal{R}}_k$ from where the process can be driven inside $\tilde{\mathcal{R}}_{k-1}$ in the presence of model uncertainty. The explicit characterization is also a necessity for the practical implementation of the MPC formulation to be presented in the next section. In this chapter, we use n-dimensional ellipsoids to mathematically express estimates of robust RTRRs at each sampling instance (note that our results are not limited to this choice of the characterization; the use of n-dimensional ellipsoids is simply to illustrate our results). The general form of the ellipsoid expression is given below:

$$\tilde{\mathcal{R}}_k = \{\mathbf{x} \mid \|\mathbf{x} - \tilde{\mathbf{c}}_k\|_{\tilde{\mathbf{P}}_k} \leq 1\} \tag{4.4}$$

where the vector $\tilde{\mathbf{c}}_k \in \mathbb{R}^n$ denotes the center of the ellipsoid, the positive-definite symmetric matrix $\tilde{\mathbf{P}}_k \in \mathbb{R}^n \times \mathbb{R}^n$ defines its size and orientation, and k indexes the batch sampling instances as before. Note that because $k = 0$ corresponds to t_f, $\tilde{\mathbf{c}}_0 = \mathbf{x}_d$ and $\tilde{\mathbf{P}}_0$ is a user-defined matrix based on the acceptable variance level of the final product quality.

To determine if a set is a valid estimate for $\tilde{\mathcal{R}}_k$ (defined by $\tilde{\mathbf{c}}_k$ and $\tilde{\mathbf{P}}_k$), we solve, for a given δ and $\mathbf{\Theta}$ (and an $\tilde{\mathcal{R}}_{k-1}$ defined by $\tilde{\mathbf{c}}_{k-1}$ and $\tilde{\mathbf{P}}_{k-1}$), the following multi-level non-linear program (NLP):

$$\min_{\mathbf{x}_0} \; J_1 = 0 \tag{4.5}$$

$$\text{subject to: } \|\mathbf{x}_0 - \tilde{\mathbf{c}}_k\|_{\tilde{\mathbf{P}}_k} \leq 1 \tag{4.6}$$

$$J_2 \geq 1 \tag{4.7}$$

$$\min_{\mathbf{u} \in \mathbf{U}} \; J_2 = \|\tilde{\mathbf{x}}(\delta) - \tilde{\mathbf{c}}_{k-1}\|_{\tilde{\mathbf{P}}_{k-1}} \tag{4.8}$$

$$\text{subject to: } \dot{\tilde{\mathbf{x}}}(t) = \mathbf{f}(\tilde{\mathbf{x}}(t), \mathbf{u}, \boldsymbol{\theta}) \tag{4.9}$$

$$\max_{\boldsymbol{\theta} \in \boldsymbol{\Theta}} \; J_3 = \|\tilde{\mathbf{x}}(\delta) - \tilde{\mathbf{c}}_{k-1}\|_{\tilde{\mathbf{P}}_{k-1}} \tag{4.10}$$

$$\text{subject to: } \dot{\tilde{\mathbf{x}}}(t) = \mathbf{f}(\tilde{\mathbf{x}}(t), \mathbf{u}, \boldsymbol{\theta}) \tag{4.11}$$

If the multi-level NLP is infeasible, we deem the estimate to be a valid robust RTRR. To understand this, consider the different levels in the multi-level NLP: for a given initial state, the two bottom most layers solve the (min-max) robust control problem, i.e., determine the input that minimizes the worst-case effect of the uncertainty. In other words, the bottom two levels compute the control action that, for the worst realization of the uncertainty, drives the state to the lowest level set of the n-dimensional ellipsoid at the next time step. The top level problem then searches over all initial conditions within the given $\tilde{\mathcal{R}}_k$ to find (if it exists) an initial condition for which the state at the next time step ends up being driven outside the robust RTRR at the next step. If the multi-level optimization problem is feasible, it implies that there are no guarantees that the process starting from the given estimate of the robust RTRR will be driven inside the robust RTRR at the next step in the presence of uncertainty even when implementing the robust control move. On the other hand, if the problem is infeasible, this implies that for every initial condition in the given robust RTRR, the process states are always contained within the robust RTRR at the next time, even in the worst-case effect of the uncertainty. An infeasible solution therefore represents that a valid robust RTRR has been found.

Remark 4.4 Note that in principle, one can add another layer in the optimization problem wherein the $\tilde{\mathbf{P}}_k$ matrix and the $\tilde{\mathbf{c}}_k$ vector are the decision variables and the objective is to maximize the 'volume' of the n-dimensional ellipsoid. Even if carried out off-line, the determination of the largest robust RTRR would become an unwieldy problem. In this chapter, we address this problem by appropriately preselecting the $\tilde{\mathbf{P}}_k$ matrix (that determines the orientation of the ellipsoid) and the $\tilde{\mathbf{c}}_k$ vector (that determines the center of the ellipsoid). In particular, the system is reverse-time integrated from all the elements in $\tilde{\mathcal{R}}_{k-1}$ (using nominal values of the uncertainty and all possible values of the inputs), and then an ellipsoid is found that best covers a subset of these points. If the multi-level optimization problem is feasible for this ellipsoid, the set is scaled down and the problem is resolved until the multi-level optimization becomes infeasible. On the other hand, if the problem is infeasible to begin with, the set is scaled up and this process is repeated until the optimization problem becomes feasible. The final ellipsoid obtained through this procedure then represents the (approximately) largest estimate of the robust RTRR (given a predecided orientation of the ellipsoid).

Remark 4.5 Note that the problem of determining the robust RTRRs cannot be addressed by extending the method of generating the RTRRs in [24] to generate point sets using all values of the uncertainty (over and above the different values of the inputs as done in [24]). The only conclusion that can be drawn for a point in such a set is that there exists a pair of input and uncertainty values such that the process can be driven to the RTRR at the next time. No guarantees can be made for the existence of an input (for any allowable value of the uncertainty) and the determination of that input (without knowing the value of the uncertain parameter) that can drive the process to the robust RTRR at the next step. This necessitates the development of the multi-level optimization-based framework for the generation of the robust RTRRs. Note also that the objective in this chapter is not to characterize the true robust RTRR (i.e., to determine all points that are contained within the robust RTRR) but to generate a workable estimate for which the existence and determination of the input to drive the process to the next robust RTRR can be guaranteed.

Remark 4.6 In [24], the necessary and sufficient conditions, namely feasibility of the RTRR-based MPC during the failure period, for safe-steering a batch were established. If the RTRR-based MPC is feasible from t^{fault} up to and including t^{repair}, the implication is that at t^{repair}, the states are contained with the RTRR at t^{repair} at which point the full control effort is recovered. Consequently, by the definition of RTRRs, the desired end-point will be reachable following t^{repair}. In the presence of uncertainty, however, the end-point reachability guarantees in the safe-steering framework are lost even if the process states at t^{repair} are contained within the RTRR at t^{repair}. This is because RTRRs as defined in [24] are generated based on nominal values of the uncertainty. Consequently, using the RTRR-based MPC formulation to preserve states in 'nominal' RTRRs during the failure period in the presence of uncertainty have no implications regarding end-point reachability and is therefore not a suitable solution to the safe-steering problem. This motivates the development of a robust MPC design for use in the safe-steering framework such that the desired control objectives in the batch can be met if the fault is recovered sufficiently fast.

4.3.2 Robust MPC Formulation

In this section, we present an MPC formulation that utilizes robust RTRRs to steer a batch system to a desired neighborhood around the end-point. Similar to the RTRR-based formulation, most of the computational burden associated with the MPC is off-line and the formulation is therefore amenable to online implementation. To this end, consider a batch process described by Eq. (4.1) for fault-free conditions (i.e., $\sigma(t) = 1$) and for which robust RTRR estimates have been characterized for a given δ, $\boldsymbol{\Theta}$, and $\mathscr{B}(\mathbf{x}_d)$. The control action at sampling instance $k = \frac{t_f - t}{\delta}$ is computed by solving the following bi-level optimization problem:

$$\min_{\mathbf{u}\in\mathbf{U}} J_{\tilde{\mathcal{R}}} = \int_t^{t+\delta} \|\tilde{\mathbf{x}}(\tau) - \mathbf{x}^*_{\text{nom}}\|_{\mathbf{Q}} + \|\Delta\mathbf{u}\|_{\mathbf{R}} \mathrm{d}t \tag{4.12}$$

$$\text{subject to: } \tilde{\mathbf{x}}(0) = \mathbf{x}(t) \tag{4.13}$$

$$\dot{\tilde{\mathbf{x}}}(\tau) = \mathbf{f}(\tilde{\mathbf{x}}(\tau), \mathbf{u}, \boldsymbol{\theta}) \tag{4.14}$$

$$\|\tilde{\mathbf{x}}(\delta) - \tilde{\mathbf{c}}_{k-1}\|_{\tilde{\mathbf{P}}_{k-1}} \leq 1 \tag{4.15}$$

$$\max_{\boldsymbol{\theta}\in\boldsymbol{\Theta}} J_\theta = \|\tilde{\mathbf{x}}(\delta) - \tilde{\mathbf{c}}_{k-1}\|_{\tilde{\mathbf{P}}_{k-1}} \tag{4.16}$$

$$\text{subject to: } \dot{\tilde{\mathbf{x}}}(\tau) = \mathbf{f}(\tilde{\mathbf{x}}(\tau), \mathbf{u}, \boldsymbol{\theta}) \tag{4.17}$$

where $\Delta \mathbf{u}$ is a vector of deviations between successive control moves, $\mathbf{Q}$ and $\mathbf{R}$ are positive-definite weighting matrices, and the objective function, $J_{\tilde{\mathcal{R}}}$, is formulated to minimize the deviations between process state trajectories and some nominal optimal state trajectories, $\mathbf{x}^*_{\text{nom}}$, and variations in the control moves. This two-tier optimization problem is formulated in a similar fashion as the bottom two levels in the robust RTRR generation algorithm of Eqs. (4.8)–(4.11). Following initialization at the current plant values Eq. (4.13), the worst-case values of the uncertainty are found from the maximization problem given by Eqs. (4.16)–(4.17). The worst-case uncertainty values are identified as those which drive the process states to the highest level set of the estimate of $\tilde{\mathcal{R}}_{k-1}$. For these worst-case values, the top level of the optimization problem searches for an input that minimizes some cost function, $J_{\tilde{\mathcal{R}}}$, and ensures the states at the next time step are contained within the robust RTRR. The methodology and algorithm used to compute the robust RTRR guarantees the feasibility of the MPC optimization problem in the fault-free scenario. Additionally, by definition, robust RTRRs take into account the requirement to drive the process to a desired end-point neighborhood. The implications on the guarantees of feasibility and driving the system to a desired end-point neighborhood are formalized below in Theorem 4.4. A sketch of a proof of this theorem is available in the Appendix.

Theorem 4.4 *Consider the constrained system of Eq. (4.1) with $\sigma = 1$ under the robust RTRR-based MPC law of Eqs. (4.12)–(4.17). If $x(t_0) \in \tilde{\mathcal{R}}(t_0)$, the MPC bi-level optimization problem remains feasible for all $t \in [t_0, t_f]$ and $x(t_f) \in B(\mathbf{x}_d)$.*

Remark 4.7 In Theorem 4.4, the condition $x(t_0) \in \tilde{\mathcal{R}}(t_0)$ guarantees the existence of an input trajectory via guaranteed feasibility of the robust RTRR-based MPC to drive the process states inside $\mathcal{B}(\mathbf{x}_d)$ in the presence of uncertainty. The condition $x(t_0) \in \tilde{\mathcal{R}}(t_0)$ is however not a necessary condition for driving the states inside $\mathcal{B}(\mathbf{x}_d)$ by batch termination even if we consider exact characterizations of true robust RTRRs in the robust RTRR-based MPC formulation. Consider the case where $x(t_0) \notin \tilde{\mathcal{R}}(t_0)$. While we cannot guarantee the existence of an input trajectory to ensure $x(t_f) \in B(\mathbf{x}_d)$, this trajectory may still exist because it might be possible to drive the state inside $\mathcal{B}(\mathbf{x}_d)$ by t_f for some realization of the uncertainty (if not for all values of the uncertainty, as required in the robust RTRR definition). The theorem, however, does establish that the robust RTRR-based MPC problem will remain initially and successively feasible and drive the process to the desired end-point neighborhood.

Remark 4.8 In the MPC formulation in Eqs. (4.12)–(4.17), the performance index Eq. (4.12) can be chosen to meet desired performance criteria or robustness objectives. For example, improving the disturbance rejection capabilities in the controller could be achieved by penalizing the Euclidean distance between the process states at sampling instance $k-1$ and the center of the $\tilde{\mathcal{R}}_{k-1}$ estimate. This objective function would tend to steer the system through the centers of the robust RTRRs, thereby lowering the possibility of any unaccounted for disturbances driving the process outside of robust RTRRs.

Remark 4.9 An important advantage of the RTRR-based predictive controller over conventional end-point-based MPC approaches is its computational efficiency while guaranteeing end-point reachability even when using the full non-linear model of the system. In the presence of uncertainty, however, it becomes important to explicitly account for the uncertainty in the RTRR-based MPC design from not only a theoretical standpoint but also from a practical standpoint. In particular, ignoring the effect of the uncertainty will lead to larger estimates of the RTRRs; therefore, if the initial conditions are within the RTRR at t_0, initial feasibility of the optimization problem can still be achieved even in the presence of uncertainty. However, this initial feasibility does not imply successive feasibility for all future time steps since large deviations in the uncertain parameters can drive the states out of the RTRR estimates at which point the control law would become infeasible. This, in turn, of course invalidates the reachability guarantees of the MPC design. Even worse, a possible scenario is that of the RTRR MPC laws remaining feasible by virtue of the large region estimates, but the uncertainty driving the process states significantly away from the the desired end-point at the last time step, resulting in a wasted batch.

Remark 4.10 One characteristic of batch processes is that the desired end-point quality, which is based on the values of quality variables at batch termination, typically remains consistent from batch to batch unless a new product is being manufactured. The main source of variation between batches is typically the initial condition in the batch as this is dictated by raw material properties, which are subject to variance depending on the source of raw materials. The robust RTRR-based MPC scheme is designed with these key properties in mind as robust RTRRs are generated for specific values of the quality variables at batch termination and also provide an explicit characterization of initial conditions for which the desired end-point quality can be met. Note that if the end-point quality in a batch is subject to change and discrete values of the other possible end-point qualities are known, robust RTRRs corresponding to all possible desired end-points can be generated beforehand and the suitable robust RTRRs can be used during controller implementation.

4.4 Robust Safe-Steering of Batch Processes

In this section, we generalize the safe-steering framework to handle model uncertainty by utilizing the robust RTRR-based MPC scheme presented in the previous section. First, we formulate the safe-steering problem and then present the robust safe-steering framework.

4.4.1 Problem Statement

For systems described by Eq. (4.1), we consider faults in the control actuator under the assumption that upon failure, the available control effort is reduced. More specifically, without loss of generality, we characterize the fault occurring in the first control actuator at a time t^{fault} which is repaired at time, t^{repair} (i.e. for $t < t^{\text{fault}}$ and $t \geq t^{\text{repair}}, \sigma(t) = 1$ while for $t^{\text{fault}} \leq t < t^{\text{repair}}, \sigma(t) = 2$) as $u^1_{\text{min,failed}} \leq u^1_2(t) \leq u^1_{\text{max,failed}} \forall t \in [t^{\text{fault}}, t^{\text{repair}})$ where u^i denotes the ith component of the input vector $\mathbf{u}$. Reduced control effort corresponds to a situation where $u^1_{\text{min,failed}} > u^1_{\text{min}}$ and $u^1_{\text{max,failed}} < u^1_{\text{max}}$. Note that in the case where a valve reverts to a completely open or shut position, we have $u^1_{\text{min,failed}} = u^1_{\text{max,failed}}$. The only inputs available for control between t^{fault} and t^{repair} are therefore $u^1_2, \ldots, u^m_2$. We define the robust safe-steering problem as the one of identifying trajectories of these functioning inputs during the fault rectification period (without requiring the value of t^{repair} or an estimate thereof to be known a-priori) in the presence of model uncertainty that will ensure the process can be driven inside the desired end-point neighborhood upon recovery of the full control effort.

4.4.2 Safe-Steering to a Desired End-Point Neighborhood

The key idea in the robust safe-steering problem is to preserve the states in robust RTRRs during the failure period by employing the robust RTRR-based MPC design. By doing so, the robust RTRR-based MPC scheme is able to drive the process to a desired end-point neighborhood following fault repair. Note also that the ability to go to a desired end-point neighborhood after fault repair is dependent on the duration of the fault. To this end, consider a batch system described by Eq. (4.1) for which the first control actuator fails at t^{fault} and is repaired at t^{repair} and the robust RTRRs for fault-free operation have been characterized for all sampling instances in the fault rectification period. We formalize the requirements for safe-steering the batch in Theorem 4.5.

Theorem 4.5 *Consider the constrained system of Eq. (4.1) with $\mathbf{x}(t_0) \in \tilde{\mathcal{R}}(t_0)$ subject to the robust RTRR-based MPC scheme given by Eqs. (4.12)–(4.17). If the MPC optimization problem remains feasible $\forall t \in [t^{\text{fault}}, t^{repair}]$, then $\mathbf{x}(t_f) \in \mathcal{B}(\mathbf{x}_d)$.*

Remark 4.11 The key idea formally expressed in Theorem 4.5 is that if a fault is repaired sufficiently fast (i.e., there exists an implementable input trajectory during the fault repair period, and one after fault repair), the robust RTRR-based MPC finds this trajectory via preserving the process states within robust RTRRs. The implication of this is that process states at t^{repair} will then belong to $\tilde{\mathcal{R}}(t^{\text{repair}})$, and according to the definition of robust RTRRs, the process can then be driven to a desired end-point neighborhood. Therefore, maintaining the system within the robust RTRRs provides a sufficient condition to ensure that the desired neighborhood can be reached upon fault recovery. In other words, the proposed robust RTRR-based MPC is able to identify the input trajectory during faulty operation (if one exists) that will enable reaching the desired end-point neighborhood upon fault recovery. In contrast, end-point-based MPC approaches can fail to find this trajectory even if it exists. The end-point MPC problem can become infeasible because it simply may not be possible to satisfy the terminal constraint with reduced control effort, which implies (appropriately truncated) infeasible solutions have to be implemented on the process. By repeatedly applying saturated versions of the infeasible solutions during the failure period, the states can be driven to an unrecoverable point from where reaching the desired end-point neighborhood is not possible even after fault recovery.

Remark 4.12 Theorem 4.5 provides sufficient conditions for fault-tolerant control in batch systems. To address the issue of necessary conditions, we note that if the fault is repaired too late, it can become impossible to preserve the system states in robust RTRRs using reduced control effort at some point between t^{fault} and t^{repair}. In this case, the states escape the robust RTRRs by t^{repair}; however, this does not necessarily imply the system states at batch termination will be outside $\mathcal{B}(\mathbf{x}_d)$. This is because the states at t^{repair} in this situation could reside in a region for which there exists a specific uncertainty realization and corresponding input trajectory that can drive the system inside the desired neighborhood. This can occur since $\mathbf{x}(t^{\text{repair}}) \in \tilde{\mathcal{R}}(t^{repair})$ is a sufficient but not necessary condition for driving the system inside $\mathcal{B}(\mathbf{x}_d)$.

Remark 4.13 The robust safe-steering framework can also be utilized to handle faults in batch systems where the batch time is flexible. To understand this, consider the case when the robust RTRR-based MPC law encounters infeasibility during the fault rectification period at sampling instance k_{infs}. In such a scenario, while there exists no control move using the limited control effort to drive the system from $\tilde{\mathcal{R}}_{k_{\text{infs}}}$ to $\tilde{\mathcal{R}}_{k_{\text{infs}}-1}$, it may be possible instead, given suitable system dynamics, to drive the system from $\tilde{\mathcal{R}}_{k_{\text{infs}}}$ to $\tilde{\mathcal{R}}_{k_{\text{infs}}+i}$ where i is a positive integer such that $k_{\text{infs}}+i$ is a sampling instance for which the robust RTRR has been generated and characterized. In such a scenario, the MPC optimization problem given by Eqs. (4.12)–(4.17) can be modified accordingly by substituting $\tilde{\mathcal{R}}_{k_{\text{infs}}+i}$ in place of $\tilde{\mathcal{R}}_{k_{\text{infs}}-i}$. Next, we revert to the original MPC law that requires the process states to be driven from $\tilde{\mathcal{R}}_k$ to $\tilde{\mathcal{R}}_{k-1}$. Consequently, in i time steps, the MPC law will again require driving the system from $\tilde{\mathcal{R}}_{k_{\text{infs}}}$ to $\tilde{\mathcal{R}}_{k_{\text{infs}}-1}$. If the fault is rectified within these i time steps, Theorem 4.4 guarantees the MPC law will now be feasible. Using this procedure maintains the process states within robust RTRRs during the failure period at the cost of an increase in the batch termination time by $i\delta$. Note that it is also possible to repeat this procedure until the fault is rectified and use different values of i each time. Also worth noting is that if no $\tilde{\mathcal{R}}_{k+i}$ exists, this points to the fact that the batch may have gone beyond recovery. In such a case, the immediate (and necessary) action of discarding the batch can be taken instead of unsuccessfully trying to reach the desired end-point neighborhood and wasting valuable material and time.

Remark 4.14 In this chapter, we consider the full state feedback problem to convey the key ideas in the robust safe-steering framework. If all system states are not measured, state observers or estimators can be incorporated into the framework such that the batch system can be safe-steered based on state estimates (see the simulation example for an illustration). A theoretical analysis of this effect however remains outside the scope of the present chapter.

4.5 Simulation Example

In this section, we first consider a fault-free environment and demonstrate the need for accounting for uncertainty. We then illustrate the robust safe-steering framework by considering an actuator failure in the process. To this end, consider a fed-batch reactor where an irreversible, first-order exothermic reaction of the form $A \rightarrow B$ takes place. The state-space model for this process takes the following form:

$$\begin{bmatrix} \dot{C}_A(t) \\ \dot{T}(t) \\ \dot{V}(t) \end{bmatrix} = \begin{bmatrix} -k_{A0} \exp\left\{\frac{E}{R}\left(\frac{1}{T_R} - \frac{1}{T(t)}\right)\right\} C_A(t) + \frac{F(t)(C_{A0}-C_A(t))}{V(t)} \\ \frac{UA(T_a(t)-T(t))}{\rho C_p V(t)} + \frac{F(t)(T_{A0}-T(t))}{V(t)} + \frac{(\Delta H)\left(-k_{A0}\exp\left\{\frac{E}{R}\left(\frac{1}{T_R}-\frac{1}{T(t)}\right)\right\}C_A(t)\right)}{\rho C_p} \\ F(t) \end{bmatrix} \tag{4.18}$$

where $C_A(t)$, $T(t)$, and $V(t)$ denote the concentration of A, reactor temperature, and volume (respectively) and constitute the state vector, $\mathbf{x}(t) = \left[C_A(t)\ T(t)\ V(t)\right]^T$. The physical meaning of the model parameters and their nominal values can be found in Table 4.1. The primary control objective considered in the simulation studies is to achieve the (arbitrarily chosen) desired end-point of $\mathbf{x}_d = \left[0.1\ 465\ 65\right]^T$ in the presence of uncertainty. The desired neighborhood around $\mathbf{x}_d$ is taken to be an ellipsoid with $\tilde{\mathbf{P}}_0 = \text{diag}\left\{10^4, 4, 11.1\bar{1}\right\}$ and $\tilde{\mathbf{c}}_0 = \mathbf{x}_d$. We consider uncertainty in the inlet temperature, T_{A0}, and the heat exchanger coefficient (UA) of $\pm 5\%$ around their

Table 4.1 Process parameters for the fed-batch reactor of Eq. (4.18)

Parameter	Description	Value	Unit
k_{A0}	Rate constant of reaction $A \rightarrow B$ at T_R	0.15	1/h
E	Activation energy of reaction $A \rightarrow B$	10^4	cal/mol
R	Universal gas constant	1.986	cal/(mol · K)
T_R	Reference temperature	290	K
C_{A0}	Concentration of A in inlet feed stream	5	mol/L
UA	Heat transfer coefficient × Area	10^4	cal/(h · K)
ρ	Solution and inlet feed density	1	kg/L
C_p	Solution and inlet feed heat capacity	65	cal/(kg · K)
T_{A0}	Temperature of inlet feed stream	293	K
ΔH	Heat of reaction $A \rightarrow B$	−4000	cal/mol

nominal values. That is, $\boldsymbol{\theta} = \left[T_{A0}\ UA\right]^T$ and the uncertainty bounds are defined by $\boldsymbol{\theta}_{\min} = \left[278.35\ 9.50 \times 10^3\right]^T$ and $\boldsymbol{\theta}_{\max} = \left[307.65\ 1.05 \times 10^4\right]^T$. The uncertainty in T_{A0} is representative of a process disturbance, whereas the heat transfer coefficient is a model parameter that is often not known precisely and varies with time due to effects of fouling. The total batch time is taken to be $t_f = 30$ min with a controller execution period of $\delta = 36$ s. The manipulated variables in the process are the inlet feed rate, F (L/h), and the heating coil temperature T_a (K), $\mathbf{u}(t) = \left[F(t)\ T_A(t)\right]^T$, with constraints, $\mathbf{u}_{\min} = \left[0\ 285\right]^T$ and $\mathbf{u}_{\max} = \left[25\ 400\right]^T$. To demonstrate the applicability of the proposed method subject to limited (noisy) measurements, we consider the case when only noisy measurements of T and V are available, $\mathbf{y}(t) = \left[T(t)\ V(t)\right]^T$, with standard deviations of 0.15 and 0.05, respectively.

To estimate C_A using online measurements of T and V, an extended Luenberger observer (ELO) is used that takes the following form:

$$\dot{\hat{\mathbf{x}}}(t) = \mathbf{f}(\hat{\mathbf{x}}(t), \mathbf{u}(t), \boldsymbol{\theta}_{\text{nom}}) + \mathbf{L}(\mathbf{y}(t) - \hat{\mathbf{y}}(t)) \tag{4.19}$$

where $\hat{\mathbf{x}}(t) = \left[\hat{C}_A(t)\ \hat{T}(t)\ \hat{V}(t)\right]^T$ and $\hat{\mathbf{y}}(t) = \left[\hat{T}(t)\ \hat{V}(t)\right]^T$ are vectors of the estimated states and outputs (respectively), the vector function, $\mathbf{f}(\cdot)$, is the right-hand side of the differential equations in Eq. (4.18), $\boldsymbol{\theta}_{\text{nom}}$ is the nominal set of parameters listed in Table 4.1, and $\mathbf{L}$ is a matrix of observer gains.

For this example, the model was successively linearized at the current estimate of the states, the computed value of the input, and the nominal set of parameters, and the gain was subsequently computed using the typical procedure used for linear dynamic systems. Specifically, the eigenvalues of the matrix, $(\mathbf{A} - \mathbf{LC})$, were placed on the left side of the complex plane where the ijth element of $\mathbf{A}$ is given by:

$$a_{i,j} = \left.\frac{\partial f_i}{\partial x_j}\right|_{\hat{\mathbf{x}}(t), \mathbf{u}(t), \boldsymbol{\theta}_{\text{nom}}} \tag{4.20}$$

4.5.1 Implementation of the Robust Reverse-Time Reachability Region-Based MPC Scheme

To demonstrate the need for accounting for uncertainty, closed-loop simulations of the fed-batch system were performed using the nominal RTRR-based MPC design in [24] and the proposed robust MPC design. First, for the given desired neighborhood, uncertainty bounds, and input constraints, nominal and robust RTRRs were generated and characterized as ellipsoids for all sampling instants using the algorithm described in the 'Generating Robust Reverse-Time Reachability Regions' section. The following objective functions were used for the MPC formulations.

$$J_{\mathcal{R}} = \|\tilde{\mathbf{x}}(\delta) - \mathbf{c}_{k-1}\|_{\mathbf{P}_{k-1}} + \int_t^{t_f} \|\Delta\mathbf{u}\|_{\mathbf{R}} \mathrm{d}t \tag{4.21}$$

$$J_{\tilde{\mathcal{R}}} = \|\tilde{\mathbf{x}}(\delta) - \tilde{\mathbf{c}}_{k-1}\|_{\tilde{\mathbf{P}}_{k-1}} + \int_t^{t_f} \|\Delta\mathbf{u}\|_{\mathbf{R}} \mathrm{d}t \tag{4.22}$$

where $J_{\mathcal{R}}$ and $J_{\tilde{\mathcal{R}}}$ denote the objective functions for the nominal and robust formulations, respectively. The penalty matrix on successive input changes/move suppression matrix was set to $\mathbf{R} = \mathrm{diag}\left\{5 \times 10^{-4}, 9 \times 10^{-4}\right\}$. The initial conditions were specified to be $\mathbf{x}_{\mathcal{R}}(0) = \left[2.648\ 275.16\ 59.883\right]^T$ and $\mathbf{x}_{\tilde{\mathcal{R}}}(0) = \left[2.65\ 279.395\ 58.483\right]^T$ for the nominal and robust MPC schemes (respectively) and verified to be within the appropriate RTRR. The initial state estimates were also chosen such that they resided within the appropriate RTRR. The deviation between the initial state estimates and the actual states were consistent for both cases. The estimates were set to $\hat{\mathbf{x}}_{\mathcal{R}}(0) = \left[2.675\ 276.786\ 59.09\right]^T$ and $\hat{\mathbf{x}}_{\tilde{\mathcal{R}}}(0) = \left[2.659\ 281.021\ 58.691\right]^T$ for the nominal and robust cases, respectively. The following set of eigenvalues were chosen for the ELO design: -0.9, -1.05, and -1.1.

The state, state estimate, and input profiles from the closed-loop simulations for both MPC designs are presented in Fig. 4.1. For the nominal case, $\mathbf{x}_{\mathcal{R}}(t_f) = \left[0.185\ 466.448\ 64.212\right]^T$, which does not reside in $\mathcal{B}(\mathbf{x}_d)$ as $\|\tilde{\mathbf{x}}(t_f) - \mathbf{c}_0\|_{\mathbf{P}_0} = 82.158$, whereas the robust MPC scheme was able to steer the system inside $\mathcal{B}(\mathbf{x}_d)$ as $\mathbf{x}_{\tilde{\mathcal{R}}}(t_f) = \left[0.0992\ 464.835\ 64.817\right]^T$ and $\|\tilde{\mathbf{x}}(t_f) - \tilde{\mathbf{c}}_0\|_{\tilde{\mathbf{P}}_0} = 0.488$. These results indicate the practical importance of explicitly accounting for uncertainty in the controller design. Using the Matlab functions, `tic` and `toc`, the average CPU time required per MPC calculation on an Intel Quad Core machine for the nominal and robust RTRR-based MPC schemes were 0.146 and 1.507 s, respectively. The higher CPU time for the robust RTRR-based design is certainly expected as the MPC optimization problem is more complex; however, the computational trade-offs are well worth the savings acquired from achieving a batch quality within the desired neighborhood.

4.5.2 Robust Safe-Steering of the Fed-Batch Process

To demonstrate the effectiveness of the proposed safe-steering framework for batch processes with model uncertainty, we consider a fault in one of the control actuators in the fed-batch process. Specifically, we consider the scenario where at $t^{\mathrm{fault}} = 0.25$ h, the actuator associated with the heating coil temperature fails. During the failure period, the heating coil can only supply limited heat to the reactor and the heating coil temperature becomes restricted to $285 \leq T_a \leq 300$ K. At $t^{\mathrm{repair}} = 0.32$ h, the fault is rectified and the full control effort is recovered. For the end-point-based predictive controller, the following objective function was used.

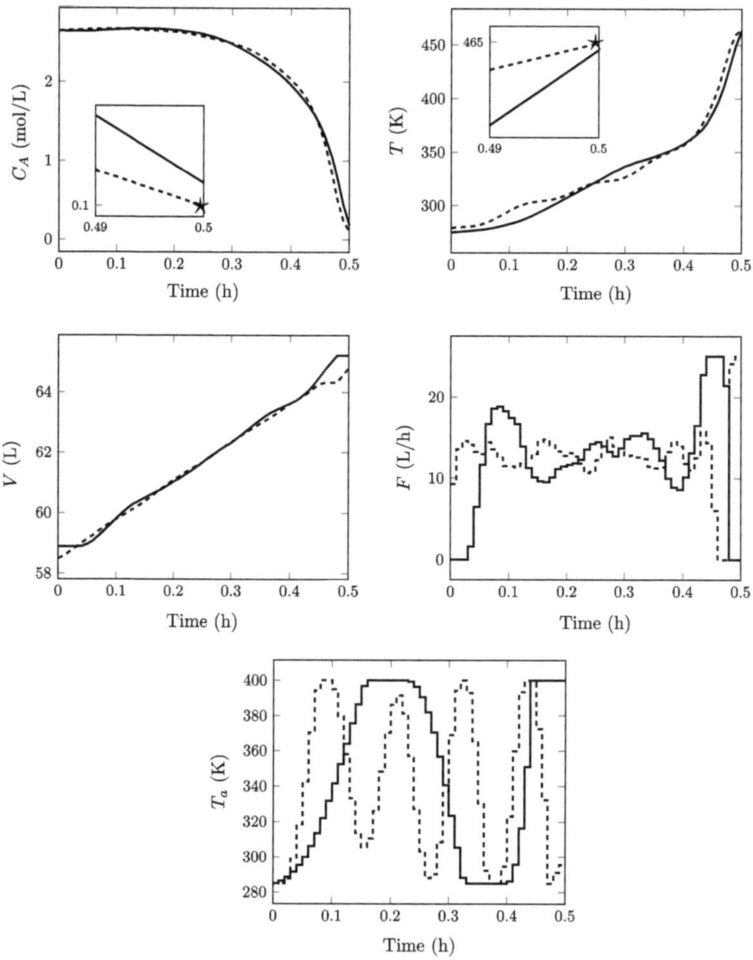

Fig. 4.1 State and input profiles of the fed-batch reactor system under the nominal (solid) and robust (dashed) RTRR-based MPC designs during a fault-free batch. The insets show the ability of the proposed robust MPC formulation to drive the states inside a desired neighborhood of $\mathbf{x}_d$. The $\star$ denotes the desired end-point variable value

$$J_{\mathcal{E}} = \|\tilde{\mathbf{x}}(t_f) - \tilde{\mathbf{c}}_0\|_{\tilde{\mathbf{P}}_0} + \int_t^{t_f} \|\Delta\mathbf{u}\|_{\mathbf{R}} \mathrm{d}t \tag{4.23}$$

The objective function in (4.22) was used for the robust RTRR-based MPC design. The move suppression matrix, initial conditions, and initial state estimates were identical in the simulations for both designs. Specifically, we set: $\mathbf{R} = \text{diag}\left\{5 \times 10^{-5}, 5 \times 10^{-6}\right\}$, $\mathbf{x}(0) = \left[2.651\ 282.712\ 58.493\right]^T$ and

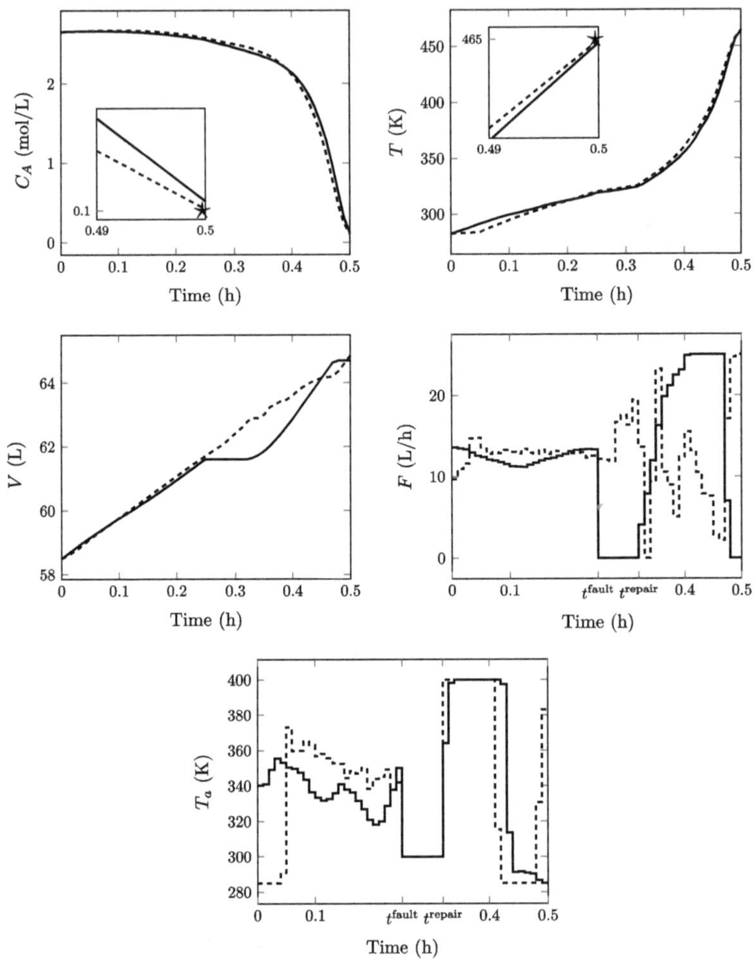

Fig. 4.2 State input profiles of the fed-batch reactor system under the end-point (solid) and robust RTRR (dashed)-based MPC designs with input failure of T_a between 0.25 and 0.32 h. The insets show the ability of the proposed safe-steering framework to drive the states inside a desired neighborhood of $\mathbf{x}_d$. The ⋆ denotes the desired end-point variable value

$\hat{\mathbf{x}}(0) = \left[2.66\ 284.337\ 58.701\right]^T$. The actual state and state estimates were verified to be within $\tilde{\mathcal{R}}(0)$. The state and input profiles for the two MPC schemes are given in Fig. 4.2. For the end-point-based MPC design, the batch is driven to the end-point of $\mathbf{x}_{\mathcal{E}}(t_f) = \left[0.121\ 464.659\ 64.686\right]^T$, which is outside of the desired neighborhood as $\|\mathbf{x}_{\mathcal{E}}(t_f) - \mathbf{c}_0\|_{\mathbf{P}_0} = 5.994$.

During the failure period, the heating coil temperature in both designs are prescribed to be 300 K while the flow rate trajectories are markedly different. Using only the flow rate, the end-point MPC design was unable to compute a trajectory that steered the system inside the desired neighborhood. Due to the repeated application of a truncated version of these poor input trajectories, the system is driven to a point by t^{repair} from where it cannot be steered inside $\mathcal{B}(\mathbf{x}_d)$ even after fault recovery. In contrast, the robust RTRR MPC design prescribed flow rates during the failure period which maintained the process states from where the batch could be driven inside the desired neighborhood upon recovery of the full control effort. The batch is ultimately steered to the end-point of $\mathbf{x}_{\tilde{\mathcal{R}}}(t_f) = \left[0.103\ 464.882\ 64.854\right]^T$, which is inside the desired neighborhood as $\|\mathbf{x}_{\tilde{\mathcal{R}}}(t_f) - \mathbf{c}_0\|_{\mathbf{P}_0} = 0.380$.

We also note the total CPU simulation time required for the robust RTRR-based MPC was 2.126 min, which was significantly shorter compared to the end-point-based MPC simulation time of 2.318 h. Closed-loop simulations of batch systems with a higher number of states or inputs would exhibit an even more substantial difference in the simulation CPU time. Moreover, with additional parameter uncertainties, wider uncertainty ranges, and the introduction of disturbances into the system, the end-point-based MPC would require more computational time because the solution at one time step would become a poorer initial guess for the next time step. On the other hand, because the robust RTRR-based MPC formulation accounts for the presence of the uncertainty and its bounds in off-line calculations (which would certainly increase), the CPU time for online control calculations would not increase significantly.

4.6 Conclusions

In this chapter, we first considered the control of batch processes subject to input constraints and model uncertainty with the objective of reaching a desired end-point neighborhood. To this end, a computationally efficient non-linear robust MPC scheme based on robust RTRRs was formulated. Prior to the MPC formulation, a multi-level optimization-based algorithm was established to generate robust RTRRs for specified uncertainty bounds, sampling period, and desired end-point neighborhood. Following the controller design, we considered the problem of uncertain batch processes subject to finite duration faults in the control actuators that cannot be handled via robust control approaches. Using the robust RTRR-based MPC as the main tool, the robust safe-steering framework was developed to address the problem of how to operate the functioning inputs during the fault repair period to ensure that the process can be driven inside a desired end-point neighborhood upon recovery of the full control effort. The computational efficiency and control performance of the robust RTRR-based MPC and its usefulness in the robust safe-steering framework were demonstrated using simulations of a fed-batch process example.

4.7 Appendix

Proof of Theorem 4.3: Consider an element $\mathbf{x}_{k-1,\text{nom}}$ in the interior of $\tilde{\mathcal{R}}_{k-1}$ (i.e., there exists a ρ^* such that $\mathcal{I}_{k-1} = \{\mathbf{y} \mid \|\mathbf{y} - \mathbf{x}_{k-1}\| \le \rho^*\} \subset \tilde{\mathcal{R}}_{k-1}$) and the point $\mathbf{x}_{k,\text{nom}}$ given by:

$$\mathbf{x}_{k,\text{nom}} = \mathbf{x}_{k-1,\text{nom}} + \int_t^{t+\delta} -\mathbf{f}(\mathbf{x}(t), \mathbf{u}_{\text{nom}}, \boldsymbol{\theta}_{\text{nom}})\mathrm{d}t \tag{4.24}$$

where δ is to be determined and $\mathbf{u}_{\text{nom}}$ and $\boldsymbol{\theta}_{\text{nom}}$ are nominal values of the input and uncertainty, respectively. It follows that

$$\mathbf{x}_{k-1,\text{nom}} = \mathbf{x}_{k,\text{nom}} + \int_t^{t+\delta} \mathbf{f}(\mathbf{x}(t), \mathbf{u}_{\text{nom}}, \boldsymbol{\theta}_{\text{nom}})\mathrm{d}t \tag{4.25}$$

Define:

$$\mathbf{x}_{k-1} = \mathbf{x}_{k,\text{nom}} + \int_t^{t+\delta} \mathbf{f}(\mathbf{x}(t), \mathbf{u}_{\text{nom}}, \boldsymbol{\theta}(t))\mathrm{d}t \tag{4.26}$$

i.e., $\mathbf{x}_{k-1}$ is the state of the system at $t + \delta$ starting at t from $\mathbf{x}_{k,\text{nom}}$ subject to the true value of the uncertain variable. From the continuity of $\mathbf{f}(\mathbf{x}, \mathbf{u}, \boldsymbol{\theta})$ on $(\mathbf{x}, \mathbf{u}, \boldsymbol{\theta})$ and that it is locally Lipschitz in $\mathbf{x}$ on $\mathcal{D} \times \mathbf{U} \times \boldsymbol{\Theta}$, the continuity of solutions of $\dot{\mathbf{x}}(t) = \mathbf{f}(\mathbf{x}(t), \mathbf{u}(t), \boldsymbol{\theta}(t))$ from $\mathbf{x}_k$ with respect to parameters (and therefore, the uncertain variable) follows (Theorem 3.5 in [33]). From the proof of Theorem 3.5 in [33], it follows that given a desired bound on the discrepancy between the evolution of the nominal and perturbed system (i.e., $\|\mathbf{x}_{k-1} - \mathbf{x}_{k-1,\text{nom}}\| \le \rho^*$), there exists a value δ^* such that if the sampling period, $\delta \le \delta^*$ then it is guaranteed that $\|\mathbf{x}_{k-1} - \mathbf{x}_{k-1,\text{nom}}\| \le \rho^*$. Therefore, $\mathbf{x}_{k,\text{nom}}$ is an element of $\tilde{\mathcal{R}}_k$ showing $\tilde{\mathcal{R}}_k \ne \emptyset$. This completes the proof of Theorem 4.3. ■

Sketch of the Proof of Theorem 4.4: The sufficiency of the condition in Theorem 4.4 can be shown by considering any $\mathbf{x}(t_0) \in \tilde{\mathcal{R}}(t_0)$. From the properties of the generation algorithm for $\tilde{\mathcal{R}}(t)$, there exists a control move that takes the states inside $\tilde{\mathcal{R}}(t + \delta)$ in a time δ. Repeating this for the duration of the batch implies that the state is driven to $\mathcal{B}(\mathbf{x}_d)$ by t_f for all possible realizations of the uncertainty. In essence, this implies that there always exists a feasible solution to the MPC law of Eqs. (4.12)–(4.17) for the entire duration of the batch. The bi-level optimization problem defining the MPC law in Eqs. (4.12)–(4.17) remains feasible for all $t \in [t_0, t_f]$ and $\mathbf{x}(t_f) \in B(\mathbf{x}_d)$. This completes the sketch of the proof of Theorem 4.4. ■

Proof of Theorem 4.5: The proof of this theorem follows from Theorem 4.4. Equating t^{repair} to t_0 results in the satisfaction of the requirements of Theorem 4.4 and therefore, the bi-level MPC optimization problem of Eqs. (4.12)–(4.17) continues to remain feasible $\forall t \in [t^{\text{repair}}, t_f]$ and $\mathbf{x}(t_f) \in \mathcal{B}(\mathbf{x}_d)$ follows. This completes the proof of Theorem 4.5. ■

References

1. Cruickshank, S.M., Daugulis, A.J., McLellan, P.J.: Dynamic modeling and optimal fed-batch feeding strategies for a two-phase partitioning bioreactor. Biotech. Bioeng. **67**, 224–233 (2000)
2. Dadebo, S.A., McAuley, K.B.: Dynamic optimization of constrained chemical engineering problems using dynamic programming. Comput. Chem. Eng. **19**, 513–525 (1995)
3. Zhang, G.P., Rohani, S.: On-line optimal control of a seeded batch cooling crystallizer. Chem. Eng. Sci. **58**, 1887–1896 (2003)
4. Soroush, M., Kravaris, C.: Optimal-design and operation of batch reactors. 1. Theoretical framework. Ind. Eng. Chem. Res. **32**, 866–881 (1993)
5. Soroush, M., Kravaris, C.: Optimal-design and operation of batch reactors. 2. A case-study. Ind. Eng. Chem. Res. **32**, 882–893 (1993)
6. Flores-Cerrillo, J., MacGregor, J.F.: Latent variable MPC for trajectory tracking in batch processes. J. Process Control **15**, 651–663 (2005)
7. Valappil, J., Georgakis, C.: State estimation and nonlinear model predictive control of end-use properties in batch reactors. In: Proceedings of the 2001 American Control Conference, vol. 2, pp. 999–1004 (2001) (Arlington, VA)
8. Alamir, M., Balloul, I.: Robust constrained control algorithm for general batch processes. Int. J. Control **72**, 1271–1287 (1999)
9. Mayne, D.Q.: Non-linear model predictive control: challenges and opportunities. In: Allgower, F., Zheng, A. (eds.) Non-Linear Model Predictive Control, pp. 23–44 (2000)
10. Crowley, T., Choi, K.Y.: Experimental studies on optimal molecular weight distribution control in a batch-free radical polymerization processes. Chem. Eng. Sci. **53**, 2769–2790 (1998)
11. Dimitratos, J., Georgakis, C., El-Aasser, M.S., Klein, A.: Dynamic modeling and state estimation for an emulsion copolymerization reactor. Comput. Chem. Eng. **13**, 21–33 (1989)
12. Kozub, D., Macgregor, J.F.: Feedback control of polymer quality in semi-batch copolymerization reactors. Chem. Eng. Sci. **47**(4), 929–942 (1992)
13. De Valliere, P., Bonvin, D.: Application of estimation techniques to batch reactors. II. Experimental studies in state and parameter estimation. Comput. Chem. Eng. **13**, 11–20 (1989)
14. Hua, X.M., Rohani, S., Jutan, A.: Cascade closed-loop optimization and control of batch reactors. Chem. Eng. Sci. **59**, 5695–5708 (2004)
15. Shi, D., Mhaskar, P., El-Farra, N.H., Christofides, P.D.: Predictive control of crystal size distribution in protein crystallization. Nanotechnology **16**(7), 562–574 (2005)
16. Shi, D., El-Farra, N.H., Li, M., Mhaskar, P., Christofides, P.D.: Predictive control of particle size distribution in particulate processes. Chem. Eng. Sci. **61**, 268–281 (2005)
17. Sheikhzadeh, M., Trifkovic, M., Rohani, S.: Real-time optimal control of an anti-solvent isothermal semi-batch crystallization process. Chem. Eng. Sci. **63**, 829–839 (2008)
18. Palanki, S., Vemuri, J.: Optimal operation of semi-batch processes with a single reaction. Int. J. Chem. React. Eng. **3** (2005)
19. Pistikopoulos, E.N., Dua, V., Bozinis, N.A., Bemporad, A., Morari, M.: On-line optimization via off-line parametric optimization tools. Comput. Chem. Eng. **26**, 175–185 (2002)
20. Huynh, N., Kazantzis, N.: Parametric optimization of digitally controlled nonlinear reactor dynamics using Zubov-like functional equations. J. Math. Chem. **38**, 499–519 (2005)
21. Mhaskar, P., El-Farra, N.H., Christofides, P.D.: Predictive control of switched nonlinear systems with scheduled mode transitions. IEEE Trans. Autom. Control **50**, 1670–1680 (2005)
22. Mhaskar, P.: Robust model predictive control design for fault-tolerant control of process systems. Ind. Eng. Chem. Res. **45**, 8565–8574 (2006)
23. Mahmood, M., Gandhi, R., Mhaskar, P.: Safe-parking of nonlinear process systems: handling uncertainty and unavailability of measurements. Chem. Eng. Sci. **63**, 5434–5446 (2008)
24. Aumi, S., Mhaskar, P.: Safe-steering of batch processes. AIChE J. **55**, 2861–2872 (2009)
25. Nomikos, P., Macgregor, J.F.: Monitoring batch processes using multiway principal component analysis. AIChE J. **40**, 1361–1375 (1994)
26. Cinar, A., Parulekar, S.J., Undey, C., Birol, G.: Batch Fermentation: Modeling: Monitoring, and Control. CRC Press, New York (2003)

27. Undey, C., Tatara, E., Williams, B.A., Birol, G., Cinar, A.: A hybrid supervisory knowledge-based system for monitoring penicillin fermentation. In: Proceedings of American Control Conference, Chicago, Illinois, vol. 6, pp. 3944–3948 (2000)
28. Undey, C., Tatara, E., Williams, B.A., Birol, G., Cinar, A.: On-line real-time monitoring of penicillin fermentation. In: International Symposium on Advanced Control of Chemical Processes, Pisa, Italy, vol. 6, pp. 243–248 (2000)
29. Undey, C., Cinar, A.: Statistical monitoring of multiphase, multistage batch processes. IEEE Control Syst. Mag. **22**, 40–52 (2002)
30. Mhaskar, P., McFall, C., Gani, A., Christofides, P.D., Davis, J.F.: Isolation and handling of actuator faults in nonlinear systems. Automatica **44**, 53–62 (2008)
31. Gandhi, R., Mhaskar, P.: Safe-parking of nonlinear process systems. Comput. Chem. Eng. **32**, 2113–2122 (2008)
32. Terwiesch, P., Agarwal, M., Rippin, D.W.T.: Batch unit optimization with imperfect modelling: a survey. J. Process Control **4**, 238–258 (1994)
33. Khalil, H.K.: Nonlinear Systems, 3rd edn. Prentice Hall, New York (2002)

Chapter 5
Transition From Batch to Continuous Operation in Bio-reactors: A Model Predictive Control Approach and Application

5.1 Introduction

Biotechnology, combined with other fields such as molecular biology, genetics, biochemistry, and chemical technology, is increasingly fostering applications in the chemical industry (e.g., environmentally friendly production of ethanol), food industry (e.g., citric acid production), and medicine (e.g., antibiotics). The microorganisms used in manufacturing these products work on a specific substrate in bio-reactors (also called fermenters) from where the final product is obtained via separation technologies. The increasing importance of bio-reactors and the presence of complex dynamic phenomenon have motivated numerous studies focusing on modeling, analysis, monitoring and control of bio-reactors (e.g., see [1–10]).

The microbial production of the biofuel ethanol, an alternative, renewable fuel with great potential, has acquired significance in recent times due to declining fossil fuels and the global strive to become energy independent. *Zymomonas mobilis* is increasingly being promoted as a suitable microorganism for the industrial production of ethanol [3, 11] in bio-reactors. The operation of the bio-reactor in continuous mode, or as a chemostat, is economically efficient from the viewpoint of production and is the most frequent means for large-scale production of biochemicals. Before the bio-reactor can be run in continuous mode, the microorganism culture needs to be grown in rich substrate conditions in a batch mode. Upon transition to continuous mode, inherent complexities of the chemical reactions that drive the bio-reactors (owing to the nonlinear dependence of growth rates on substrate concentrations and bio-reactor conditions such as temperatures and pH) result in a host of dynamic characteristics including unstable steady states and oscillations [11]. Furthermore, under continuous operation, the process has to be operated in the presence of constraints which originate from expression of performance criteria (requiring product concentration, for instance, to be above a prescribed value) or from physical limitations on the control actuators (such as valves or pumps). The presence of constraints, however, limits the set of initial conditions starting from where closed-loop stability can be achieved and this fact needs to be accounted for in not only determining

P. Mhaskar et al., *Modeling and Control of Batch Processes*,
Advances in Industrial Control, https://doi.org/10.1007/978-3-030-04140-3_5

the appropriate switching time from batch to continuous mode, but in the choice of control design under continuous operation as well.

One control method well suited to handling constraints within an optimal control framework is model predictive control (MPC). In MPC, a control trajectory is computed by solving an optimization problem that minimizes a prescribed performance objective over a finite horizon (predicted using a process model) while satisfying constraints on the input and state variables. To imbue the control design with the desirable features of feedback control, only the first 'piece' of the control trajectory is implemented on the system, and the control trajectory is recomputed using the next available state measurement. Several research efforts have focused on the design and stability characterization of model predictive controllers (see, e.g., [12–17] for an excellent review). In these predictive control approaches, explicit stability constraints are incorporated which require the process state to go to the desired equilibrium point or some neighborhood of the equilibrium point by the end of the horizon. The stability guarantees, however, are based on an assumption of initial feasibility of the optimization problem. More recently, Lyapunov-based predictive controllers were designed that guarantee feasibility from an explicitly characterized set of initial conditions subject to input [18], input and state constraints [19] and uncertainty [20] (see [21] for further results and references). The extensive results in model predictive control notwithstanding, application of model predictive control approaches to control of bio-reactors have received limited attention, and the existing applications (e.g., see [22]) do not explicitly account for the process nonlinearity or provide an explicit characterization of the set of initial conditions starting from where the bio-reactor can be stabilized.

In contrast to continuous operation of bio-reactors, where the control objective is stabilization, the natural control objective during the batch run (preparatory to the operation of a bio-reactor in continuous mode) is achieving desired targets at the end of the batch. When the product is generated using only the batch mode, the desired product properties are utilized to formulate targets at the end of the batch, or in describing optimal trajectory criteria. Several research efforts have focused on developing control designs that accomplish this objective (see, e.g., [23–25]). However, when the process is being run in batch mode only to 'prepare' for an eventual transition to continuous mode of operation, then it is necessary to account for the stability properties under continuous operation in formulating the operation criteria for the batch mode. Transitioning from batch mode to continuous mode at an inappropriate time, or not accounting for the presence of constraints under continuous operation may result in instability and loss of product under continuous operation. Motivated by these considerations, this chapter considers the production of ethanol in bio-reactors and addresses the problem of determining the time of transition from batch to continuous operation in a way that ensures stability under continuous mode of operation subject to input constraints. To this end, first a nonlinear model predictive control design is utilized that stabilizes the bio-reactor under continuous mode of operation. The key property of the Lyapunov-based model predictive control design is that it provides guaranteed stability from an explicitly characterized set of initial conditions. Additional constraints are incorporated in the predictive control design that expand on the set of initial conditions that can be stabilized by control designs that only require the value of the Lyapunov function to decay. Then, the explicit characterization of the set of stabilizable initial conditions under continuous mode of operation is used in designing the batch mode of operation. Specifically, a predictive control approach is used in determining the batch duration that results in stabilizable values of the state variables at the end of the batch run. An application of the proposed method is demonstrated on an ethanol production process. A dynamic model of the bio-reactor for producing ethanol, developed in [11], is used to demonstrate the application of the proposed method.

5.2 Process Description and Modeling

A model describing the fermentation of glucose to ethanol by *Zymomonas mobilis* in a continuously stirred fermenter can be described by the following set of four differential equations [11]:

$$
\begin{aligned}
\frac{d[C_s]}{dt} &= -\frac{1}{Y_{SX}}\left\{\frac{C_s C_e}{(K_s + C_s)}\right\} - m_S C_X + D(C_{SO} - C_S)\\
\frac{d[C_X]}{dt} &= \left\{\frac{C_s C_e}{(K_s + C_s)}\right\} + D(C_{XO} - C_X)\\
\frac{d[C_e]}{dt} &= (k_1 - k_2 C_p + k_3 C_p^2)\left\{\frac{C_s C_e}{(K_s + C_s)}\right\} + D(C_{eO} - C_e)\\
\frac{d[C_P]}{dt} &= \frac{1}{Y_{PX}}\left\{\frac{C_s C_e}{(K_s + C_s)}\right\} + m_P C_X + D(C_{PO} - C_P)
\end{aligned}
\tag{5.1}
$$

where C_s, C_X, C_e, and C_P denote the substrate (glucose), biomass, inhibitor and product (ethanol) concentrations in the reactor, respectively, D denotes the dilution rate, and C_{S0} denotes the inlet substrate concentration. The cellular component, C_e, is an internal key component that is required for biomass (cell) growth and product formation [11]. The inhibitory effect of ethanol is modeled not through a direct influence on the specific growth rate but indirectly through this component, C_e. The values of the parameters, taken from [11], are given in Table 5.1. The evolution of the bio-reactor during the batch mode of operation is simulated by setting $D = 0$. Under continuous operation, the open-loop system behaves in an oscillatory fashion (i.e., the system possesses an open-loop unstable steady state surrounded by a limit cycle). The control objective is to stabilize the bio-reactor by using the inlet glucose concentration C_{S0} as the manipulated input subject to the constraints $100\ g/l \leq C_{S0} \leq 180\ g/l$, and by appropriately executing the transition from batch to continuous operation.

To facilitate controller design and presentation of our results in a compact form, we first write the model of Eq. (5.1) in a deviation (from the operating unstable steady-state) variable form, by defining $x = [x_1\ x_2\ x_3\ x_4]^T$ where $x_1 = C_s - C_{s,\,nom}$, $x_2 = C_X - C_{X,\,nom}$, $x_3 = C_e - C_{e,\,nom}$, $x_4 = C_P - C_{P,\,nom}$, $u = C_{S0} - C_{S0,\,nom}$ and obtain a continuous-time nonlinear system with the following state-space description:

$$\dot{x}(t) = f_k(x(t)) + g_k(x(t))u_k \tag{5.2}$$

$$|u_k| \leq u_k^{max} \tag{5.3}$$

where $x(t) \in \mathbb{R}^4$ denotes the vector of process state variables and $u(t) \in [-u^{max}, u^{max}] \subset \mathbb{R}$ denotes the constrained manipulated input associated with the kth mode of operation. $k(t)$, which takes values in the finite index set K, represents a discrete state that indexes the vector fields $f(\cdot)$, $g(\cdot)$ as well as the manipulated input $u(\cdot)$, and represents the mode of operation ($k = 1$ implies batch mode and $k = 2$ implies continuous mode). The notation $L_f h$ denotes the standard Lie derivative of a scalar function $h(\cdot)$ with respect to the vector function $f(\cdot)$. The explicit form of the vector fields $f_k(x(t))$ and $g_k(x(t))$ can be obtained by comparing Eqs. (5.1) and (5.3) and is omitted for brevity. Switching between the two modes occurs at a time T_{batch} determined by the supervisor.

The presence of constraints on the manipulated input implies that for the continuous mode, only a finite set of initial conditions (the so-called null controllable region) can be stabilized to the desired equilibrium point. Specific control designs typically stabilize a subset of the null controllable region, and knowledge of this set can be utilized in appropriately transiting from batch to continuous mode of operation. In the remainder of the chapter, we first present a predictive control design that allows for an explicit characterization of its stability region. That is, it allows for determining, before implementation, the set of initial conditions that can be stabilized subject to input constrains. This characterization is utilized in determining the appropriate time for transition from the batch mode to continuous mode of operation in a way that closed-loop stability is guaranteed. To simplify the presentation of our results, we will focus only on the state feedback problem where measurements of all process states are available for all times.

5.3 Predictive Control Design for Continuous Operation

In this section, the main steps involved in the predictive controller design are outlined. These include: (a) formulating appropriate stability constraints in the control problem and (b) obtaining an explicit characterization of the set of initial conditions starting from where the closed-loop system is guaranteed to be stable. The control objective under continuous operation is to stabilize the process of Eq. (5.3) in the presence of constraints on the manipulated input.

(a) *Predictive controller formulation*

We present a Lyapunov-based predictive controller design that allows for an explicit characterization of the set of initial conditions starting from where the predictive controller is guaranteed to be feasible (and therefore stabilizing) [18, 19].

Additionally, the predictive controller utilized in this chapter incorporates constraints that go beyond those employed in [18, 19] that only require the Lyapunov function value to decay, thereby enhancing the set of initial conditions starting from where closed-loop stability is achieved. In this chapter, we demonstrate an application of the predictive controller formulation given by:

$$u = argmin\{J(x, t, u(\cdot))|u(\cdot) \in S\} \tag{5.4}$$

$$s.t. \;\; \dot{x} = f_2(x) + g_2(x)u \tag{5.5}$$

$$\dot{V}(x(\tau)) \le -\epsilon^* \;\; \forall\, \tau \in [t, t+\Delta) \; if \;\; V(x(t)) > \delta^{'} \tag{5.6}$$

$$V(x(\tau)) \le \delta^{'} \; \forall\, \tau \in [t, t+\Delta) \;\; if \;\; V(x(t)) \le \delta^{'} \tag{5.7}$$

$$\begin{aligned} u^*(x(t+\Delta)) &\le u^*(x(t)) \;\; if \;\; V(x(t)) > c^{max} \; and \; u^*(x(t)) > 0 \\ u^*(x(t+\Delta)) &\le u^{max} \;\; if \;\; V(x(t)) > c^{max} \; and \; u^*(x(t)) = 0 \end{aligned} \tag{5.8}$$

where $u^*(x(t))$ denotes the smallest nonnegative number for which $L_{f_2}V(x(t)) \le |L_{g_2}V(x)(t))|u^*$. Note that $\dot{V}(x(t)) = L_{f_2}V(x) + L_{g_2}V(x)u$, therefore if $L_{f_2}V(x(t)) \le |L_{g_2}V(x)(t))|u^*$, it implies that for $u = -u^*sgn(L_{g_2}V(x))$ where $sgn(x) = 1$ if $x \ge 0$ and $sgn(x) = -1$ if $x < 0$, $\dot{V}(x(t)) = L_{f_2}V(x) + L_{g_2}V(x)u \le 0$. $u^*(x(t))$ thus represents the smallest magnitude of the control action that can make the Lyapunov function derivative negative. $S = S(t, T)$ is the family of piecewise continuous functions (functions continuous from the right), with period Δ, mapping $[t, t+T]$ into U and T is the horizon. Equation (5.5) is the model describing the time evolution of the state x under continuous operation, V is a control Lyapunov function (CLF), and $\delta^{'}$, $\epsilon^* > 0$ are design parameters. A control $u(\cdot)$ in S is characterized by the sequence $\{u[j]\}$ where $u[j] := u(j\Delta)$ and satisfies $u(t) = u[j]$ for all $t \in [j\Delta, (j+1)\Delta)$. The performance index is given by

$$J(x, t, u(\cdot)) = \int_t^{t+T} \left[\|x^u(s; x, t)\|_Q^2 + \|u(s)\|_R^2\right] ds + vV(x(t+\Delta)) \tag{5.9}$$

where Q is a positive semi-definite symmetric matrix, R is a strictly positive definite symmetric matrix, and $v > 0$. $x^u(s; x, t)$ denotes the solution of Eq. (5.3), due to control u, with initial state x at time t. The minimizing control $u^0(\cdot) \in S$ is then applied to the process over the interval $[t, t+\Delta)$ and the procedure is repeated indefinitely.

The meaning and implication of the constraints of Eqs. (5.6)–(5.8) are as follows: The constraint of Eq. (5.6) requires the control action to enforce a decay in the value of the Lyapunov function value over the next time interval; because of the discrete nature (implement and hold) of the control action, such decay may not be achievable for all state values and is only requested to drive the process to a desired neighborhood of the origin defined by $V(x) \le \delta^{'}$. Once the process reaches the desired neighborhood of the origin, the constraint of Eq. (5.7) prevents the state from escaping that neighborhood. For initial conditions within an invariant set Ω (as defined below), a continuous decay in the Lyapunov function value is sufficient to drive the state to the desired neighborhood of the origin (see [18, 19] for a rigorous formulation of the stability properties and the proof). For initial conditions outside the set Ω, the constraint of Eq. (5.8) essentially asks for the control action to be computed such that for the process state at the next time instant, negative definiteness of $\dot{V}$ can be achieved by using less control action than was achievable before (in the case that some control action was required to achieve negative definiteness of $\dot{V}$), or that negative definiteness of $\dot{V}$ can be achieved using some allowable control action (in the case that no control action was required to achieve negative definiteness of $\dot{V}$). To understand this better, recall that the evolution of the Lyapunov function is described by the equation $\dot{V} = L_f V + L_g V u$. When $L_f V \geq 0$, some minimum control effort is required to ensure negative definiteness of $\dot{V}$. In such cases, when $L_g V \neq 0$, the control action can be utilized to achieve negative definiteness of $\dot{V}$, and the smallest magnitude of the control action that can achieve negative definiteness of $\dot{V}$ is denoted by u^*. (The Lyapunov function being a CLF precludes the possibility that $L_f V > 0$ and $L_g V = 0$.) If $L_f V < 0$, this means that even if no control action were to be implemented, the Lyapunov function would decay 'by itself', yielding $u^* = 0$. In this case, the constraint simply requires that at the next time instance, negative definiteness of $\dot{V}$ be achievable using a control action that is within the constraints.

(b) *Characterization of stability properties*

For the predictive controller above, using an auxiliary control design of the form ([26], see also [27, 28]):

$$u_{bc} = -\left(\frac{\alpha(x) + \sqrt{(\alpha(x))^2 + (u_{max}\beta(x))^4}}{(\beta(x))^2\left[1 + \sqrt{1 + (u_{max}\beta(x))^2}\right]}\right)(L_g V)^T \tag{5.10}$$

when $L_g V \neq 0$ and $u_{bc} = 0$ when $L_g V = 0$, where $\alpha(x) = L_f V + (\rho\|x\|)$, $\beta(x) = \|(L_G V)^T\|$ and $\rho > 0$, it can be shown that given the desired neighborhood of the equilibrium point (characterized by δ'), there exist parameters Δ (the implement and hold time), and ϵ (the requested rate of decay of the Lyapunov function) that the optimization problem is guaranteed to be feasible from an explicitly characterized set of initial conditions denoted by $\Omega := \{x \in \mathbb{R}^n | V(x) \leq c^{max}, L_f V(x) \leq \|L_g V\| u^{max}\}$ (see [18] for the proof). Furthermore, for initial conditions inside Ω, the control action prescribed by the control law of Eq. (5.10) can be used as an initial guess to the optimization problem (it can be shown that this initial guess is guaranteed to be feasible). For initial conditions outside Ω, the constraint of Eq. (5.8), if feasible, serves to expand on the set of initial conditions starting from where closed-loop stability is achieved. Specifically, the satisfaction of the constraint of Eq. (5.8) could result in stabilization of initial conditions outside Ω, which would otherwise not be stabilized in the absence of the constraint of Eq. (5.8).

Remark 5.1 In contrast to other predictive control designs, where the stability of the closed-loop system depends on an assumption of initial feasibility of the optimization problem, the predictive controller of Eqs. (5.4)–(5.8) is guaranteed to be feasible (and therefore stabilizing) from an explicitly characterized set of initial conditions. This key property is achieved due to (1) requesting a decay in the value of the Lyapunov function only during the next hold time (instead of requiring the Lyapunov function to decay at the end of the horizon, or for the state to go to some invariant neighborhood by the end of the horizon) and (2) using an auxiliary bounded control approach [26–28] to ascertain that a feasible solution to the optimization problem exists. Specifically, not requiring the states to go to some neighborhood by the end of the horizon takes away the dependence of feasibility of the constraint on having a 'large enough' horizon (while still ensuring closed-loop stability). The auxiliary control approaches serve to determine the set of initial conditions starting from where negative definiteness of the Lyapunov function derivative can be guaranteed using some allowable control action, thereby guaranteeing feasibility of the optimization problem.

Remark 5.2 Note that the problem of designing control laws that guarantee stability in the presence of input constraints has been extensively studied within the framework of nonlinear control (see, for example, [26, 28–30]). The bounded controller design proposed in [28] (inspired by the results on bounded control in [26]) is an example of a Lyapunov-based controller design that (1) guarantees stability in the presence of constraints and (2) allows for an explicit characterization of the closed-loop stability region. Other control designs that allow for an explicit characterization of the stability region can be used to provide a feasible initial guess to the optimization problem as well. Furthermore, referring to the control law of Eq. (5.10) (and this holds for other Lyapunov-based control laws as well), it is important to note that a general procedure for the construction of CLFs for nonlinear systems of the form of Eq. (5.3) is currently not available. Yet, for several classes of nonlinear systems that arise commonly in the modeling of engineering applications, it is possible to use suitable approximations (e.g., [31]) or exploit system structure to construct CLFs. For example, for feedback linearizable systems, quadratic Lyapunov functions can be chosen as candidate CLFs and can be made CLFs with appropriate choice of the function parameters based on the process parameters (see, for example, [32]). Also, for nonlinear systems in strict-feedback form, backstepping techniques can be employed for the construction of CLFs [32].

Remark 5.3 Note that if one were to use a bounded control design to achieve stability and characterize the stability region, one would also get the stability guarantees from an explicitly characterized set of initial conditions. The bounded control designs, however, do not possess a framework for incorporating performance considerations in computing the control action. In contrast, the predictive control approach allows for incorporating constraints (to achieve stability) while also specifying performance considerations through the objective function. More importantly, the bounded controllers are designed to only achieve negative definiteness of $\dot{V}$. For initial conditions within Ω, decreasing value of the Lyapunov function ensures that the process stays within Ω and is stabilized. For initial conditions outside Ω, while initially a decrease in the Lyapunov function value may be achieved, the state trajectory may drift out of the set of initial conditions from where $\dot{V} < 0$ is achievable, and once again, the bounded control designs do not have a mechanism to prevent this from happening. The predictive control approach allows for specifying meaningful stability constraints for initial conditions outside Ω, which, if feasible, drive the state trajectory so that successive decay in the value of the Lyapunov function is achievable, thereby expanding on the set of initial conditions from where closed-loop stability is achievable.

5.4 Determining Batch to Continuous Transition Time

The characterization of the set of initial conditions starting from where closed-loop stability can be achieved under continuous operation provides a natural target for batch operation. Depending on the specific bio-reactor in consideration, the value of the process states at the end of the batch run can be influenced by various parameters such as the initial conditions for the batch run, process conditions during the batch run (such as temperature or pH), and the batch duration itself. When it is possible to influence the process trajectory via a manipulated input over the batch duration, such a manipulated input can be utilized in driving the state trajectory at the end of the batch run inside the stability region. Furthermore, when such a manipulated input is available, the state trajectory during the batch operation can be made to follow a desired or optimum trajectory. In the absence of such a manipulated input, as in the case with the bio-reactor under consideration, the choice of the batch duration can serve as an important factor in managing the transition to continuous operation. The control design in the present chapter recognizes that the product at the end of the batch will be the starting point for the continuous operation and thereby formulates appropriate conditions that determine the batch termination time. In this sense, the proposed approach serves to facilitate the operation of the bio-reactor in continuous mode. Note also that an estimate of the batch duration allows for process operators to prepare for the eventual transition to continuous operation. The algorithm that is used in determining the switch from batch to continuous operation is as follows:

$$T_{batch} = argmin\{V(x(t))\} \tag{5.11}$$

$$\begin{aligned} s.t. \ \dot{x} &= f_1(x) \\ x(0) &= x_0 \end{aligned} \tag{5.12}$$

where T_{batch} is the batch time, x_0 is the initial value of the process states, $V(x)$ is the Lyapunov function used in designing the controller under continuous operation, and Eq. (5.12) is the model describing the evolution of the state x under batch operation.

Remark 5.4 For the given initial conditions for the batch run, the optimization problem of Eqs. (5.11)–(5.12) chooses a time T_{batch} where the state trajectory goes closest to the desired equilibrium point. The use of the Lyapunov function as a closeness measure has important implications in the ability to achieve a stabilizing transition to continuous mode of operation. Specifically, since the stability region under continuous operation is characterized by a level curve of the Lyapunov function, the choice of the batch duration for which the value of the Lyapunov function is minimum naturally means that such a batch duration time, if exists, will be chosen such that the state at the time of the transition resides in the stability region under continuous operation.

Remark 5.5 Note that for the specific bio-reactor considered in this chapter, no manipulated input variables are available during the batch operation. For a given initial condition and operating conditions, the solution to Eqs. (5.11)–(5.12) only determines (not guarantees) whether the state trajectory during batch operation passes through the stability region under continuous operation. Note also that while the predictive controller under continuous operation provides guarantees for all initial conditions in the stability region, it is also possible for the predictive controller to stabilize initial conditions outside the stability region. Therefore, the algorithm of Eqs. (5.11)–(5.12) does not impose a constraint on the states at the end of the batch to be within the stability region under continuous operation (to avoid feasibility issues) but only requires determining the time during the evolution of the batch that the state profile goes through a minimum in the Lyapunov function value. Furthermore, avoiding undesirable scenarios like washout is already accounted for in the stability region characterization. Specifically, if a certain initial condition under continuous operation leads to washout, such initial conditions are naturally excluded from the stability region.

Remark 5.6 Note that in principle, one could have simply monitored the evolution of the state trajectory and initialized the transition to continuous mode of operation once the process state enters the stability region under continuous mode. In most bio-reactors, however, the transition from batch to continuous mode needs to be done under sterile conditions and accordingly needs preparation time. The estimate of the batch transition time provides the operators with a time window to prepare for the transition to continuous operation. Furthermore, the calculation can be repeated midway (or more frequently) through the first estimate and corrected if necessary to account for model inaccuracies and disturbances.

Remark 5.7 While in this chapter we have assumed continuous measurement of the process state variables to demonstrate the key idea in the controller design and determining the time of transition, the same approach can be utilized in handling the case when not all state variables are measured. For the specific bio-reactor under consideration, it may also entail expansion of the process model to include continuously measurable quantities such as dissolved oxygen or evolved carbon dioxide and to use these measurements, together with a state estimation scheme, to estimate the process states. Uncertainty can be handled by utilizing a robust model predictive control approach of the form in [20]. A rigorous analysis of the output feedback problem and uncertainty, however, remains outside the scope of the current chapter.

Table 5.1 Parameter values and units

$K_s = 0.5$ kg/m^3
$k_1 = 16$ h^{-1}
$k_2 = 0.497$ m^{3}kg^{-1}h^{-1}
$k_3 = 0.00383$ m^{6}kg^{-2}h^{-1}
$m_s = 2.16$ kg/kg/h
$m_P = 1.1$ kg/kg/h
$Y_{sx} = 0.0244498$ kg/kg
$Y_{px} = 0.0526315$ kg/kg
$D = 0.03$ h^{-1}
$C_{x0} = 0.0$ kg/m^3
$C_{e0} = 0.0$ kg/m^3
$C_{p0} = 0.0$ kg/m^3
$C_{s0} = 140.0$ kg/m^3
$C_{s,s} = 21.27$ kg/m^3
$C_{x,x} = 1.052$ kg/m^3
$C_{e,s} = 0.0323$ kg/m^3
$C_{p,s} = 58.5413$ kg/m^3

5.5 Simulation Results

Several simulation runs were carried out to evaluate the effectiveness of the proposed predictive controller under continuous operation as well as determining appropriately the transition time from the batch mode to continuous mode. Note that under the operating conditions listed in Table 5.1, the open-loop system behaves in an oscillatory fashion (i.e., the system possesses an open-loop unstable steady state surrounded by a stable limit cycle). First, the implementation of the proposed predictive controller to continuous operation of the bio-reactor is demonstrated. A quadratic Lyapunov function of the form $V(x) = x^T Px$ was chosen in the predictive controller design and the set of initial conditions starting from where $\dot{V} < 0$ is achievable (the set Π), and the invariant set Ω within Π (defined by $V(x) \leq 13.1$) was computed. The P matrix was obtained by solving the Riccati inequality using the A and B matrices resulting from the linearization of the process around the desired equilibrium point. For this specific process, choosing a P matrix this way ensures that at least locally around the equilibrium point, there exists a compact set Π for which it is possible to achieve negative definiteness of $\dot{V}$. Specifically, for such a choice of P, the Lyapunov function is a control Lyapunov function locally around the equilibrium point. The Lyapunov function being a control Lyapunov function serves to ensure that whenever $L_g V = 0$ (i.e., the control action cannot be used to enforce negative definiteness of the Lyapunov function derivative), then for those states $L_f V < 0$ (i.e., the system, for such values of the states, is inherently 'stable' in the sense that the

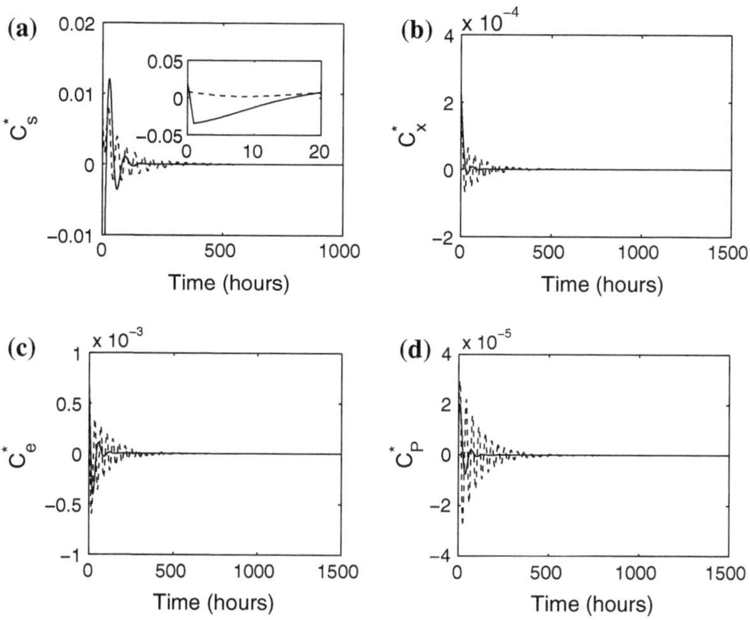

Fig. 5.1 Evolution of closed-loop dimensionless **a** glucose, **b** biomass, **c** inhibitor, and **d** product concentrations starting from an initial condition inside (dashed lines) and outside (solid lines) the guaranteed stability region

value of the Lyapunov function derivative is negative). For this process, Lyapunov functions of the form $\alpha_1 x_1^2 + \alpha_2 x_2^2 + \alpha_3 x_3^2 + \alpha_4 x_4^2$ with $\alpha_i > 0$ were also tested. An estimate of the stability region could not be obtained, because for every choice of $\alpha_i > 0$ tested, there would always be states for which $L_g V = 0$ and $L_f V > 0$ (i.e., the candidate Lyapunov function did not qualify as a control Lyapunov function). The Lyapunov function obtained using the P matrix as described above provides an estimate (albeit possibly a conservative one) of the stability region. The weighting matrices in the predictive controller were chosen as $Q = P$ and $R = rI_u$ with $r = 0$ when $V(x) > c^{max}$ and $r = 0.01$ when $V(x) \leq c^{max}$, where I_u is the identity matrix of appropriate dimension. Such a choice of the objective function enforces as fast a decay of the Lyapunov function as possible for initial conditions outside the guaranteed stability region. Once the state enters the stability region, the penalty on the control action ensures smooth control action.

We first demonstrate the implementation of the predictive controller for an initial condition inside Ω. From an initial condition, $C_s(0) = 21.438\ \frac{\text{kg}}{\text{m}^3}$, $C_x(0) = 1.0517\ \frac{\text{kg}}{\text{m}^3}$, $C_e(0) = 0.0323\ \frac{\text{kg}}{\text{m}^3}$, $C_p(0) = 58.5413\ \frac{\text{kg}}{\text{m}^3}$, for which $V_{c_1} = 12.54 < 13.1 = c^{max}$, the predictive control achieves stabilization of the closed-loop system as expected. The evolution of the process states is shown by the dashed lines in Fig. 5.1, and the corresponding manipulated input and Lyapunov function evolution is shown by the dashed lines in Fig. 5.2. Note that the stability region esti-

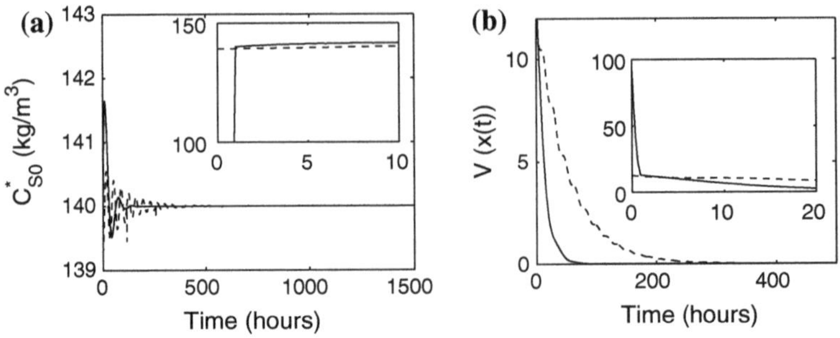

Fig. 5.2 Evolution of the **a** inlet glucose concentration and **b** Lyapunov function value starting from an initial condition inside (dashed lines) and outside (solid lines) the guaranteed stability region

mate for this process ends up being fairly conservative due to the restriction on finding an invariant set Ω within the set of initial conditions from where $\dot{V} < 0$ is achievable. To demonstrate the enhancement in the set of initial conditions that can be stabilized, we consider an initial condition $C_s(0) = 21.73\ \frac{\text{kg}}{\text{m}^3}$, $C_x(0) = 1.0519\ \frac{\text{kg}}{\text{m}^3}$, $C_e(0) = 0.0323\ \frac{\text{kg}}{\text{m}^3}$, $C_p(0) = 58.5414\ \frac{\text{kg}}{\text{m}^3}$, for which $V_{c_1} = 94.0209$. This initial state is outside the stability region (i.e., $V(x(0)) > c^{max}$); however, it resides in the set of initial conditions for which $\dot{V} < 0$ is achievable. The predictive controller of Eqs. (5.4)–(5.8) is implemented and it computes a control action such that not only $\dot{V} < 0$ but also that $\dot{V} < 0$ for the future state can be achieved using lesser available control action, and the controller is able to stabilize the closed-loop system (solid lines in Figs. 5.1–5.2). The inset in Fig. 5.2 illustrates the large control action initially taken to bring the process state closer to the desired equilibrium point.

Having demonstrated the implementation of the predictive controller under continuous operation, we now utilize the stability region characterization in determining the transition from batch to continuous mode of operation. Specifically, for an initial condition (for the batch mode) of $C_s(0) = 37.03\ \frac{\text{kg}}{\text{m}^3}$, $C_x(0) = 0.945\ \frac{\text{kg}}{\text{m}^3}$, $C_e(0) = 0.006\ \frac{\text{kg}}{\text{m}^3}$, $C_p(0) = 50.77\ \frac{\text{kg}}{\text{m}^3}$, with an arbitrarily selected batch duration of $T_{batch} = 10$ h, it is seen that closed-loop stability is not achieved upon transition to continuous mode. Specifically, after transition to continuous mode of operation, the process state is far from the stability region under continuous operation, and the stability constraints in the predictive controller are simply not feasible. If the constraints are removed to yield a feasible solution, closed-loop stability is not achieved (see dashed lines in Figs. 5.3 and 5.4). However, when the batch transition time is picked according to the proposed method, it results in a $T_{batch} = 5.3$ h, and the process state at the end of the batch run resides in the stability region under continuous operation. Figure 5.4b shows the evolution of the Lyapunov function value. Under batch operation, the value of the Lyapunov function goes through a minima as the state trajectory passes 'closest' to the desired equilibrium point. When allowed to evolve further under batch operation, the states move away from the desired equilibrium point, and in this case, so far away that stabilization is not achieved upon transition-

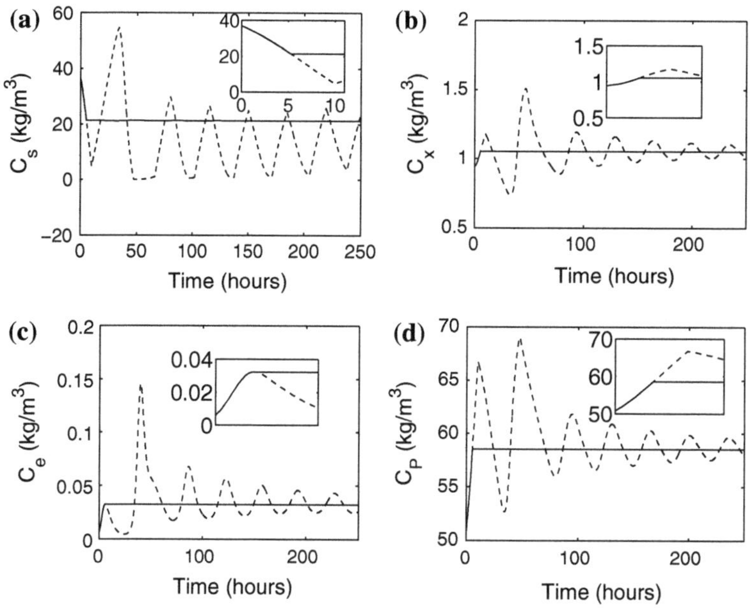

Fig. 5.3 Evolution of closed-loop dimensionless **a** glucose, **b** biomass, **c** inhibitor, and **d** product concentrations following an arbitrarily decided transition from batch to continuous mode at $T_{batch} = 10\,\text{h}$ (dashed lines) and with $T_{batch} = 5.3\,\text{h}$ determined using the proposed algorithm (solid lines)

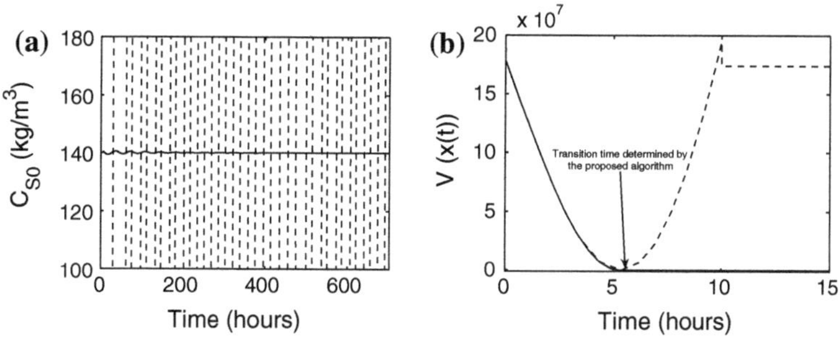

Fig. 5.4 Evolution of the **a** inlet glucose concentration and **b** Lyapunov function value following an arbitrarily decided transition from batch to continuous mode at $T_{batch} = 10\,\text{h}$ (dashed lines) and with $T_{batch} = 5.3\,\text{h}$ determined using the proposed algorithm (solid lines)

ing to continuous mode. In contrast, the algorithm of Eqs. (5.11)–(5.12) determines the time of transition to be 5.3 h (see solid line in Fig. 5.4b) which corresponds to the time that the Lyapunov function value is the smallest. Transitioning to continuous mode of operation at $t = 5.3\,\text{h}$ results in closed-loop stability (see solid lines in Figs. 5.3 and 5.4).

5.6 Conclusions

This chapter considered the problem of determining the transition of ethanol producing bio-reactors from batch to continuous operation and subsequent control subject to constraints and performance considerations. A Lyapunov-based nonlinear model predictive controller was first utilized that stabilizes the bio-reactor under continuous mode of operation. The key idea in the predictive controller is the formulation of appropriate stability constraints that allow an explicit characterization of the set of initial conditions from where feasibility of the optimization problem, and hence, closed-loop stability is guaranteed. The predictive controller also expanded on the set of initial conditions that can be stabilized by control designs that only require the value of the Lyapunov function to decay. Then, the explicit characterization of the set of stabilizable initial conditions was used in determining the appropriate time for which the reactor must be run in batch mode. Specifically, the predictive control approach was utilized in determining the appropriate batch length that achieves stabilizable values of the state variables at the end of the batch. Closed-loop simulations demonstrated the ability of the predictive controller to achieve closed-loop stability and the necessity of determining appropriately the time of switching from batch to continuous operation.

References

1. Mantzaris, N.V., Daoutidis, P.: Cell population balance modeling and control in continuous bioreactors. J. Process Control **14**, 775–784 (2004)
2. Hjortso, M., Nielsen, J.: A conceptual model of autonomous oscillations in microbial cultures. Chem. Eng. Sci. **49**, 1083–1095 (1994)
3. McLellan, P.J., Daugulis, A.J., Li, J.H.: The incidence of oscillatory behavior in the continuous fermentation of zymomonas mobilis. Biotech. Prog. **15**, 667–680 (1999)
4. Namjoshi, A., Kienle, A., Ramkrishna, D.: Steady-state multiplicity in bioreactors: bifurcation analysis of cybernetic models. Chem. Eng. Sci. **58**, 793–800 (2003)
5. Garhyan, P., Mahecha-Botero, A., Elnashaie, S.S.E.H.: Complex bifurcation/chaotic behavior of acetylcholinesterase and cholineacetyltransferase enzymes system. Appl. Math. Model. **9**, 824–853 (2006)
6. Dochain, D., Babary, J.P., Tallmaamar, N.: Modeling and adaptive-control of nonlinear distributed parameter bio-reactors via orthogonal collocation. Automatica **28**, 873–883 (1992)
7. Gudi, R.D., Shah, S.L., Gray, M.R.: Adaptive multirate state and parameter-estimation strategies with application to a bioreactor. AIChE J. **41**, 2451–2464 (1995)
8. Ryhiner, G., Dunn, I.J., Heinzle, E., Rohani, S.: Adaptive online optimal-control of bio-reactors-application to anaerobic degradation. J. Biotech. **22**, 89–105 (1992)
9. Thatipamala, R., Hill, G.A., Rohani, S.: On-line state estimation and adaptive optimization using state equations for continuous production of bioethanol. J. Biotech. **48**, 179–190 (1996)
10. Panjapornpon, C., Soroush, M., Seider, W.D.: Model-based controller design for unstable, non-minimum-phase, nonlinear processes. Ind. Eng. Chem. Res. **45**, 2758–2768 (2006)
11. Mahecha-Botero, A., Garhyan, P., Elnashaie, S.S.E.H.: Bifurcation, stabilization, and ethanol productivity enhancement for a membrane fermentor. Math. Comput. Model. **41**, 391–406 (2005)

12. Keerthi, S., Gilbert, E.: Optimal, infinite horizon feedback laws for a general class of constrained discrete time systems. J. Optim. Theory Appl. **57**, 265–293 (1988)
13. Muske, K.R., Rawlings, J.B.: Model predictive control with linear-models. AIChE J. **39**, 262–287 (1993)
14. Valluri, S., Soroush, M., Nikravesh, M.: Shortest-prediction-horizon non-linear model-predictive control. Chem. Eng. Sci. **53**, 273–292 (1998)
15. Kazantzis, N., Kravaris, C.: Energy-predictive control: a new synthesis approach for nonlinear process control. Chem. Eng. Sci. **54**, 1697–1709 (1999)
16. Langson, W., Chryssochoos, I., Rakovic, S.V., Mayne, D.Q.: Robust model predictive control using tubes. Automatica **40**, 125–133 (2004)
17. Mayne, D.Q., Rawlings, J.B., Rao, C.V., Scokaert, P.O.M.: Constrained model predictive control: stability and optimality. Automatica **36**, 789–814 (2000)
18. Mhaskar, P., El-Farra, N.H., Christofides, P.D.: Predictive control of switched nonlinear systems with scheduled mode transitions. IEEE Trans. Autom. Control **50**, 1670–1680 (2005)
19. Mhaskar, P., El-Farra, N.H., Christofides, P.D.: Stabilization of nonlinear systems with state and control constraints using Lyapunov-based predictive control. Syst. Control Lett. **55**, 650–659 (2006)
20. Mhaskar, P.: Robust model predictive control design for fault-tolerant control of process systems. Ind. Eng. Chem. Res. **45**, 8565–8574 (2006)
21. Christofides, P.D., El-Farra, N.H.: Control of Nonlinear and Hybrid Process Systems: Designs for Uncertainty, Constraints and Time-Delays
22. Ramaswamy, S., Cutright, T.J., Qammar, H.K.: Control of a continuous bioreactor using model predictive control. Process Biochem. **40**, 2763–2770 (2005)
23. Srinivasan, B., Bonvin, D., Visser, E., Palanki, S.: Dynamic optimization of batch processes-II. role of measurements in handling uncertainty. Comput. Chem. Eng. **27**, 27–44 (2003)
24. Shi, D., Mhaskar, P., El-Farra, N., Christofides, P.: Predictive control of crystal size distribution in protein crystallization. Nanotechnology **16**(7), 562–574 (2005)
25. Shi, D., El-Farra, N.H., Li, M., Mhaskar, P., Christofides, P.D.: Predictive control of particle size distribution in particulate processes. Chem. Eng. Sci. **61**, 268–281 (2005)
26. Lin, Y., Sontag, E.D.: A universal formula for stabilization with bounded controls. Syst. Control Lett. **16**, 393–397 (1991)
27. El-Farra, N.H., Christofides, P.D.: Integrating robustness, optimality, and constraints in control of nonlinear processes. Chem. Eng. Sci. **56**, 1841–1868 (2001)
28. El-Farra, N.H., Christofides, P.D.: Bounded robust control of constrained multivariable nonlinear processes. Chem. Eng. Sci. **58**, 3025–3047 (2003)
29. Teel, A.: Global stabilization and restricted tracking for multiple integrators with bounded controls. Syst. Control Lett. **18**, 165–171 (1992)
30. Liberzon, D., Sontag, E.D., Wang, Y.: Universal construction of feedback laws achieving ISS and integral-ISS disturbance attenuation. Syst. Control Lett. **46**, 111–127 (2002)
31. Dubljevic, S., Kazantzis, N.: A new lyapunov design approach for nonlinear systems based on zubov's method. Automatica **38**, 1999–2007 (2002)
32. Freeman, R.A., Kokotovic, P.V.: Robust Nonlinear Control Design: State-Space and Lyapunov Techniques. Birkhauser, Boston (1996)

Part III
Integrating Multi-model Dynamics with PLS Based Approaches

Chapter 6
Integrating Data-Based Modeling and Nonlinear Control Tools for Batch Process Control

6.1 Introduction

In the first part of the book, control designs were presented and illustrated using models that utilized mechanistic information. The high expenses associated with every batch dictate the need for the development of dedicated modeling tools for batch processes that minimize wasted batches in the model development process and yet provide a model that captures the essential nonlinear and complex nature of the process.

In response to this, input–output-based multi-model approaches, such as piecewise affine (PWA) models, have been developed in which multiple linear models are used to capture local dynamics in the data and then combined, yielding a single model capable of describing the major nonlinearities. Note, however, that PWA models are inherently discontinuous since the identification of the correct cluster is a discrete decision. In the Takagi, Sugeno, and Kang (TSK) modeling approach, the state-input space is not partitioned using clustering algorithms, but instead, each input variable is individually partitioned using fuzzy rule sets thereby ignoring state-input interactions, and a weighted combination of the models (based on so-called membership functions) is used for prediction.

The availability of measurements (beyond the designated inputs and outputs) and possible correlations among the data has motivated the application of latent variable modeling methods, particularly partial least squares (PLS) regression, to identify dynamic batch process models, albeit limited by the assumption of a linear relationship between the variables. In the approaches to incorporate nonlinear relationships, the model's predictive capability depends on the choice of the nonlinear mapping (quadratic functions, neural networks, etc.). Another example of a modeling technique (where the clustering is done essentially along time points, or appropriate markers, in the batch duration) is that using principal component analysis (PCA) models, where the process is described via deviations from a nominal process trajectory [1].

P. Mhaskar et al., *Modeling and Control of Batch Processes*,
Advances in Industrial Control, https://doi.org/10.1007/978-3-030-04140-3_6

Existing batch control approaches (model-based or otherwise) specify the desired end-point quality either indirectly, through first determining a process trajectory that terminates at the desired end-point and then using classical or advanced control tools (such as model predictive control (MPC)) to track the desired trajectory, or directly in an end-point-based MPC framework. The inapplicability of step test-based tuning techniques in batch processes negatively impacts the control performance of trajectory tracking approaches using proportional–integral–derivative controllers. In trajectory tracking predictive control approaches [1–8], with modeling errors and process noise, a process trajectory deemed optimal in off-line calculations may be significantly suboptimal, or even infeasible in online implementation. In end-point-based MPC, a dynamic optimization (DO) problem that incorporates the final product quality specifications (in the objective function and/or constraints) is solved at each sampling instance until batch termination. In these approaches, since the input trajectory from the current time to the end of the batch has to be optimized, the size of the optimization problem can become intractable specifically during the early stages in the batch. Input parametrization techniques [9, 10] constitute a fairly recently developed method for making nonlinear dynamic optimization problems computationally tractable for real-time application. The impact of plant-model mismatch and disturbances, however, remains significant in this approach (owing, in part, to use of approximate first-principles models with parametric errors and unmodeled disturbances).

The majority of the existing work addressing faults in batch systems focuses primarily on fault detection and isolation while the design of explicit fault-tolerant control structures (FTCS) for batch systems has received limited attention. Existing FTCS for batch processes are mostly robust control designs that treat faults as disturbances. However, upon fault occurrence in a batch system, the final product quality may become unreachable if the fault is not repaired sufficiently fast. Additionally, implementing an input trajectory prescribed by controllers with limited fault-tolerant properties can drive the system states to a point from where the final quality is not reachable even if the fault is repaired. In response to these issues, we developed a control and safe-steering framework [11] that utilized a first-principles model and presented a computationally efficient MPC design that addressed the problem of determining how to utilize functioning inputs during fault rectification to enable desired product properties reachability following fault repair. The proposed design represents a computationally efficient framework that is amenable for integration with appropriately derived data-based models for control of batch processes.

Motivated by the above considerations, this chapter considers the problem of designing an integrated framework seamlessly merging data-based models with nonlinear control tools for control of batch processes. The rest of this manuscript is organized as follows: First, the class of processes considered is presented followed by reviews of the auto-regressive exogenous (ARX)-type modeling approach, PLS regression, and the clustering algorithm used in the proposed multi-model approach. Next, a framework is presented for developing a data-based model for batch processes that makes use of all existing measurement and captures the nonlinear nature of the batch. In this modeling approach, a database of previous batch trajectories is initially clustered into a number of operating regions, a weighting scheme is then devised for the training data that can also be utilized to appropriately weight the local models given an initial condition and input, and finally, local linear models are estimated simultaneously using partial least squares (PLS) regression. The resulting model is then incorporated within the RTRR-based MPC and safe-steering framework. Specifically, a methodology and algorithm are presented to generate RTRRs using the data-based model for their subsequent use within a RTRR-based MPC design for batch processes. Then, simulation results of a fed-batch reactor system and a nylon-6,6 batch polymerization system, subject to sensor noise, disturbances, and time-varying parameters, are presented to demonstrate the effectiveness and applicability of the proposed modeling and control approach in the presence of limited measurements. Finally, we summarize our results.

6.2 Preliminaries

In this section, we first describe the class of batch processes considered. Then, we give an overview of auto-regressive eXogenous (ARX)-type modeling, a popular technique for developing linear input–output models of dynamic systems, and then illustrate how multivariate regression tools, such as partial least squares (PLS) regression, can be used within the ARX-type modeling framework to utilize all available measurement data (beyond just the input/output data). Then, we review fuzzy c-means clustering and the concept of reverse-time reachability regions (RTRRs).

6.2.1 Process Description

We consider batch process systems subject to input constraints and failures described by:

$$\begin{aligned} \dot{\boldsymbol{x}} &= \boldsymbol{f}(\boldsymbol{x}, \boldsymbol{u}_\sigma) \\ \boldsymbol{y} &= \boldsymbol{g}(\boldsymbol{x}) \\ t &\in [t_0, t_f], \boldsymbol{u}_\sigma(\cdot) \in \mathcal{U}_\sigma, \boldsymbol{x}(t_0) = \boldsymbol{x}_0 \end{aligned} \tag{6.1}$$

where $\boldsymbol{x} \in \mathbb{R}^n$ denotes the vector of state variables, $\boldsymbol{y} \in \mathbb{R}^p$ denotes the vector of measurable output variables, and $\boldsymbol{u}_\sigma \in \mathbb{R}^m$ denotes the vector of constrained manipulated inputs, taking values in a nonempty convex subset $\mathcal{U}_\sigma$ of $\mathbb{R}^m$, where $\mathcal{U}_\sigma = \{\boldsymbol{u} \in \mathbb{R}^m \mid \boldsymbol{u}_{\mathsf{min},\sigma} \leq \boldsymbol{u} \leq \boldsymbol{u}_{\mathsf{max},\sigma}\}$, where $\boldsymbol{u}_{\mathsf{min},\sigma}, \boldsymbol{u}_{\mathsf{max},\sigma} \in \mathbb{R}^m$ denote the constraints on the manipulated inputs. The times t_0 and t_f denote the initial time and batch termination times, respectively. The variable, $\sigma \in \{1, 2\}$, is a discrete variable that indexes fault-free and faulty operation with $\sigma = 1$ signifying fault-free operation and $\sigma = 2$ signifying faulty operation. The fault scenarios considered in this chapter involve actuator failure resulting in reduced available control effort for a finite duration of time lasting from the time of fault occurrence, t^{fault}, to the time of fault repair, t^{repair}. Throughout the manuscript, we assume that for any $\boldsymbol{u}(\cdot) \in \mathcal{U}_\sigma$, the solution of the batch system of Eq. (6.1) exists and is continuous for all $t \in [t_0, t_f]$.

6.2.2 Auto-regressive Exogenous (ARX) Models

In the ARX-type modeling approach, the process outputs at a specific sampling instance are assumed to depend linearly on the previous process conditions (defined by the process outputs and inputs). Mathematically, ARX models are defined as:

$$\boldsymbol{y}(k) = \sum_{i=1}^{n_y} \mathbf{A}_i \boldsymbol{y}^{\mathrm{T}}(k-i) + \sum_{j=1}^{n_u} \mathbf{B}_i \boldsymbol{u}^{\mathrm{T}}(k-j) + \boldsymbol{v}(k) \tag{6.2}$$

where $\boldsymbol{y}(k) \in \mathbb{R}^p$ and $\boldsymbol{u}(k) \in \mathbb{R}^m$ are the process output and input vectors at sampling instant k (respectively), $\mathbf{A}_i \in \mathbb{R}^{p \times p}$ and $\mathbf{B}_i \in \mathbb{R}^{p \times m}$ denote the model coefficient matrices, $\boldsymbol{v}(k) \in \mathbb{R}^p$ is the noise vector, and n_y and n_u denote the (maximum) number of time lags in the outputs and inputs (respectively) and define the order of the ARX model. For outputs which do not require the maximum number of lags, the appropriate elements in $\mathbf{A}_i$ and $\mathbf{B}_i$ can be set to zero. In the case of full state measurements, $n_y = n_u = 1$ is a natural choice, which then results in the state-space system described by:

$$\boldsymbol{x}(k) = \mathbf{A}\boldsymbol{x}(k-1) + \mathbf{B}\boldsymbol{u}(k-1) \tag{6.3}$$

To facilitate the estimation of the model coefficient matrices, Eq. (6.2) can be rewritten in matrix form as follows (for a given n_y and n_u):

$$\boldsymbol{y}(k) = \boldsymbol{\beta}\bar{\boldsymbol{x}}(k) + \boldsymbol{v}(k) \tag{6.4}$$

where $\boldsymbol{\beta} = \left[\mathbf{A}_1 \cdots \mathbf{A}_{n_y}\ \mathbf{B}_1 \cdots \mathbf{B}_{n_u}\right]$ (appropriately zero padded) collects the model coefficient matrices and $\bar{\boldsymbol{x}}^{\mathrm{T}}(k) = \left[\boldsymbol{y}(k-1) \cdots \boldsymbol{y}(k-n_y)\ \boldsymbol{u}(k-1) \cdots \boldsymbol{u}(k-n_u)\right]$ is a vector of lagged concatenated outputs and inputs. Given plant data, a response matrix, $\mathbf{Y}$, and regressor matrix, $\bar{\mathbf{X}}$, can be constructed corresponding to $\boldsymbol{y}(k)$ and $\bar{\boldsymbol{x}}(k)$ (respectively) in Eq. (6.4) by sorting the data sample-wise, and the model parameters can be estimated using partial least squares (PLS) regression to account for any colinearity and correlation in the data due to the variables describing the same underlying phenomena in the process or because of the data being collected under closed-loop conditions [12–14].

Mathematically, PLS regression consists of decomposing $\bar{\mathbf{X}}$ and $\mathbf{Y}$ as the sum of the outer products of a score and loading vector:

$$\bar{\mathbf{X}} = \sum_{j=1}^{A} \boldsymbol{t}_j \boldsymbol{p}_j^{\mathrm{T}} + \mathbf{E} = \mathbf{T}\mathbf{P}^{\mathrm{T}} + \mathbf{E} \tag{6.5}$$

$$\mathbf{Y} = \sum_{j=1}^{A} \boldsymbol{r}_j \boldsymbol{q}_j^{\mathrm{T}} + \mathbf{F} = \mathbf{R}\mathbf{Q}^{\mathrm{T}} + \mathbf{F} \tag{6.6}$$

where $\boldsymbol{t}_j$ and $\boldsymbol{r}_j$ are the input and output scores representing the projections of the variables in $\bar{\mathbf{X}}$ and $\mathbf{Y}$ on the subspaces, $\boldsymbol{p}_j$ and $\boldsymbol{q}_j$ define the orientation of the corresponding subspaces, the matrices, $\mathbf{T}$, $\mathbf{P}$, $\mathbf{R}$, and $\mathbf{Q}$, simply contain their corresponding vectors, and $\mathbf{E}$ and $\mathbf{F}$ denote residual matrices. Because it is desired to obtain a useful relationship between the original data matrices, $\bar{\mathbf{X}}$ and $\mathbf{Y}$, the two matrices are linked by an inner relation between their scores of the form:

$$\boldsymbol{r}_j = \boldsymbol{b}_j \boldsymbol{t}_j + \boldsymbol{e}_j \tag{6.7}$$

where $\boldsymbol{b}_j$ are the coefficients of the inner relationship and $\boldsymbol{e}_j$ are the residuals. In common PLS algorithms, such as Nonlinear Iterative Partial Least Squares (NIPALS), the subspace orientation and scores for both matrices are determined simultaneously so as to maximize the correlation between $\bar{\mathbf{X}}$ and $\mathbf{Y}$ and therefore obtain the optimal fit for the inner relationship [15, 16]. The final result from PLS regression is a linear model between $\bar{\mathbf{X}}$ and $\mathbf{Y}$ where the coefficients are functions of the scores and loadings from the matrix decompositions:

$$\mathbf{Y} = \bar{\mathbf{X}} \boldsymbol{\beta}_{\mathsf{PLS}} + \mathbf{G} \tag{6.8}$$

where $\mathbf{G}$ denotes the residuals and $\boldsymbol{\beta}_{\mathsf{PLS}} = f(\mathbf{P}, \mathbf{T}, \mathbf{Q})$.

6.2.3 Fuzzy c-Means Clustering

Prior to developing multiple dynamic PLS models, an important step in the proposed multi-model approach is to first locate the operating points around which the individual local linear models will be identified. One approach to find this set of operating points is to partition the historical batch database into a number of clusters (i.e., a group of points in the database that are mathematically similar). Subsequently, a corresponding linear model can be identified for each cluster. For the current chapter, we employ fuzzy c-means clustering [17] to partition the database (the results are not limited to this particular type of clustering; any other meaningful clustering algorithm can be used instead). Assuming full state measurements in a given batch database (see the simulation example for the case of limited availability of measurements), let $\bar{\mathbf{X}}^{\mathrm{T}} = \left[\bar{\boldsymbol{x}}_1 \cdots \bar{\boldsymbol{x}}_i \cdots \bar{\boldsymbol{x}}_N\right]$ be a matrix of $n + m$ rows and N columns where each column is a different instance of $\left[\boldsymbol{x}^{\mathrm{T}}(k)\ \boldsymbol{u}^{\mathrm{T}}(k)\right]$ (vertically concatenated states and inputs at sampling instant k). The state-input space in $\bar{\mathbf{X}}^{\mathrm{T}}$ can be partitioned into

L different clusters using fuzzy clustering, which assigns each sample, $\bar{\boldsymbol{x}}_i$, a degree of belonging to a cluster $\ell \in [1, L]$. The partition information can be represented by an L by N membership matrix, $\mathbf{U} = \{\mu_{\ell,i}\}$, where each row contains the membership information for ℓth cluster for all N points. In fuzzy clustering, the elements in $\mathbf{U}$ must satisfy the following conditions [18]:

$$\mu_{\ell,i} \in [0, 1], \quad \text{for } 1 \leq \ell \leq L \quad 1 \leq i \leq N \tag{6.9}$$

$$\sum_{\ell=1}^{L} \mu_{\ell,i} = 1, \quad \text{for } 1 \leq i \leq N \tag{6.10}$$

$$0 < \sum_{i=1}^{N} \mu_{\ell,i} < N, \quad \text{for } 1 \leq \ell \leq L \tag{6.11}$$

Equation (6.10) requires that the total membership of each observation (which ranges from 0 to 1 (Eq. (6.9))) equals one. The cluster centers (based on minimizing the total variance in the data from cluster centers) are obtained by minimizing the following objective function (the so-called c-means functional) [17, 19]:

$$J = \sum_{i=1}^{N} \sum_{\ell=1}^{L} \left(\mu_{\ell,i}\right)^f \|\bar{\boldsymbol{x}}_i^{\mathrm{T}} - \boldsymbol{c}_\ell\|^2 \tag{6.12}$$

where $\boldsymbol{c}_\ell \in \mathbb{R}^{n+m}$ denote the cluster center vectors, which have to be determined. The weighting exponent parameter, f, determines the fuzziness of the resulting clusters with $f = 1$ implying hard, nonoverlapping partitions. For this chapter (as is typically the case), we choose $f = 2$. The partition matrix elements, $\mu_{\ell,i}$, and cluster centers, $\boldsymbol{c}_\ell$, that minimize the objective function and satisfy the constraints in Eqs. (6.9)–(6.11) have been shown to be (for $f = 2$) [17, 19]:

$$\mu_{\ell,i} = \frac{\|\bar{\boldsymbol{x}}_i^{\mathrm{T}} - \boldsymbol{c}_\ell\|^{-2}}{\sum_{\ell=1}^{L} \|\bar{\boldsymbol{x}}_i^{\mathrm{T}} - \boldsymbol{c}_\ell\|^{-2}} \tag{6.13}$$

and

$$\boldsymbol{c}_\ell = \frac{\sum_{i=1}^{N} \mu_{i,\ell}^2 \bar{\boldsymbol{x}}_i^{\mathrm{T}}}{\sum_{i=1}^{N} \mu_{i,\ell}^2} \tag{6.14}$$

where $\|\bar{\boldsymbol{x}}_i^{\mathrm{T}} - \boldsymbol{c}_\ell\|$ denotes the Euclidean distance between point i and the ℓth cluster center. From Eq. (6.14), it can be seen that the center point of each cluster is also the mean of all the points, weighted by their membership degrees. In view of this, in fuzzy clustering, the degree of $\bar{\boldsymbol{x}}_i$ belonging to cluster ℓ is taken to be inversely proportional to the squared distance between the point and cluster center, $\boldsymbol{c}_\ell$ (i.e., $\mu_{\ell,i} \propto \|\bar{\boldsymbol{x}}_i^{\mathrm{T}} - \boldsymbol{c}_\ell\|^{-2}$), which is then normalized across all clusters.

Remark 6.1 In the fuzzy clustering done for this chapter, points that are mathematically 'similar' according to the Euclidean 2-norm are clustered, resulting in (overlapping) $n + m$-dimensional spherical clusters. To account for the different variances in the directions of the coordinate axes of $\bar{\mathbf{X}}$, each variable can be scaled to unit variance through dividing by its standard deviation. This is equivalent to weighting the norm used in the clustering algorithm by an appropriate diagonal matrix (i.e., a diagonal matrix comprising the inverse variances of each variable). To accommodate nonspherical clusters, extensions of fuzzy c-means clustering that consider different norms or different shapes can be utilized as well [20, 21].

Remark 6.2 The number of clusters in fuzzy c-means clustering is the most important parameter in the algorithm. Well-defined criteria (based on the cluster geometry) to iteratively refine the number of clusters have been presented [22–24]. To evaluate the goodness of the final fuzzy partitions, many validation measures have also been introduced [24, 25]. Among these, the Xie-Beni index [25], which is a ratio of the total within-cluster variance to the separation of the cluster centers (and therefore should be minimal for the best partition), has been found to perform well in practice.

Remark 6.3 In the absence of full state measurements, the columns in $\bar{\mathbf{X}}^{\mathrm{T}}$ are replaced by concatenated lagged outputs and inputs: $\left[\boldsymbol{y}(k-1) \cdots \boldsymbol{y}(k-n_y)\boldsymbol{u}(k-1)\right]$ $\left[\cdots \boldsymbol{u}(k-n_u)\right]^{\mathrm{T}}$. In this case, the dimension of the space required to be clustered is $n_y + n_u + p + m$, which may be prohibitively large for computation. In this chapter, this high dimensionality problem was addressed by first decomposing the $\bar{\mathbf{X}}$ matrix using principal component analysis (PCA) and subsequently clustering the resulting latent variable (or score) space. The resulting loading matrix from PCA, $\mathbf{P}$, can be used to relate the original cluster space variables (i.e., the lagged inputs and outputs) to the latent variables according to $\mathbf{T} = \bar{\mathbf{X}}\mathbf{P}$, where $\mathbf{T}$ denotes the projections of each row in $\bar{\mathbf{X}}$ onto the subspace (i.e., the scores). Note that this result follows from Eq. (6.5) and the fact that $\mathbf{P}$ is orthonormal. Typically, a much lower number of principal components (compared to $n_y + n_u + p + m$) is required to be retained to completely characterize $\bar{\mathbf{X}}$ since $\bar{\mathbf{X}}$ can include many lagged variables and therefore correlations among the columns.

6.2.4 Reverse-Time Reachability Regions

The reverse-time reachability region (RTRR) of a batch system at time, t, $\mathcal{R}(t)$, was defined as the set of process states from where a batch process could be steered to a desired end-point, $\boldsymbol{x}_{\mathrm{des}}$, by the end of the batch while satisfying the input constraints [11]. The definition of the discrete time version of this set is reproduced below:

Definition 6.1 For the batch process described by Eq. (6.1) with sampling period δ, the RTRR at time $t = t_{\mathrm{f}} - q\delta$, indexed by q, is the set [11]:

$$\mathcal{R}_q = \{\boldsymbol{x}_0 \mid \boldsymbol{x}(t) = \boldsymbol{x}_0,\ \dot{\boldsymbol{x}} = \boldsymbol{f}(\boldsymbol{x}, \boldsymbol{u})\ \exists\ \boldsymbol{u} = \{\boldsymbol{u}[i]\} \in \mathcal{U}\ \forall i = 1, \ldots, k \text{ where } \boldsymbol{u}[i] = \boldsymbol{u}(i\delta) \\ \text{and satisfies } \boldsymbol{u}(t) = \boldsymbol{u}[i]\ \forall t \in [i\delta, (i+1)\delta),\ \text{such that } \boldsymbol{x}(t_{\text{f}}) = \boldsymbol{x}_{\text{des}}\} \tag{6.15}$$

An algorithm to generate estimates of RTRRs as point sets at each sampling instance consists of successive off-line integrations of the reverse-time model of the system (i.e., $\dot{\boldsymbol{x}} = -\boldsymbol{f}(\boldsymbol{x}, \boldsymbol{u})$) [11]. However, this algorithm requires a first-principles model of the process, which may not always be available. Another contribution of the present chapter is the modification of the RTRR generation algorithm to use a data-based model of the process.

6.3 Integrating Data-Based Modeling Methods with Nonlinear Control Tools

In this section, we first propose a general multi-model approach for modeling batch systems applicable using full state measurements or limited output measurements and work through the underlying details. Then, for the case of full state measurements, we present an optimization-based algorithm to sequentially characterize RTRR estimates using the resultant data-based model and then formulate a predictive controller design using these sets.

6.3.1 *Multi-model Approach*

In presenting the multi-model approach, we first assume full state measurements and address the case of limited measurements through a remark (see Remark 6.7). In the proposed multi-model approach, a necessary preprocessing step is to cluster the batch database into L clusters as described earlier. Assuming this step has been completed, the basic idea in the proposed approach is to identify several local linear models around the cluster center points and then combine them with appropriate weights to describe the global nonlinear behavior. For the individual linear models, we employ the state-space models in Eq. (6.3), which are identified using PLS regression. Mathematically, this idea is expressed by the following model:

$$\boldsymbol{x}(k) = \sum_{\ell=1}^{L} w_\ell(k)\hat{\boldsymbol{\beta}}_\ell \left[\boldsymbol{x}^{\mathrm{T}}(k-1)\ \boldsymbol{u}^{\mathrm{T}}(k-1)\right] \tag{6.16}$$

where w_ℓ is the (normalized) weight given to model ℓ of L total models and $\hat{\boldsymbol{\beta}}_\ell$ is a matrix of model coefficients (with appropriate dimensions). Note that if the weights are determined independently (from estimating the individual model parameters),

Eq. (6.16) becomes linear in $\hat{\boldsymbol{\beta}}_\ell$ and the system identification problem reduces to a PLS regression problem.

Intuitively, from the process description in Eq. (6.1), the weight placed on a local linear model should depend on the proximity of the states and input to the center point of the set of points for which the local linear model is identified, captured in the normalized fuzzy clustering membership function (Eq. (6.13)). For instance, if a state-input combination nearly coincides with a specific cluster center point, the local linear model corresponding to that cluster should be given most of the weight. This is consistent with Eq. (6.13) as the membership function value corresponding to that cluster will be close to 1, and while for the remaining clusters, the membership function will be near 0.

Since all the local models can potentially contribute during prediction, it is important to identify the local linear PLS models simultaneously. To facilitate this regression, the process data can be arranged in the following linear model form [26]:

$$\mathbf{Y} = \left[\mathbf{U}_1 \otimes \bar{\mathbf{X}} \cdots \mathbf{U}_\ell \otimes \bar{\mathbf{X}} \cdots \mathbf{U}_L \otimes \bar{\mathbf{X}}\right]\left[\hat{\boldsymbol{\beta}}_1 \cdots \hat{\boldsymbol{\beta}}_\ell \cdots \hat{\boldsymbol{\beta}}_L\right]^{\mathrm{T}} \tag{6.17}$$

where $\otimes$ denotes element by element multiplication and $\mathbf{U}_\ell$ is defined as:

$$\mathbf{U}_\ell = \begin{bmatrix} \mu_{\ell,1} & \cdots & \mu_{\ell,1} \\ \vdots & \cdots & \vdots \\ \mu_{\ell,N} & \cdots & \mu_{\ell,N} \end{bmatrix} \tag{6.18}$$

Note that $\mathbf{U}_\ell$ has the same dimensions as $\bar{\mathbf{X}}$. Since Eq. (6.17) is linear in the parameters, $\hat{\boldsymbol{\beta}}_\ell$, PLS regression can be performed to simultaneously identify the local linear models. In summary, the proposed multi-model approach unifies the concepts of ARX modeling, PCA/PLS techniques, fuzzy c-means clustering, and multiple linear models inspired from a 'state-space' representation, where the clustering is done based on states/outputs and inputs, the model parameters are identified using the strength of PCA/PLS techniques, and the nonlinearity of the process is captured via the use of clustering techniques in combination with local linear models.

Remark 6.4 The proposed modeling method addresses some of the key issues with existing modeling approaches for batch systems. In contrast to PWA modeling in which a crisp clustering algorithm such as k-means is typically used and the model selection is a discrete decision, the current multi-model approach employs fuzzy clustering and allows weighting the different models appropriately using a continuous weighting function. This becomes particularly important during periods of transition in the process when it is evolving from one operating region to another. In this case, as multiple models will be weighted appropriately, the information from several different models can be used, resulting in more accurate predictions. In multi-model approaches using crisp clustering algorithms such as PWA modeling, only samples belonging to a specific cluster can contribute in determining the model and no surrounding information is used. That is, artificial boundaries are established for the clusters and their corresponding models, which can lead to abrupt changes in predictions during transitional periods. Furthermore, the presence of discrete model selection in the PWA framework also negatively impacts their use in optimization-based control designs by requiring solutions of optimization problems that include continuous (the control action) as well as discrete (the choice of the model) variables.

Remark 6.5 Note that the key difference between the present approach and the latent variable modeling methods [1] is the parameter used in the clustering. The latent variable modeling approaches rely on the batch trajectories being ‘similar’ such that the effect of state and input evolution on the process dynamics can be captured through ‘time’ or ‘phase’ markers. In contrast, the present chapter explicitly accounts for the dependence of the dynamics on the process states and inputs, thereby allowing more variation in the batch trajectories, essentially leading to a richer model. The latent variable modeling [1] also leads to an inherently linear PCA model, whereas the weighting in the proposed modeling approach allows capturing nonlinearity in the data/process. Furthermore, when using the PCA model to predict the outputs given a set of control moves, a missing data algorithm is used and the outputs are computed implicitly (not explicitly as in the proposed approach) such that they maintain the average correlation structure as in the batch database [1]. However, in many cases, the correlation structure determined by the PCA model may no longer hold for a new initial condition due to the strong nonlinearities in the system and/or there are significant changes in the correlation structure as the batch proceeds. In contrast, if states/inputs close to the initial condition were encountered (albeit say at a different time point in historical batch trajectories), the present approach will be able to use that information to predict meaningful process evolution for the new batch.

Remark 6.6 While the present approach has the ability to capture the different phases that a batch goes through due to the inherent nonlinearity of the dynamics, it does not explicitly account for time-varying dynamics or where the dynamics exhibit a switched nature (i.e., phases where particular reagents are added only during the phase). While explicit incorporation of these conditions remains outside the scope of the present chapter, the efficacy of the proposed method to handle time-varying parameters is demonstrated in the simulation example.

Remark 6.7 For the case of limited output measurements, the matrix $\bar{\mathbf{X}}$ (or its PCA decomposition) as described in Remark 6.3 is required to be clustered. Thus, in order to compute the model weights (w_ℓ) using Eq. (6.13), the $\bar{\boldsymbol{x}}$ vector in Eq. (6.13) is constituted of lagged outputs and inputs (or their scores as described in Remark 6.3). This implies the lagged outputs and inputs are used to infer the current operating conditions as opposed to the current states and inputs. The proposed model in the case of limited measurements takes the following form:

$$\hat{\boldsymbol{y}}(k) = \sum_{\ell=1}^{L} w_\ell(k)\hat{\boldsymbol{\beta}}_\ell \left[\boldsymbol{y}^{\mathrm{T}}(k-1) \cdots \boldsymbol{y}^{\mathrm{T}}(k-n_y)\; \boldsymbol{u}^{\mathrm{T}}(k-1) \cdots \boldsymbol{u}^{\mathrm{T}}(k-n_u)\right] \tag{6.19}$$

This model can be readily incorporated in conventional output-based predictive controllers for batch systems. In this chapter, we utilize the output model form in designing a trajectory tracking predictive controller for a nylon-6,6 batch polymerization process.

6.3.2 Reverse-Time Reachability Region Generation Using the Data-Based Model

In this section, assuming full state measurements, we present a methodology to generate RTRRs using the data-based modeling approach developed earlier. Due to unavoidable discrepancies between a process and its empirical model, instead of considering exact reachability to a desired end-point, we consider reachability to a desired end-point neighborhood, $\mathcal{B}(\boldsymbol{x}_{\text{des}})$. We define a data-based version of a RTRR as the set of states from where the data-based model of the process can be driven inside the desired end-point neighborhood by the end of the batch. Denoting this set at sampling instant q as $\hat{\mathcal{R}}_q$, the definition of an empirical RTRR is stated as:

Definition 6.2 For the data-based model of the form in Eq. (6.16), the empirical RTRR at time $t = t_{\text{f}} - q\delta$, indexed by q, is the set:

$$\hat{\mathcal{R}}_q = \{\hat{\boldsymbol{x}}_0 \mid \boldsymbol{x}(0) = \hat{\boldsymbol{x}}_0,\ \boldsymbol{x}(k) = \sum_{\ell=1}^{L} w_\ell(k)\hat{\boldsymbol{\beta}}_\ell \left[\boldsymbol{x}^{\mathrm{T}}(k-1)\ \boldsymbol{u}^{\mathrm{T}}(k-1)\right]$$
$$\text{for } k = 1, \ldots, q\ \exists\ \boldsymbol{u}(k) \in \mathcal{U}$$
$$\forall\, k = 0, \ldots, q-1 \text{ such that } \boldsymbol{x}(q) \in \mathcal{B}(\boldsymbol{x}_{\text{des}})\} \tag{6.20}$$

In formulating an empirical RTRR-based predictive control design, an explicit characterization of these sets is required. In this chapter, we use ellipsoids to mathematically express empirical RTRRs at each sampling instance. The general form of the ellipsoid expression is given below:

$$\hat{\mathcal{R}}_q \approx \left\{\boldsymbol{x} \mid (\boldsymbol{x} - \boldsymbol{c}_q)^{\mathrm{T}}\mathbf{W}_q(\boldsymbol{x} - \boldsymbol{c}_q) \leq 1\right\} \tag{6.21}$$

where the vector $\boldsymbol{c}_q \in \mathbb{R}^n$ denotes the center of the ellipsoid, the positive-definite symmetric matrix $\mathbf{W}_q \in \mathbb{R}^{n\times n}$ defines its size and orientation, and q indexes the batch sampling instances as before. We note that because $q = 0$ corresponds to t_{f}, $\boldsymbol{c}_0 = \boldsymbol{x}_{\text{des}}$ and $\mathbf{W}_0$ is a user-defined matrix based on the acceptable variance level of the final product quality. An equivalent representation of the ellipsoid was used in this chapter in which the ellipsoid is expressed as the image of a unit ball under an affine transformation. That is, consider the unit ball, $S(0, 1) = \left\{\boldsymbol{x} \mid \boldsymbol{x}^{\mathrm{T}}\boldsymbol{x} \leq 1\right\}$ in $\mathbb{R}^n$ and the affine transformation $T(\boldsymbol{x}) = \mathbf{H}\boldsymbol{x} + \boldsymbol{d}$ where $\mathbf{H} \in \mathbb{R}^{n\times n}$ is a nonsingular matrix and $\boldsymbol{d} \in \mathbb{R}^n$. Applying the affine transformation to a point on the unit ball, we have

$z = \mathbf{H}\boldsymbol{x} + \boldsymbol{d}$, which implies $\boldsymbol{x} = \mathbf{H}^{-1}(z - \boldsymbol{d})$. An ellipsoid can then be expressed through an affine transformation of the unit ball:

$$T(S(0,1)) = \left\{z \mid \left(\mathbf{H}^{-1}(z-\boldsymbol{d})\right)^{\mathrm{T}}\left(\mathbf{H}^{-1}(z-\boldsymbol{d})\right) \leq 1\right\} = \left\{z \mid (z-\boldsymbol{d})^{\mathrm{T}}\mathbf{V}^{-1}(z-\boldsymbol{d}) \leq 1\right\} \tag{6.22}$$

where $\mathbf{V} = \mathbf{H}\mathbf{H}^{\mathrm{T}} \in \mathbb{R}^{n\times n}$ is a positive-definite symmetric matrix. Thus, from Eq. (6.22), defining $\mathbf{H}_q$ and $\boldsymbol{d}_q$ is equivalent to defining the ellipsoid parameters $\mathbf{W}_q$ and $\boldsymbol{c}_q$ in Eq. (6.21).

Starting at $q = 1$, an explicitly characterized estimate of $\hat{\mathcal{R}}_q$ is identified from where the model states can be driven inside $\hat{\mathcal{R}}_{q-1}$. This procedure is repeated until an empirical RTRR is identified for every sampling instant in the batch. Given the RTRR ellipsoid parameters at $q-1$ and I (predetermined) points (generated from a uniform distribution) on the surface of a unit ball denoted by $\left\{\boldsymbol{x}_{\mathsf{ub}}^{(1)}, \ldots, \boldsymbol{x}_{\mathsf{ub}}^{(i)}, \ldots, \boldsymbol{x}_{\mathsf{ub}}^{(I)}\right\}$, the following NLP is solved to determine the ellipsoid parameters, $\mathbf{H}_q$ and $\boldsymbol{d}_q$ (and therefore $\mathbf{W}_q$ and $\boldsymbol{c}_q$):

$$\max_{\mathbf{H}_q, \boldsymbol{d}_q, \boldsymbol{u}^{(i)} \in \mathcal{U}} \quad \det \mathbf{H}_q \tag{6.23}$$

$$\text{subject to: } \boldsymbol{x}^{(i)} = \mathbf{H}_q \boldsymbol{x}_{\mathsf{ub}}^{(i)} + \boldsymbol{d}_q \;\; \forall\, i = 1, \ldots, I \tag{6.24}$$

$$\boldsymbol{x}_{\mathsf{nxt}} = \sum_{\ell=1}^{L} w_\ell \hat{\boldsymbol{\beta}}_\ell \left[\boldsymbol{x}^{(i)} \; \boldsymbol{u}^{(i)}\right] \tag{6.25}$$

$$w_\ell = \frac{\| \left[\boldsymbol{x}^{(i)} \; \boldsymbol{u}^{(i)}\right]^{\mathrm{T}} - \boldsymbol{c}_\ell \|^{-2}}{\sum_{\ell=1}^{L} \| \left[\boldsymbol{x}^{(i)} \; \boldsymbol{u}^{(i)}\right]^{\mathrm{T}} - \boldsymbol{c}_\ell \|^{-2}} \tag{6.26}$$

$$(\boldsymbol{x}_{\mathsf{nxt}} - \boldsymbol{c}_{q-1})^{\mathrm{T}} \mathbf{W}_{q-1} (\boldsymbol{x}_{\mathsf{nxt}} - \boldsymbol{c}_{q-1}) \leq 1 \tag{6.27}$$

$$\mathbf{H}_q = \mathbf{L}_q \mathbf{L}_q^{\mathrm{T}} \tag{6.28}$$

The independent decision variables in this NLP are the ellipsoid parameters ($\mathbf{H}_q$ and $\boldsymbol{d}_q$) and I control moves (corresponding to the I initial conditions on the surface of the ellipsoid). The NLP is formulated to maximize the volume of the current RTRR ellipsoid while ensuring for I uniformly distributed points on the surface of the ellipsoid, there exists a control action (as prescribed by a predictive controller using the data-based model) that can drive the ellipsoid surface point inside the next RTRR. Equation (6.24) represents the affine transformation of the I unit ball points into the ellipsoid surface points. Equation (6.28) represents the Cholesky decomposition of $\mathbf{H}_q$, where $\mathbf{L}_q \in \mathbb{R}^{n\times n}$ is a lower triangular matrix, and ensures $\mathbf{H}_q$ is positive-definite and symmetric. Note that ascertaining the feasibility of the optimization problem for the I surface points does not guarantee the feasibility of all points on the surface, or for that matter, for the internal points. While the nonlinear (and nonconvex) nature of the optimization problem prevents such guarantees, in practice this conclusion can be reached by choosing a sufficiently large I. To ensure that the I chosen is 'sufficiently large', in this chapter, I was increased until changes in the solution were below a predefined tolerance. To further verify that a control action exists to drive the states inside the next RTRR for the internal points of the ellipsoid, an appropriately defined NLP [27] is solved.

6.3.3 Empirical Reverse-Time Reachability Region-Based MPC Formulation

Consider a batch system described by Eq. (6.1) for fault-free conditions and for which empirical RTRR estimates have been characterized for a given δ and $\mathcal{B}(\boldsymbol{x}_{\text{des}})$. The control action at sampling instance $q = (t_{\text{f}} - t)/\delta$ is computed by solving the following NLP:

$$\min_{\boldsymbol{u}(k)\in\mathcal{U}} J_{\mathcal{R}} = \sum_{k=1}^{P} \alpha(\hat{\boldsymbol{x}}(k) - \boldsymbol{c}_{q-k})^{\text{T}}\mathbf{W}_{q-k}(\hat{\boldsymbol{x}}(k) - \boldsymbol{c}_{q-k}) + \gamma(\Delta\boldsymbol{u}(k)^{\text{T}}\mathbf{R}\Delta\boldsymbol{u}(k)) \tag{6.29}$$

$$\text{subject to: } \hat{\boldsymbol{x}}(0) = \boldsymbol{x}(t) \tag{6.30}$$

$$\hat{\boldsymbol{x}}(k+1) = \sum_{\ell=1}^{L} w_{\ell}(k)\hat{\boldsymbol{\beta}}_{\ell}\left[\boldsymbol{x}^{\text{T}}(k)\ \boldsymbol{u}^{\text{T}}(k)\right] \tag{6.31}$$

$$w_{\ell} = \frac{\|\left[\boldsymbol{x}^{\text{T}}(k)\ \boldsymbol{u}^{\text{T}}(k)\right] - \boldsymbol{c}_{\ell}\|^{-2}}{\sum_{i=1}^{L}\|\left[\boldsymbol{x}^{\text{T}}(k)\ \boldsymbol{u}^{\text{T}}(k)\right]^{\text{T}} - \boldsymbol{c}_{i}\|^{-2}} \tag{6.32}$$

where $\Delta\boldsymbol{u}(k) \in \mathbb{R}^{m}$ denotes a vector of differences between the current and previous input values across the horizon, $\mathbf{R}$ is a positive-definite weighting matrix, and the objective function, $J_{\mathcal{R}}$, is formulated to minimize variations in the control moves and maintain the process states inside the empirical RTRRs over the prediction horizon. The relative importance of the two terms in $J_{\mathcal{R}}$ can be traded off using appropriate values for the weights, α and γ.

Remark 6.8 The predictive model in the MPC formulation, specifically the nonlinear weighting function, makes this optimization problem a NLP, which can potentially be too computationally expensive for real-time application. However, this nonlinearity, while capturing the process dynamics much better than a single linear model, is likely to be much less severe compared to nonlinearities typically found in first-principles deterministic models. Consequently, the optimization problem should remain efficiently solvable even for moderate values of P.

Remark 6.9 Note that due to the unavoidable plant-model mismatch, the proposed MPC formulation does not offer any guarantees regarding the reachability of the process inside the desired end-point neighborhood. In particular, even if one were to impose a constraint in the MPC requiring the states to go to the next RTRR, the feasibility of the constraint (guaranteed if the current states are in the corresponding RTRR) would not guarantee that the process state would be inside the RTRR at the next step. Yet, the determination of the RTRRs, specifically the ellipsoid matrices, provides useful weighting matrices to penalize state deviations to enforce the states to never significantly diverge from conditions where the desired end-point neighborhood can be reached. This also results in important fault-tolerant characteristics as discussed next.

Remark 6.10 In order to generate empirical RTRRs using the data-based model, the database must include state measurements. If the number of states is known and a deterministic model of the system is available but overly complex for use in any online applications (such as state estimation or model-based control), the states of the system can be back calculated off-line from the database measurements using a variety of state estimation tools such as a moving horizon estimator, extended Kalman filter, or the unscented Kalman filter. The resulting states can be used to populate the database, and a state-space model of the form shown in Eq. (6.16) can be developed. This model will capture the nonlinearity in the batch and is more amenable to online applications and usable for generating empirical RTRR estimates (and a corresponding RTRR-based MPC design).

Remark 6.11 Note that if a fault occurs during the batch operation, in the absence of any knowledge of the fault repair time, the only meaningful control objective is to take control action at the current time such that if full control effort were to be restored at the next time step, reachability to the desired end-point neighborhood can be achieved. The RTRR-based MPC, by preserving the states in the reachability regions during the fault repair period, implements exactly this control objective. In contrast, end-point based predictive controllers try to achieve a (potentially) inherently un-achievable objective- that of driving the process to the desired end-point neighborhood subject to the reduced control effort, and in doing so could drive the process to a point from where the desired end-point neighborhood is not reachable even after fault repair.

6.4 Simulation Results

In this section, two simulation examples are presented. The first one illustrates the details of the proposed modeling and control design subject to varying initial conditions, noisy measurements, and time-varying uncertainty, and the second demonstrates an application subject to lack of full state information.

6.4.1 Fed-Batch Reactor

In this section, a data-based model of a fed-batch system is extracted from an artificially generated historical database using the proposed modeling methodology. Then, the resulting model is utilized in the RTRR-based predictive controller. To this end, consider a fed-batch reactor system where an irreversible series of reactions of the form $2\mathsf{A} \xrightarrow{k_1} \mathsf{B} \xrightarrow{k_2} 3\mathsf{C}$ take place. The state-space model has been derived using standard modeling assumptions and using typical parameters values for this system [28]. The state vector is $\boldsymbol{x} = \left[C_{\mathsf{A}}\ C_{\mathsf{B}}\ C_{\mathsf{C}}\ T\ V\right]^{\mathrm{T}}$ where C_{A}, C_{B}, and C_{C} denote the concentrations of species A, B, and C (respectively) and T and V denote the reactor temperature and volume (respectively). The manipulated inputs are the inlet feed rate, F (L/h), and heating coil temperature, T_{hx} (K), $\boldsymbol{u} = \left[F\ T_{\mathsf{hx}}\right]^{\mathrm{T}}$, with constraints $\boldsymbol{u}_{\mathsf{min}} = \left[0\ 288\right]^T$ and $\boldsymbol{u}_{\mathsf{max}} = \left[20\ 360\right]^T$. The primary control objective considered is to drive the process states inside a (arbitrarily chosen) desired end-point around $\boldsymbol{x}_{\mathsf{des}} = \left[2.752\ 1.601\ 0.8422\ 365.756\ 112.425\right]^T$. The desired neighborhood is taken to be an ellipsoid with $\mathbf{W}_0 = \mathrm{diag}\left\{25, 400, 100, 0.04, 1\right\}$ and $\boldsymbol{c}_0 = \boldsymbol{x}_{\mathsf{des}}$. The batch termination time, t_{f}, is taken to be 1 h with a sampling period, δ, of 0.025 h.

Data-Based Model Development

A database consisting of forty batches was generated for the fed-batch reactor system using the state-space model. Ten batches were set aside as the validation data set. With two available inputs, reference trajectories of C_{B} and T (see Fig. 6.1) were chosen to be tracked by manipulating F and T_{hx} (respectively) using two PI controllers. Both PI controllers were tightly tuned for one set of initial conditions and fixed for the remaining 39 batches. The criteria used to tune the PI controllers were the integral of time-weighted absolute error (ITAE) and a reasonably smooth input trajectory.

For a more realistic representation of plant data, sensor noise, disturbances, and a time-varying parameter were also considered. The range of initial conditions and sensor noise levels are summarized in Table 6.1 along with the PI tuning parameters. To simulate disturbances, T^{in} was stochastically varied throughout the duration

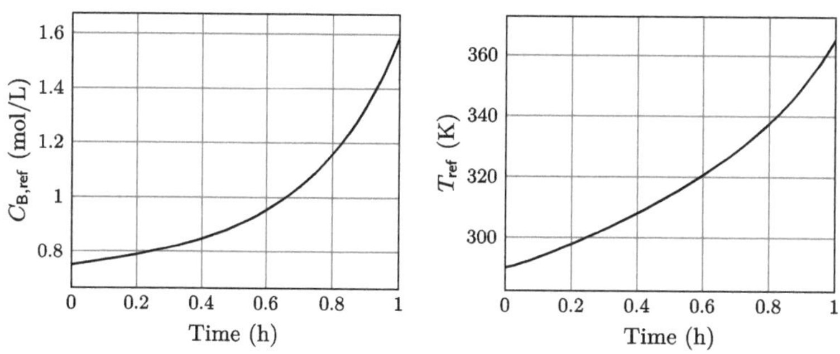

Fig. 6.1 Nominal trajectories of C_{B} and T tracked using two PI controllers to generate the batch database for the fed-batch system

Table 6.1 Simulation parameters used for database generation for the fed-batch system

(a) Initial condition ranges and sensor noise levels

	Initial condition range	Sensor noise
C_A	4.861–5.106 mol/L	±2.2% of original signal
C_B	0.692–0.801 mol/L	±2.8% of original signal
C_C	0.446–0.539 mol/L	±2.1% of original signal
T	280.931–300.057 K	0.10 standard deviation
V	97.952–101.959 L	0.10 standard deviation

(b) PI controller parameters

	Loop 1	Loop 2
Controlled variable	C_B	T
Manipulated variable	F	T_{hx}
Proportional gain	6.10	−0.035
Integral time (h)	0.80	0.01

of each batch around its nominal value in the range 295–305 K. The time-varying parameter was chosen to be the heat exchanger coefficient, UA. At the start of each batch, UA was assigned a value in the range 28,620–30,349 cal/(h · K) and then decreased exponentially to simulate fouling.

With forty time steps in each batch and thirty batches, the operating region space required to be partitioned consisted of 1200 observations. To find the optimum number of clusters, L was varied from $L = 10$ to $L = 100$. For each clustering choice, PLS models were fit for a range of principal components, the trajectories of the validation batches were predicted using the PLS models, and the root-mean-squared error (RMSE) in the predictions was tabulated. The lowest RMSE in the predictions was obtained with $L = 20$ clusters and 142 principal components. Note that from the data arrangement used to compute the PLS models (Eq. (6.17)), the maximum number of principal components is 152. Generally, retaining a high number of principal components produced very low residuals, but the predictive capabilities of the model are reduced due to overfitting. This is because with an excessive number of principal components, the model begins to fit the random noise element in the data. A trade-off between low residuals and high predictive power was found using 142 principal components.

For a specific initial condition and a set of input trajectories in the database, Fig. 6.2 shows the model output from the local linear models together with the corresponding database trajectories. Similar fits were observed for the remaining 29 different initial conditions and input trajectories. Figure 6.3 illustrates the predictive capabilities of the model for an initial condition and input trajectories which are not in the training data, but in the validation data set. The temperature and volume ranges in the figures are significantly larger compared to the ranges for the concentrations because their initial values (and values for the duration of the batch) are an order of magnitude greater than all the concentrations. As a result, the prediction error for the concen-

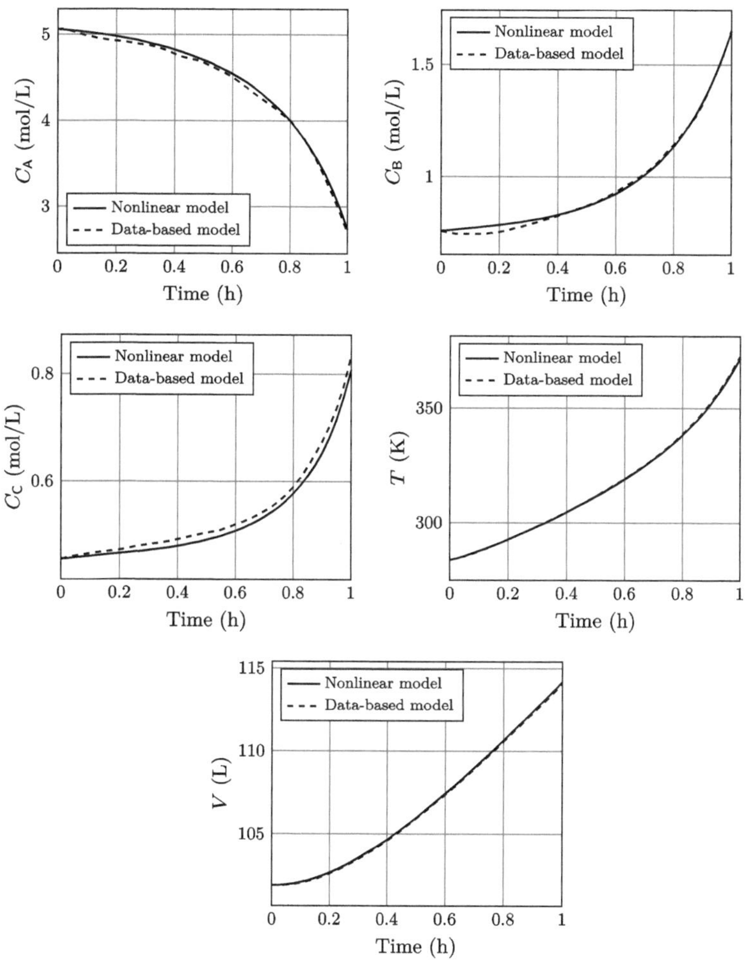

Fig. 6.2 Output from using multiple local linear models and the corresponding batch trajectory in the training data for the fed-batch system

trations are more noticeable compared to the temperature and volume. Overall, the multi-model approach was able to capture the major nonlinearities in the database.

MPC Implementation Using the Data-Based Model

In this section, the proposed RTRR-based tracking MPC design is implemented on the fed-batch reactor system and the control performance is compared with that under PI control. The performance was measured by computing the level set of the desired end-point neighborhood for the final values of the states. A value of less than one indicated the final states were within the desired neighborhood (see Eq. (6.21)). Closed-loop simulations were performed for ten different initial conditions that were not in the training or validation data set. All initial conditions were also verified to be within the

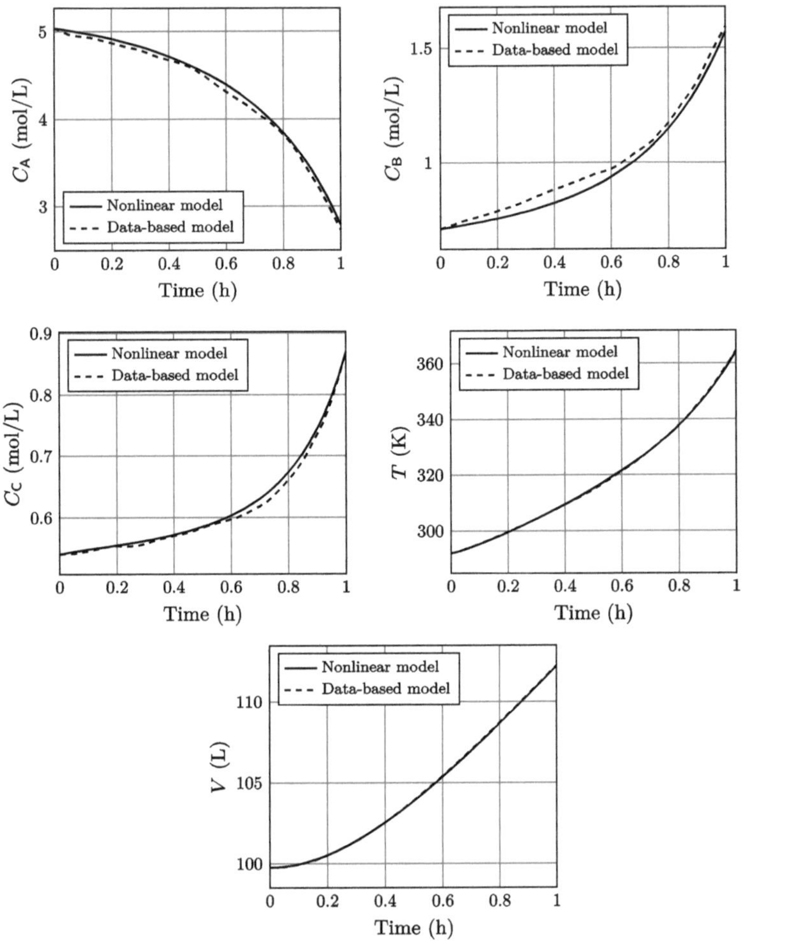

Fig. 6.3 Output from using multiple local linear models and the corresponding batch trajectory in the validation data set, demonstrating the good prediction capability of the proposed modeling method for the fed-batch system

empirical RTRR at the initial time. Sensor noise, a stochastic disturbance in T^{in}, and a time-varying UA parameter were also considered. Similar to the PI tuning, the RTRR-based predictive controller was tuned once for a specific set of initial conditions and left unchanged for the remainder of the simulations to avoid confounding the results with tuning. For RTRR-based MPC, the following values were picked for the tuning parameters: $\mathbf{R} = \text{diag}\{0.01, 0.005\}$ and $P = 18$. The final level sets obtained from the simulations are shown in Table 6.2. The RTRR-based MPC design was able to drive the system inside $\mathcal{B}(\boldsymbol{x}_{\text{des}})$ for all ten initial conditions, whereas the PI controller failed in more than half of the cases. On average, the final MPC and PI level sets were 0.359 and 2.153 (respectively). In Fig. 6.4, a representative set of closed-loop

Table 6.2 Final level sets of $\mathcal{B}(\boldsymbol{x}_{\text{des}})$ achieved using a PI controller and the RTRR-based predictive controller for the fed-batch system

Initial condition:	1	2	3	4	5	6	7	8	9	10
Final PI level set:	2.479	7.220	0.436	1.972	2.233	0.808	0.682	4.045	0.463	1.190
Final MPC level set:	0.700	0.939	0.237	0.932	0.160	0.164	0.0263	0.0883	0.152	0.190

profiles is presented (initial condition #10). To demonstrate that the MPC problem remains efficiently solvable despite being nonlinear, we note that with $P = 18$, the longest CPU time required time to solve the MPC problem was 0.452 s (using GAMS with IPOPT as the solver on Intel Quad Core machine).

To demonstrate the fault tolerance of the RTRR-based controller, we consider faults in both of the control actuators and compare the performance of the RTRR-based MPC design with PI control. Specifically, starting from $\boldsymbol{x}(0) = \left[4.977\ 0.763\ 0.539\ 289.485\ 100.456\right]^{\text{T}}$, we consider the scenario where at $t^{\text{fault}} = 0.25\,\text{h}$, the actuators associated with the flow rate and heating coil fail and the maximum flow rate and heating coil temperature are reduced to $\boldsymbol{u}_{\max} = \left[10\ 310\right]^T$ (from $\left[20\ 360\right]^T$). At $t^{\text{repair}} = 0.45\,\text{h}$, the fault is rectified and full control effort is recovered. Note that during the failure period, the prediction horizon of the RTRR-based MPC was reduced from $P = 18$ to $P = 1$, to avoid having to assume the failure situation any longer than necessary (in computing the control action). For the PI controller, the batch is driven to the end-point of $\boldsymbol{x}(t_{\text{f}}) = \left[3.647\ 1.237\ 0.712\ 339.084\ 110.763\right]^T$, which is well outside of $\mathcal{B}(\boldsymbol{x}_{\text{des}})$ resulting in a final level set value of 104.393. On the other hand, the RTRR-based MPC drives the process to $\boldsymbol{x}(t_{\text{f}}) = \left[2.810\ 1.588\ 0.867\ 365.605\ 112.148\right]^T$ which corresponds to a final level set of 0.173. The closed-loop profiles for this case are shown in Fig. 6.5. The PI controller prescribes the heat exchanger temperature to remain saturated during the failure period, whereas the RTRR prescribes more meaningful temperature toward the latter stages of the fault. As a result, the RTRR-based controller is able to recover after the fault and essentially begins to track the nominal state trajectories which terminate at the desired end-point.

6.4.2 Nylon-6,6 Batch Polymerization

In this section, we apply the data-based modeling methodology on a complex nonlinear batch polymerization process to extract models for the key measurable process variables. Subsequently, we employ the model in a trajectory tracking predictive controller and compare the tracking performance with a PI controller. For our studies, we utilize the mathematical model of nylon-6,6 polymerization wherein the polymer is produced by the amidation of adipic acid and hexamethylenediamine (HMD) in a batch autoclave reactor with a steam jacket for providing the heat needed for vaporiza-

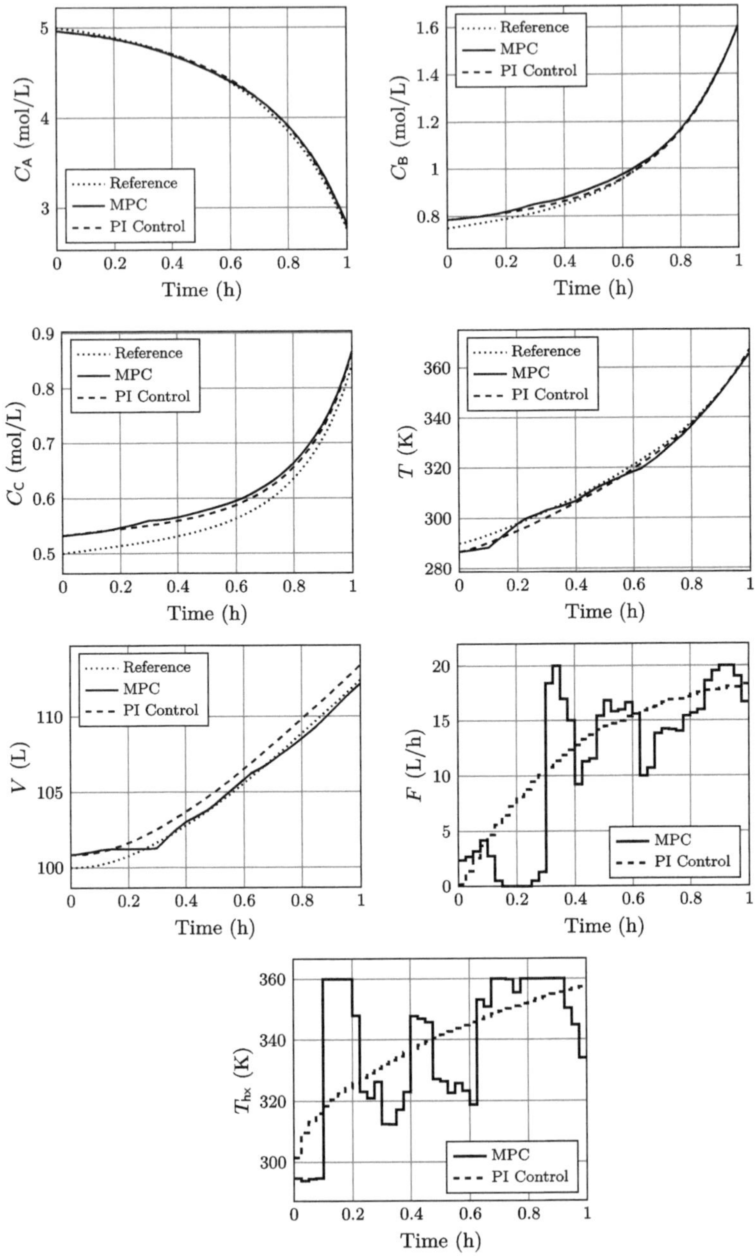

Fig. 6.4 Representative state and input profiles of the fed-batch reactor system under PI control and RTRR-based MPC with no input failures. The nominal set of state trajectories that terminate at the desired end-point is also shown

Table 6.3 Tracking performance of the PI controller and the proposed MPC formulation for the nylon-6,6 batch polymerization system

Initial condition:	1	2	3	4	5	6	7	8	9	10
PI T ITAE:	9.10	10.18	3.33	4.95	12.99	6.99	13.23	3.14	14.63	4.91
MPC T ITAE:	2.18	1.51	1.33	1.76	2.98	1.21	2.72	1.19	1.90	1.89
PI P ITAE:	3.08	3.46	5.02	2.71	1.93	10.29	5.18	1.20	1.43	4.61
MPC P ITAE:	1.01	1.22	1.47	2.03	1.43	1.04	1.57	0.860	1.31	1.58

tion (and reaction) and a valve for venting vaporized water [29]. The reaction model, modeling assumptions (and their explanations), parameter values, and kinetic relationships are available in the existing literature [29, 30] and omitted here for brevity. The final state-space model of the process consists of nine coupled ordinary differential equations (ODEs) with the state vector comprised of the molar amounts of each functional group and evaporated HMD, the reaction medium mass, temperature, and volume, and reactor pressure. The final model takes the general form shown in Eq. (6.1) with the outputs, $\boldsymbol{y}$, being the reaction mixture temperature, T (K), and reactor pressure, P (psi). The manipulated inputs are the steam jacket pressure, P_j (psi), and vent rate, v (kg/h), $\boldsymbol{u} = \left[P_j \; v\right]^{\mathrm{T}}$, constrained between $\boldsymbol{u}_{\mathrm{min}} = \left[700 \; 0\right]^{\mathrm{T}}$ and $\boldsymbol{u}_{\mathrm{max}} = \left[1800 \; 2000\right]^{\mathrm{T}}$. The duration of the batch is $t_{\mathrm{f}} = 3\,\mathrm{h}$ with a sampling period of 60 s.

For this chapter, we focus specifically on tracking trajectories of the reaction medium temperature, T, and reactor pressure, P, by manipulating the steam jacket pressure, P_j and vent rate, v. We assume reference trajectories for T and P, denoted by T_{ref} and P_{ref} (respectively), have been identified appropriately in some fashion, and Fig. 6.6 presents these specific trajectories.

Data-Based Model Development

For the nylon-6,6 system, the pressure dynamics were significantly faster than the temperature dynamics, leading to a weak correlation between the two outputs. Consequently, individual linear models were computed for the outputs as opposed to a PLS model which predicts both outputs. To develop the data-based models for the two outputs, a database of previous batches was first generated. To this end, the deterministic nylon-6,6 polymerization model [29] was simulated fifteen times from different initial conditions (five batches were reserved as the validation data set). As in the fed-batch case, the set of reference temperature and pressure profiles presented in Fig. 6.6 were tracked reasonably well using two PI controllers to generate the database. For the PI loop-pairing, the vent rate was used to track the reactor pressure while the steam jacket pressure was used to track the temperature. Both PI controllers were tightly tuned (using ITAE and smooth input trajectories as the criteria) for one set of initial conditions and fixed for the remaining batches.

The following parameters were required to be specified for the modeling approach: the lag structure (n_y and n_u) and the number of clusters, L. To understand the identification procedure, consider just the temperature model. The temperature model was identified by first generating a series of different models as follows: First, n_y and

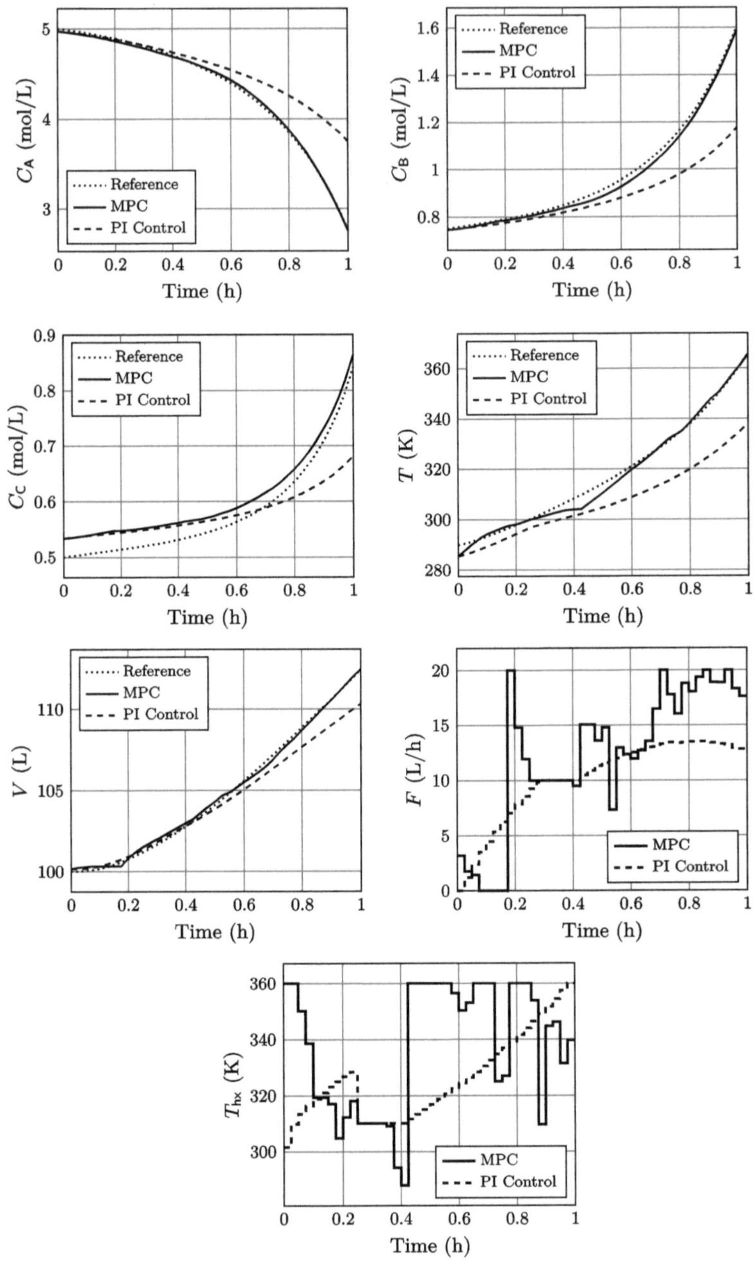

Fig. 6.5 Representative state and input profiles of the fed-batch reactor system under PI control and RTRR-based MPC with input failures between 0.25 and 0.45 h. The nominal set of state trajectories that terminates at the desired end-point is also shown. The profiles demonstrate the fault-tolerant characteristic of the RTRR-based MPC design

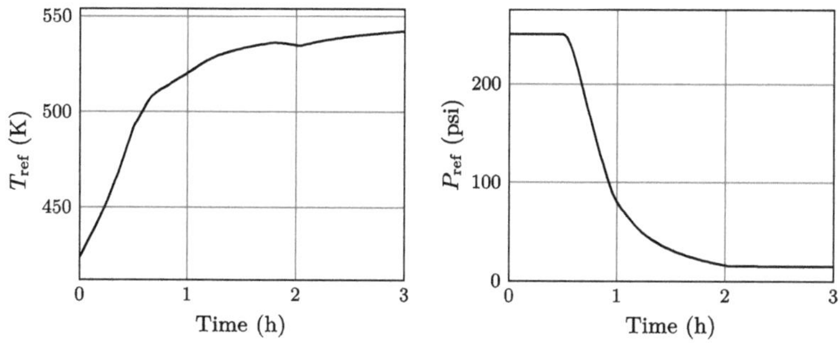

Fig. 6.6 Nominal trajectories of T and P tracked using two PI controllers to generate the batch database for the nylon-6,6 batch polymerization system

n_u were specified, thus setting the dimensions of $\bar{\mathbf{X}}$. Next, $\bar{\mathbf{X}}$ was decomposed using PCA, the resulting score space was clustered for $L = 2$ to $L = 20$, and local linear models were identified for each choice of L via principal component regression (PCR) [15]. The quality of each model was evaluated by its RMSE in predicting the five validation batches. This was repeated for several choices for the lag structure. For the temperature model, $L = 5$ and lags of 1, 0, 1, and 1 in the temperature, reactor pressure, steam jacket pressure, and vent (respectively) yielded the lowest RMSE. Note that zero lags for the reactor pressure indicate that the reactor pressure was not used in predicting the reaction mixture temperature. A similar procedure was repeated for the pressure model, and the final lag structure was found to be 0, 1, 0, and 1 for the temperature, reactor pressure, steam jacket pressure, and vent rate (respectively) for which $L = 1$. These results were expected for the pressure model since they correspond to a linear dynamic model between the pressure and vent rate, which is consistent with the state-space model. In Fig. 6.7, we compare the output of the nonlinear model with the model output from the data-based model for a set of initial conditions that was in the original training data. In Fig. 6.8, we consider initial conditions which were in the validation data set. Overall, the multi-model approach captured the nonlinear nature of the batch and provided relatively reliable predictions.

MPC Implementation Using the Data-Based Model

A predictive controller for tracking reference temperature and pressure profiles for the nylon-6,6 polymerization process is presented in this section. The control action at each sampling instance can be computed by solving the optimization problem below.

$$\min_{u_{\min} \le u(k) \le u_{\max}} J = \sum_{k=1}^{P} \|\hat{y}(k) - y_{\text{ref}}(k)\|_{\mathbf{Q}} + \|\Delta u(k)\|_{\mathbf{R}} \tag{6.33}$$

$$\text{subject to: Equation (6.19)} \tag{6.34}$$

$$\hat{y}(0) = y(t) \tag{6.35}$$

where $\mathbf{Q}$ and $\mathbf{R}$ are positive-definite matrices, and P is the prediction horizon.

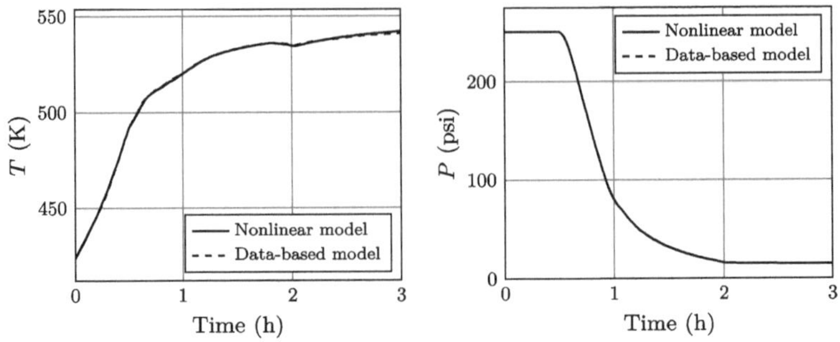

Fig. 6.7 Output from using multiple local linear models and the corresponding batch trajectory in the training data for the nylon-6,6 batch polymerization system

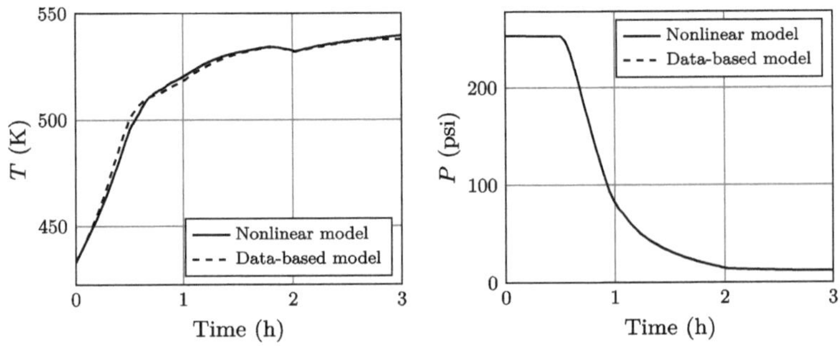

Fig. 6.8 Output from using multiple local linear models and the corresponding batch trajectory in the validation data set for the nylon-6,6 batch polymerization system, demonstrating the good prediction capability of the proposed modeling method

Closed-loop simulations for ten new initial conditions (all from within the range of initial conditions in the training data) were performed using the proposed trajectory tracking MPC design, and the performance was compared against a PI controller. The tuning parameters used for the proposed MPC were: $\mathbf{Q} = \text{diag}\{2.75, 27.5\}$, $\mathbf{R} = \text{diag}\{0.02, 0.02\}$, and $P = 12$.

On average, the proposed MPC controller offered a significant improvement of approximately 78 and 65% for temperature and pressure tracking (See Table 6.3 respectively). In all simulations, the proposed predictive controller outperformed the PI controller. A representative set of closed-loop simulation results is presented in Fig. 6.9. In this case, the ITAEs for the proposed predictive controller improved on the PI controller by 77 and 26%. Overall, the simulation results clearly demonstrate the advantages of implementing the proposed trajectory tracking predictive controller over PI control.

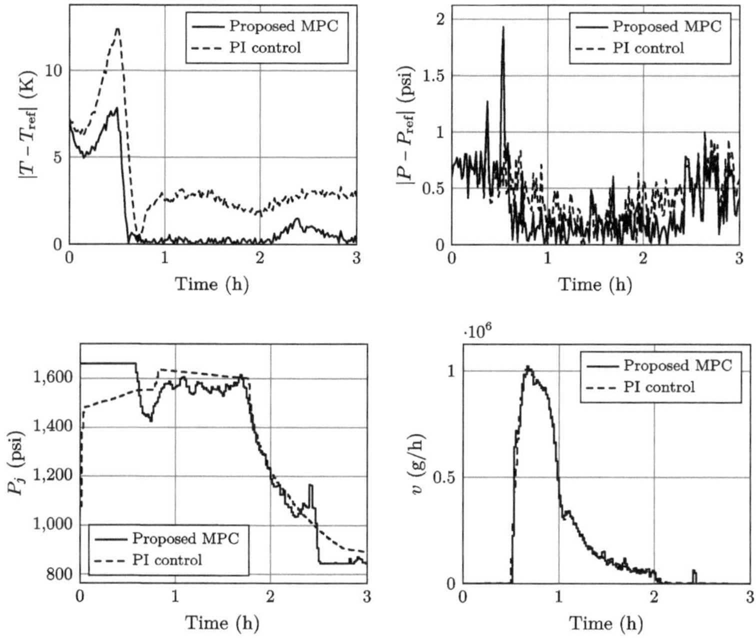

Fig. 6.9 Representative output tracking error and input profiles of the nylon-6,6 batch polymerization system under PI control and the proposed trajectory tracking MPC design

6.5 Conclusions

In this chapter, we addressed the problem of uniting empirical modeling approaches with nonlinear control tools for control of batch systems. In the proposed approach, we exploited the availability of historical batch data, the simplicity of local linear models, the data extraction capabilities of PLS/PCR, and the use of appropriate clustering and weighting techniques in conjunction with multiple models to capture the nonlinear nature of a batch process. The resulting model from this approach was employed to generate empirical RTRRs, which were subsequently incorporated in a predictive control design. The efficacy of the RTRR-based MPC design and superior performance, as well as fault-handling ability, with respect to PI control was demonstrated through a fed-batch simulation example. The data-based modeling approach was also applied to derive the models for use in a trajectory tracking MPC design for a nylon-6,6 batch polymerization process with limited measurements. The tracking performance of the proposed trajectory tracking controller was demonstrated to be significantly superior to PI control.

References

1. Flores-Cerrillo, J., MacGregor, J.F.: Latent variable MPC for trajectory tracking in batch processes. J. Proc. Control **15**, 651–663 (2005)
2. Soroush, M., Kravaris, C.: Optimal-design and operation of batch reactors. 1. Theoretical framework. Ind. Eng. Chem. Res. **32**, 866–881 (1993)
3. Soroush, M., Kravaris, C.: Optimal-design and operation of batch reactors. 2. A case-study. Ind. Eng. Chem. Res. **32**, 882–893 (1993)
4. Cruickshank, S.M., Daugulis, A.J., McLellan, P.J.: Dynamic modeling and optimal fed-batch feeding strategies for a two-phase partitioning bioreactor. Biotech. Bioeng. **67**, 224–233 (2000)
5. Zhang, G.P., Rohani, S.: On-line optimal control of a seeded batch cooling crystallizer. Chem. Eng. Sci. **58**, 1887–1896 (2003)
6. Corriou, J.P., Rohani, S.: A new look at optimal control of a batch crystallizer. AIChE J. **54**(12), 3188–3206 (2008)
7. Trifkovic, M., Sheikhzadeh, M., Rohani, S.: Multivariable real-time optimal control of a cooling and antisolvent semibatch crystallization process. AIChE J. **55**(10), 2591–2602 (2009)
8. Bhat, S.A., Huang, B.: Preferential crystallization: multi-objective optimization framework. AIChE J. **55**(2), 383–395 (2009)
9. Palanki, S., Vemuri, J.: Optimal operation of semi-batch processes with a single reaction. Int. J. Chem. React. Eng. **3**, 1219 (2005)
10. Pistikopoulos, E.N., Dua, V., Bozinis, N.A., Bemporad, A., Morari, M.: On-line optimization via off-line parametric optimization tools. Comput. Chem. Eng. **26**, 175–185 (2002)
11. Aumi, S., Mhaskar, P.: Safe-steering of batch processes. AIChE J. **55**, 2861–2872 (2009)
12. Lin, W.L., Qin, S.J., Ljung, L.: On consistency of closed-loop subspace identification with innovation estimation. In: IEEE Conference on Decision and Control (CDC), pp. 2195–2200 (2004)
13. Huang, B., Ding, X.S., Qin, S.J.: Closed-loop subspace identification: an orthogonal projection approach. J. Process Control **15**(1), 53–66 (2005)
14. Wang, J., Qin, S.J.: Closed-loop subspace identification using the parity space. Automatica **42**(2), 315–320 (2006)
15. Geladi, P., Kowalski, B.: Partial least-squares regression: a tutorial. Anal. Chim. Acta **185**, 1–17 (1986)
16. Höskuldsson, A.: PLS regression methods. J. Chemom. **2**, 211–228 (1988)
17. Seber, G.A.F.: Multivariate Observations. Wiley, New York (1984)
18. Ruspini, E.H.: Numerical methods for fuzzy clustering. Inf. Sci. **2**(3), 319–350 (1970)
19. Bezdek, J.: Pattern Recognition with Fuzzy Objective Function Algorithms. Kluwer Academic Publishers, Norwell (1981)
20. Gustafson, D.E., Kessel, W.C.: Fuzzy clustering with a fuzzy covariance matrix. In: Proceedings of the IEEE Conference on Decision and Control (CDC), pp. 761–766 (1979)
21. Bezdek, J., Coray, C., Gunderson, R., Watson, J.: Detection and characterization of cluster substructure I. Linear structure: fuzzy *c*-lines. SIAM J. Appl. Math. **40**(2), 339–357 (1981)
22. Krishnapuram, R., Freg, C.: Fitting an unknown number of lines and planes to image data through compatible cluster merging. Pattern Recognit. **25**(4), 385–400 (1992)
23. Frigui, H., Krishnapuram, R.: Clustering by competitive agglomeration. Pattern Recognit. **30**(7), 1109–1119 (1997)
24. Gath, I., Geva, A.: Unsupervised optimal fuzzy clustering. IEEE Trans. Pattern Anal. Mach. Intell. **11**, 773–780 (1989)
25. Xie, X.L., Beni, G.: A validity measure for fuzzy clustering. IEEE Trans. Pattern Anal. Mach. Intell. **13**, 841–847 (1991)
26. Yen, J., Wang, L., Gillespie, C.: Improving the interpretability of TSK fuzzy models by combining global learning and local learning. IEEE Trans. Fuzzy Syst. **6**(4), 530–537 (1998)
27. Yen, J., Wang, L., Gillespie, C.: Robust model predictive control and fault-handling of batch processes. AIChE J. **57**, 1796–1808 (2011)

28. Fogler, H.S.: Elements of Chemical Reaction Engineering, 4th edn. pp. 625–627. Prentice Hall, Upper Saddle River (2006)
29. Russell, S.A., Robertson, D.G., Lee, J.H., Ogunnaike, B.A.: Control of product quality for batch nylon-6,6 autoclaves. Chem. Eng. Sci. **53**(21), 3685–3702 (1998)
30. Joly, M., Pinto, J.: Optimal control of product quality for batch nylon-6,6 autoclaves. Chem. Eng. J. **97**, 87–101 (2004)

Chapter 7
An Adaptive Data-Based Modeling Approach for Predictive Control of Batch Systems

7.1 Introduction

In the previous chapter, a modeling approach was presented which utilized multi-models for predicting batch dynamics. To enable improved disturbance rejection, one contribution of the present chapter is the generalization of the modeling approach of [1] to incorporate the ability to capture time-varying parameters/use information available from new operating conditions immediately (instead of waiting until the end of the batch to update the model). Adding an adaptive element to model-based control designs wherein the model's parameters are updated online at each sampling instant has been a popular method for improving closed-loop performance [2]. Other approaches have focussed on ensuring the ability to compute online the manipulated input trajectory by exploiting the structure of the optimal solution [3–6].

The recursive least squares (RLS) algorithm (a linear estimator) is most commonly used for updating empirical model parameters due to the model's assumed linear form (with respect to the parameters). One drawback of conventional RLS algorithms is an inherent assumption of static/stationary model parameters. The RLS framework can straightforwardly accommodate a random walk model for handling time-varying parameters, but the more popular method to accomplish this has been to discount past data in an exponentially weighted manner with a forgetting factor. The use of a forgetting factor (or, in fact, any mechanism to account for time-varying dynamics) is particularly important for maintaining the validity of empirical models of batch processes. In particular, because of using a linear model (or appropriate combinations of linear models) to describe the inherently nonlinear process, new operating conditions can be encountered during a batch around which the model is highly inaccurate or completely invalid, specifically if such conditions were absent from the 'training' data set. By updating the model more aggressively based on the current operating conditions, the newly encountered local dynamics can be modeled to some extent, helping to preserve the model's accuracy and validity.

Motivated by the above considerations, in this chapter, we add online learning ability to our previously developed modeling approach [1] and use the resulting

P. Mhaskar et al., *Modeling and Control of Batch Processes*,
Advances in Industrial Control, https://doi.org/10.1007/978-3-030-04140-3_7

adaptive model in a trajectory tracking predictive controller. Two algorithms are used for adapting the model: (1) the standard RLS algorithm with a forgetting factor and (2) a probabilistic RLS (PRLS) algorithm (also with a forgetting factor) specifically developed for the modeling approach. The rest of this manuscript is organized as follows: We begin by presenting the class of processes that is considered and then review the multi-model approach. Next, we add online learning ability to the multi-model framework. We begin by showing how the standard RLS algorithm can be applied in a straightforward manner. This is followed by the development of a PRLS estimator for each local model that takes the model's probability of being representative of the current plant dynamics into account during the update. Simulation results of a nylon-6,6 batch polymerization process (subject to limited measurements, sensor noise, and disturbances in the initial conditions) are then presented. We initially extract a process model from an artificial database of previous batches and then make it adaptive using RLS and PRLS. Subsequently, we demonstrate the improved closed-loop performance achieved from using the adaptive model (over a nonadaptive model) in a trajectory tracking predictive controller. Finally, we summarize our results.

7.2 Preliminaries

In this section, we first describe the class of batch processes considered. Then, we give an overview of the data-based modeling approach for batch processes developed in our earlier work [1].

7.2.1 *Process Description*

We consider batch processes subject to input constraints and uncertainties described by:

$$\begin{aligned} \dot{\boldsymbol{x}} &= \boldsymbol{f}(\boldsymbol{x}, \boldsymbol{u}, \boldsymbol{\theta}) \\ \boldsymbol{y} &= \boldsymbol{g}(\boldsymbol{x}) + \boldsymbol{v} \\ t &\in [t_0, t_f], \boldsymbol{u}(\cdot) \in \mathcal{U}, \boldsymbol{\theta}(\cdot) \in \Theta, \boldsymbol{x}(t_0) = \boldsymbol{x}_0 \end{aligned} \tag{7.1}$$

The vectors, $\boldsymbol{x} \in \mathbb{R}^n$ and $\boldsymbol{y} \in \mathbb{R}^p$, denote the state variables and noise corrupted output variables (respectively), $\boldsymbol{v} \in \mathbb{R}^p$ is a zero-mean, normally distributed measurement noise vector, and $\boldsymbol{u} \in \mathbb{R}^m$ is the vector of constrained manipulated inputs, taking values in a nonempty, convex set, $\mathcal{U} \triangleq \{\boldsymbol{u} \mid \boldsymbol{u}_{\min} \leq \boldsymbol{u} \leq \boldsymbol{u}_{\max}\} \subset \mathbb{R}^m$ where $\boldsymbol{u}_{\min}$ and $\boldsymbol{u}_{\max}$ define the minimum and maximum (respectively) allowable input values. The vector, $\boldsymbol{\theta} \in \mathbb{R}^q$, is constituted of bounded, possibly time-varying uncertain variables, taking values in the nonempty, compact set, $\Theta \triangleq \{\boldsymbol{\theta} \mid \boldsymbol{\theta}_{\min} \leq \boldsymbol{\theta} \leq \boldsymbol{\theta}_{\max}\} \subset \mathbb{R}^q$

where $\boldsymbol{\theta}_{\min}$ and $\boldsymbol{\theta}_{\min}$ denote the minimum and maximum (respectively) allowable uncertainty realizations. The times, t_0 and t_f, denote the initial time and batch termination times, respectively. Throughout the manuscript, we assume that for any input trajectory, $\boldsymbol{u}(\cdot) \in \mathcal{U}$, the solution of the batch system exists (with initial conditions, $\boldsymbol{x}_0$) and is continuous for all $t \in [t_0, t_f]$.

7.2.2 Multi-model Data-Driven Modeling for Batch Systems

In this section, we review a multi-model, data-based modeling approach for batch processes [1]. Mathematically, the model for the process output (for the sake of clarity, we present the equation describing the evolution of a single process output; the method, however, is applicable to the case of multiple process outputs) takes the form of a weighted combination of L auto-regressive exogenous (ARX) models as shown below in (7.2):

$$\begin{aligned}\hat{y}(k) &= \sum_{\ell=1}^{L} w_\ell(k)\bar{\boldsymbol{x}}(k)\hat{\boldsymbol{\beta}}_\ell \\ &= \sum_{\ell=1}^{L} w_\ell(k)\left[\boldsymbol{y}'(k-1)\ \ldots\ \boldsymbol{y}'(k-n_y)\ \boldsymbol{u}'(k-1)\ \ldots\ \boldsymbol{u}'(k-n_u)\right]\hat{\boldsymbol{\beta}}_\ell\end{aligned} \tag{7.2}$$

where $w_\ell(k)$ is model ℓ's weight at sampling instant, k, and $\hat{\boldsymbol{\beta}}$ defines the ℓth local model. The vector, $\bar{\boldsymbol{x}}(k)$, consists of lagged concatenated outputs and inputs where the scalars, n_y and n_u, denote the number of lags in the outputs and inputs (respectively). For notational simplicity, we have assumed the same number of lags, n_y and n_u, for each output and input variable (respectively) and the same lag structure for all L models. These assumptions can be readily relaxed. Different criteria can be used to select the 'optimum' lag structure with the general objective being to achieve low prediction error with the minimum number of model parameters (i.e., lags), which prevents against overfitting and maintains model simplicity [7].

Using the following definitions,

$$\begin{aligned}\hat{\boldsymbol{\beta}} &\triangleq \left[\hat{\boldsymbol{\beta}}_1 \ldots \hat{\boldsymbol{\beta}}_\ell \ldots \hat{\boldsymbol{\beta}}_L\right]' \\ \boldsymbol{h}(k) &\triangleq \left[w_1(k)\bar{\boldsymbol{x}}(k) \ldots w_\ell(k)\bar{\boldsymbol{x}}(k) \ldots w_L(k)\bar{\boldsymbol{x}}(k)\right]\end{aligned}$$

Equation (7.2) can be rewritten in the common least squares vector form:

$$\hat{y}(k) = \boldsymbol{h}(k)\hat{\boldsymbol{\beta}} \tag{7.3}$$

The model identification procedure is an initial clustering step followed by solving a linear regression problem. In the first step, for a given lag structure, a matrix, $\bar{\mathbf{X}}$, corresponding to $\bar{\boldsymbol{x}}(k)$, is generated by sorting the plant data sample-wise, and then, $\bar{\mathbf{X}}$ (or its equivalent latent variable space—see the remarks below) is clustered into L clusters using fuzzy c-means clustering. In fuzzy c-means clustering, points that are mathematically 'similar' according to the Euclidean 2-norm are clustered into overlapping spherical clusters with corresponding center points [8]. Each cluster represents a region in the $\bar{\mathbf{X}}$ space where an associated model has the highest degree of validity, and in this way, the cluster's center point represents the linearization point of its associated model. Using the results of the clustering step (the cluster center points), the weights, $w_\ell(\cdot)$, for the training data can be computed prior to the model coefficients (to be discussed shortly). Consequently, the $\boldsymbol{h}(k)$ vector in (7.3) is completely specified for the training data. Thus, a regressor matrix corresponding to $\boldsymbol{h}(k)$, $\mathbf{H}(k)$, can be constructed, and the local linear models, $\hat{\boldsymbol{\beta}}$, are computable using linear regression.

Intuitively, from (7.1), the model weights, $w_\ell(\cdot)$, should depend on the current values of the states, inputs, and uncertainty realizations since they define the system dynamics through $\boldsymbol{f}(\cdot)$. In other words, the local models should be weighted according to the current process conditions. In the absence of state measurements, the vector of lagged outputs and inputs, $\bar{\boldsymbol{x}}(k)$, can be used to infer the current process conditions, and each model's weight can be assigned based on the proximity of the operating conditions to its center point. For instance, denoting model ℓ's center point as $\boldsymbol{c}_\ell$, its weight should be inversely proportional to the squared distance between $\bar{\boldsymbol{x}}(k)$ and $\boldsymbol{c}_\ell$:

$$w_\ell(k) \propto \|\bar{\boldsymbol{x}}'(k) - \boldsymbol{c}_\ell\|^{-2}$$

Normalizing this expression over all clusters yields the following normalized weighting function:

$$w_\ell(k) = \frac{\|\bar{\boldsymbol{x}}'(k) - \boldsymbol{c}_\ell\|^{-2}}{\sum_{i=1}^{L} \|\bar{\boldsymbol{x}}'(k) - \boldsymbol{c}_i\|^{-2}}$$

This weighting function is identical to the fuzzy c-means clustering *membership* function, implying the weights for the training data are computed as a by-product of the clustering step.

The number of clusters is an important parameter in this approach. To evaluate the goodness of the final fuzzy partitions, many validation measures have also been proposed with the most popular being the Xie-Beni index [9], which is a ratio of the total within-cluster variance to the separation of the cluster centers (and therefore should be minimal for the best partition). In this chapter, we iteratively refined the number of clusters based on how well an independent validation data set was predicted. Thus, there was a balancing of the number of clusters and prediction error.

Remark 7.1 In performing the linear regression to estimate the model coefficients, ordinary least squares (OLS) regression may be unsuitable since the columns in the regressor matrix may be highly correlated and/or colinear as a result of lagging process variables, leading to rank deficiency and imprecise model parameter estimates. Moreover, the process variables can be cross-correlated (correlated with each other and other variables at different sampling instants) since the data may be collected under closed-loop conditions [10, 11], which introduces relationships between the process inputs and previous outputs. Possible colinearities/correlations in the regressor matrix that are problematic in OLS regression can be handled by using latent variable tools such as principal component regression (PCR) or partial least squares regression (PLS) when the process has several correlated output variables [12].

Remark 7.2 From the definition of $\bar{\mathbf{X}}$, the dimension of the space required to be clustered is $(p \times n_y) + (m \times n_u)$, which can be prohibitively high for the required computations. Additionally, as previously discussed, the columns of $\bar{\mathbf{X}}$ can be correlated. The dimensionality and correlation problems can be handled by first decomposing $\bar{\mathbf{X}}$ using principal component analysis (PCA) [12] and clustering the resulting score/latent variable space, $\mathbf{T}$, instead of $\bar{\mathbf{X}}$. The loading matrix from PCA, $\mathbf{P}$, can be used to relate the measurable variables to the scores by $\mathbf{T} = \bar{\mathbf{X}}\mathbf{P}$. The score space typically has a much lower dimension and better numerical properties (i.e., orthogonality), making the clustering computations more numerically stable and computationally tractable. Note that having chosen to represent the nonlinear dynamics with local linear model, PCA becomes a natural choice to sift through extraneous data and identify the core dimensionality (states) of the system.

7.3 Online Estimation of the Data-Based Model Parameters

The parameters of the data-based model (7.2) are the cluster center points (which define the weighting function) and the linear model coefficients. One way to adapt this model is where both sets of parameters are updated at every sampling instant using the plant measurements. In this chapter, we look at a subset of this problem by assuming that the originally computed cluster center points hold for the current batch. Of course, if the new batch encounters significantly new set of conditions, the cluster points can always be updated before updating the parameters. Once finished, the data from the new batch are added to the existing database, and the cluster centers along with the ARX model coefficients are updated. That is, the model coefficients and cluster centers are updated in an intra- and inter-batch fashion, respectively.

In this section, we address the problem of how to recursively update the ARX model coefficients during a batch as measurements become available. We start by demonstrating that the standard RLS algorithm is a natural solution to this problem and brings attention to a few issues regarding its implementation. Next, we develop a probabilistic RLS (PRLS) algorithm specific to the multi-model approach that adopts a localized, probabilistic approach to the model updates.

7.3.1 Recursive Least Squares (RLS) Parameter Estimation

Consider the scenario where OLS regression or PCR has been used to estimate the model coefficients and their covariances, denoted henceforth by $\hat{\boldsymbol{\beta}}(0)$ and $\mathbf{P}(0)$ (respectively). The model coefficient estimates can be updated based on the error between the predicted output and the plant measurement by re-arranging the regression solution in a recursive form, yielding the standard recursive least squares (RLS) algorithm [13]. The updated estimates given by RLS are equivalent to those that would be obtained from OLS/PCR after appending the response and regressor matrices with new rows corresponding to the current measurement. In this way, the old training data is given just as much importance as the new data. With limited training data (as is the case for batch processes), operating conditions can be encountered during a new batch which are not modeled by the original linear models. Under these circumstances, it is desirable to update the models more aggressively based on the new plant data.

One extension of RLS to cover this situation is to replace the conventional least squares criterion with one having time-varying weighting of the data as shown below:

$$V(\hat{\boldsymbol{\beta}}, N) = \frac{1}{2\sigma_y^2} \sum_{i=1}^{N} \lambda^{N-i} \left(y(i) - \boldsymbol{h}(i)\hat{\boldsymbol{\beta}} \right)^2 \tag{7.4}$$

where N is the number of observations, σ_y^2 is the measurement noise variance, and the scalar, $0 < \lambda \leq 1$, is a *forgetting factor*. With the inclusion of λ, the most recent data is given unit weight while data that is n time units old is weighted down by λ^n. The effective memory length of the data (i.e., the number of observations used at any sampling instant for the update), N_t, is given by [13]:

$$N_t = \frac{1}{1 - \lambda} \tag{7.5}$$

The set of recursive equations shown in (7.6)–(7.9) for computing $\hat{\boldsymbol{\beta}}(k)$ and its covariance, $\mathbf{P}(k)$, can be derived for the least squares criterion in (7.4) [13].

$$v(k) = y(k) - \boldsymbol{h}(k)\hat{\boldsymbol{\beta}}(k-1) \tag{7.6}$$

$$\hat{\boldsymbol{\beta}}(k) = \hat{\boldsymbol{\beta}}(k-1) + \boldsymbol{k}(k)v(k) \tag{7.7}$$

$$\boldsymbol{k}(k) = \frac{\mathbf{P}(k-1)\boldsymbol{h}'(k)}{\boldsymbol{h}(k)\mathbf{P}(k-1)\boldsymbol{h}'(k) + \lambda\sigma_y^2} \tag{7.8}$$

$$\mathbf{P}(k) = \frac{1}{\lambda} \left[\mathbf{I} - \boldsymbol{k}(k)\boldsymbol{h}(k)\right] \mathbf{P}(k-1) \tag{7.9}$$

Note that by starting from a time-varying criterion, the inherent assumption of static model parameters associated with RLS without a forgetting factor is removed. This becomes important if the process has time-varying uncertainties. In the above set of recursive equations, $v(k)$ is the prediction error or innovation, $y(k)$ is the plant measurement, and $\boldsymbol{k}(k)$ is a gain vector. The estimate and covariance can be initialized at the solution of the original regression problem, $\hat{\boldsymbol{\beta}}(0)$ and $\mathbf{P}(0)$. For $\lambda = 1$, (7.7)–(7.9) reduce to the standard RLS algorithm and the effective memory length of the data is ∞, signifying all the data is used with equal weight. Additionally, (7.9) indicates that the estimate variance always decreases as the batch progresses for $\lambda = 1$. This can be understood as the level of uncertainty in the estimates decreasing with time as more information is being used for the estimation.

When $\lambda < 1$, as new data becomes available, old data is continuously discounted. While this behavior is necessary for adapting more aggressively based on new data, it becomes problematic when there is no information in the new data, which is referred to as periods of low excitation. During low excitation periods, data loss occurs since old data is discounted while the new data has no information. Consequently, the covariance matrix elements can increase unboundedly and become ill-conditioned, leading to the estimates 'blowing-up' or 'bursting.' In a closed-loop environment, periods of low excitation can occur when the controller begins to track constant or slowly varying set-points during which the inputs and outputs show little variation between successive sampling instants and no information is obtainable from the plant data while the old (useful) data continues to be discounted. A common practice to keep the plant persistently excited has been to add small dither signals on top of the inputs that do not significantly affect the closed-loop behavior.

Low excitation periods are less common in batch processes compared to continuous processes. Recall that the set-point trajectories for batch systems are typically computed off-line or regenerated periodically during the batch by solving a (dynamic) optimization problem. The set-point trajectories tend to vary throughout the batch duration while covering a wide range of operating conditions. The inputs are continually adjusted in response, keeping the process more or less persistently excited. The flexibility to use complex, time-varying set-point trajectories is, in fact, one of the main reasons for the popularity of batch processing. Nevertheless, there may be periods when the set-point trajectories plateau specifically during the initial and/or finishing stages of the batch. There have been many techniques proposed to handle the effects of low excitation [13]. Two approaches that require limited modifications to the RLS framework include using a variable forgetting factor [14] such that the information content in the filter remains constant or the constant trace algorithm [13] wherein the covariance matrix is explicitly bounded by scaling it at each iteration such that its trace is constant.

7.3.2 Probabilistic Recursive Least Squares (PRLS) Parameter Estimation

The standard RLS algorithm with a forgetting factor in (7.6)–(7.9) simultaneously updates all of the local ARX model coefficients based on the prediction error. Suppose at a given sampling instant, model ℓ is a substantially better representation of the plant dynamics compared to the other models. In this situation, model ℓ should be updated to a larger extent than the other models once the measurement becomes available. However, with the standard RLS algorithm, models known not to be a good representation of the current plant dynamics can be unnecessarily updated. This motivates the idea of adopting a more local update approach. To this end, we develop the probabilistic recursive least squares (PRLS) algorithm for each model that takes the probability of the plant measurement originating from the different models into account.

Denote the original model coefficient estimates and their covariance for the ℓth model as $\hat{\boldsymbol{\beta}}_\ell(0)$ and $\mathbf{P}_\ell(0)$ (respectively). In a local update approach, L estimators are operated independently of each other (i.e., each estimator only considers its corresponding model). As a result, the following L events are mutually exclusive in each estimator:

$$\mathcal{E}_\ell : y(k) \text{ originated from plant dynamics representable by model } \ell, \qquad \ell = 1, \ldots L \tag{7.10}$$

For the development of the PRLS equations, we first introduce the following assumptions.

Assumption 1 There is negligible prediction error from the originally identified model such that:

$$\mathbf{E}\left[y(k)\right] = \hat{y}(k) \tag{7.11}$$

where $\mathbf{E}\left[\cdot\right]$ denotes the expected value and $\hat{y}(k)$ is the model prediction of the plant measurement, $y(k)$, from (7.2).

Assumption 2 At a given sampling instant, the plant dynamics are representable by a linear combination of the L local linear models in (7.2). Thus, the events in (7.10) are exhaustive (in addition to being mutually exclusive in each estimator):

$$\sum_{\ell=1}^{L} \Pr\left\{\mathcal{E}_\ell\right\} = 1$$

where $\Pr\left\{\cdot\right\}$ denotes the probability of an event.

As an initial step, we compute the posterior probability of a plant measurement originating from model ℓ using a Bayesian approach. This is formalized below in a theorem.

Theorem 1 *Consider L estimators (corresponding to the L models) operating independently and subject to the assumption above. The posterior probability, $\gamma_\ell(k)$, that the plant measurement, $y(k)$, originated from plant dynamics representable by model ℓ is given by:*

$$\gamma_\ell(k) = \frac{\mathcal{N}\left(y(k);\ \hat{y}_\ell(k), \sigma_\ell^2(k)\right) \|\bar{\boldsymbol{x}}'(k) - \mathbf{c}_\ell\|^{-2}}{\sum_{i=1}^{L} \mathcal{N}\left(y(k);\ \hat{y}_i(k), \sigma_i^2(k)\right) \|\bar{\boldsymbol{x}}'(k) - \mathbf{c}_i\|^{-2}} \tag{7.12}$$

where $\mathcal{N}\left(x;\ \mu, \sigma^2\right)$ represents the value of x on a normal distribution with mean, μ, and variance, σ^2, and $\hat{y}_\ell(k)$ and $\sigma_\ell^2(k)$ are given by:

$$\hat{y}_\ell(k) = \bar{\boldsymbol{x}}(k)\hat{\boldsymbol{\beta}}_\ell(k-1) \tag{7.13}$$

$$\sigma_\ell^2(k) = \bar{\boldsymbol{x}}(k)\mathbf{P}_\ell(k-1)\bar{\boldsymbol{x}}'(k) + \sigma_y^2 \tag{7.14}$$

□

Proof: Using the total probability theorem with respect to the events in (7.10), the expected value of the plant measurement at sampling instant k, given measurements up to $k-1$, can be written as:

$$\mathbf{E}\left[y(k) \mid Y^{(k-1)}\right] = \sum_{\ell=1}^{L} \mathbf{E}\left[y(k) \mid \mathcal{E}_\ell, Y^{(k-1)}\right] \Pr\left\{\mathcal{E}_\ell \mid Y^{(k-1)}\right\} \tag{7.15}$$

where $Y^{(k-1)}$ denotes the measurement sequence up to $k-1$. We have used the fact that the events in (7.10) are mutually exclusive and exhaustive. Note that this follows from the total probability theorem, which, in general, can be stated as:

$$\mathbf{E}\left[x\right] = \sum_{i=1}^{L} \mathbf{E}\left[x \mid \mathcal{E}_i\right] \Pr\left\{\mathcal{E}_i\right\}$$

where x is a continuous random variable and $\{\mathcal{E}_1, \ldots, \mathcal{E}_2\}$ are mutually exclusive and exhaustive events.

With no prediction error (the first assumption), the expected value of the plant measurement conditioned on the event that the ℓth model represents the plant dynamics follows from (7.2) with $w_\ell = 1$ and $w_i = 0$ for $i \neq \ell$ where $i \in [1, L]$.

$$\mathbf{E}\left[y(k) \mid \mathcal{E}_\ell, Y^{(k-1)}\right] = \bar{\boldsymbol{x}}(k)\hat{\boldsymbol{\beta}}_\ell(k-1)$$

Using this result and (7.11), (7.15) can be expressed as:

$$\hat{y}(k) = \sum_{\ell=1}^{L} \bar{\boldsymbol{x}}(k)\hat{\boldsymbol{\beta}}_\ell(k-1)\Pr\left\{\mathcal{E}_\ell \mid Y^{(k-1)}\right\} \tag{7.16}$$

Comparing (7.2) and (7.16), we have:

$$\Pr\left\{\mathcal{E}_\ell \mid Y^{(k-1)}\right\} = w_\ell(k) = \frac{\|\bar{\boldsymbol{x}}'(k) - \mathbf{c}_\ell\|^{-2}}{\sum_{i=1}^{L} \|\bar{\boldsymbol{x}}'(k) - \mathbf{c}_i\|^{-2}} \tag{7.17}$$

Thus, the prior probability of $\mathcal{E}_\ell$ (based on information up to $k-1$) can be estimated with model ℓ's weight (i.e., its membership function value). The prior probability can be corrected once the plant measurement becomes available using Bayes' rule. Denoting this posterior probability as $\gamma_\ell(k)$, the application of Bayes' rule yields:

$$\gamma_\ell(k) \triangleq \Pr\left\{\mathcal{E}_\ell \mid y(k), Y^{(k-1)}\right\} = \frac{p\left(y(k) \mid Y^{(k-1)}, \mathcal{E}_\ell\right) \Pr\left\{\mathcal{E}_\ell \mid Y^{(k-1)}\right\}}{p\left(y(k) \mid Y^{(k-1)}\right)}$$
$$= \frac{p\left(y(k) \mid Y^{(k-1)}, \mathcal{E}_\ell\right) \|\bar{\boldsymbol{x}}'(k) - \boldsymbol{c}_\ell\|^{-2}}{p\left(y(k) \mid Y^{(k-1)}\right) \sum_{i=1}^{L} \|\bar{\boldsymbol{x}}'(k) - \boldsymbol{c}_i\|^{-2}} \tag{7.18}$$

where $Y^{(k)}$ has been partitioned as $\left\{y(k), Y^{(k-1)}\right\}$ and $p\left(y(k) \mid Y^{(k-1)}, \mathcal{E}_\ell\right)$ is the prior likelihood of the measurement conditioned on $\mathcal{E}_\ell$. This conditional likelihood is simply the measurement's height on the probability density function (pdf) of model ℓ's prediction. With zero-mean Gaussian measurement noise, the pdf of the prediction is a normal distribution with mean, $\hat{y}_\ell(k)$, and variance, $\sigma_\ell^2(k)$. These pdf parameters are given by (7.13) and (7.14), respectively. The mean is simply the prediction from only using model ℓ (i.e., the pdf is centered at the model's prediction). The variance equation follows from the assumed linear model form, and its derivation procedure is identical to the one for the innovation covariance in the Kalman filter. Thus, $p\left(y(k) \mid Y^{(k-1)}, \mathcal{E}_i\right)$ is given by:

$$p\left(y(k) \mid Y^{(k-1)}, \mathcal{E}_i\right) = \mathcal{N}\left(y(k);\ \hat{y}_\ell(k), \sigma_\ell^2(k)\right) \tag{7.19}$$

The denominator in (7.18) can be rewritten by invoking the total probability theorem with respect to the events in (7.10) and using the results in (7.17) and (7.19):

$$p\left(y(k) \mid Y^{(k-1)}\right) = \sum_{i=1}^{L} p\left(y(k) \mid Y^{(k-1)}, \mathcal{E}_i\right) \Pr\left\{\mathcal{E}_i \mid Y^{(k-1)}\right\}$$
$$= \frac{\sum_{i=1}^{L} \mathcal{N}\left(y(k);\ \hat{y}_i(k), \sigma_i^2(k)\right) \|\bar{\boldsymbol{x}}'(k) - \boldsymbol{c}_\ell\|^{-2}}{\sum_{i=1}^{L} \|\bar{\boldsymbol{x}}'(k) - \boldsymbol{c}_i\|^{-2}} \tag{7.20}$$

Substituting (7.19) and (7.20) yields the final expression for the posterior probability in the theorem, completing the proof. □

Remark 7.3 A standard assumption of zero-mean, normally distributed measurement/sensor noise is made for the result in the theorem, but the key idea of the theorem holds even when this is not the case. If the noise distribution's parameters are known, the likelihood expressions derived in (7.19) and (7.20) can be modified accordingly. In the absence of any information about the noise characteristics, the assumption of zero-mean, normally distributed noise is a meaningful way to determine the likelihoods that are required for the result in the theorem.

Remark 7.4 In this theorem, we use the model weights (obtained from the membership function) as the prior probabilities when applying Bayes' rule. To rigorously show the equivalence between the prior probabilities and model weights, an assumption of negligible prediction error was made. In practice, this assumption will of course not hold because of unavoidable plant-model mismatch. Hence, the weights essentially represent an *estimate* of the prior probabilities with the quality of the estimate depending on the model quality. Note also that the prior probabilities are corrected to some extent once the measurement becomes available from Bayes' rule, and the level of plant-model mismatch can also be reduced as the model is continually updated/adapted online. As presented in the sequel, if the initial model built using the multi-model approach is of good quality, the result of this theorem provides a mechanism to update the individual models within the RLS framework in a probabilistic sense. This is formalized below in the next theorem.

Theorem 2 *Consider the problem of probabilistically updating each model's coefficients (using independent estimators) at sampling instant k when we have* $\mathbf{y}(\cdot)$ *and* $\boldsymbol{u}(\cdot)$ *up to k and* $k-1$*, respectively. The coefficients for model* ℓ *that minimize the time-varying least squares criterion in (7.4) in a probabilistic sense at k are given by:*

$$v_\ell(k) = y(k) - \bar{\boldsymbol{x}}(k)\hat{\boldsymbol{\beta}}_\ell(k-1) \tag{7.21}$$

$$\hat{\boldsymbol{\beta}}_\ell(k) = \hat{\boldsymbol{\beta}}_\ell(k-1) + \boldsymbol{k}_\ell(k)\gamma_\ell(k)v_\ell(k) \tag{7.22}$$

$$\boldsymbol{k}_\ell(k) = \frac{\mathbf{P}_\ell(k-1)\bar{\boldsymbol{x}}'(k)}{\bar{\boldsymbol{x}}(k)\mathbf{P}_\ell(k-1)\bar{\boldsymbol{x}}'(k) + \lambda\sigma_y^2} \tag{7.23}$$

$$\mathbf{P}_\ell(k) = \mathbf{P}_\ell(k-1)\left[1 - \gamma_\ell(k)\right] + \frac{\gamma_\ell(k)}{\lambda}\left[\mathbf{I} - \boldsymbol{k}_\ell(k)\bar{\boldsymbol{x}}(k)\right]\mathbf{P}_\ell(k-1) + \tilde{\mathbf{P}}_\ell(k) \tag{7.24}$$

where $\tilde{\mathbf{P}}_\ell(k) = \boldsymbol{k}_\ell(k)\left[\gamma_\ell(k)v_\ell^2(k) - \gamma_\ell(k)^2 v_\ell^2(k)\right]\boldsymbol{k}_\ell(k)'$. □

Proof: Note that this proof is a modification of the probabilistic data association filter derivation [15]. Given the plant measurement, $y(k)$, the following complementary, mutually exclusive, and exhaustive events are possible in the ℓth estimator:

$$\begin{aligned} &\mathcal{E}_\ell : y(k) \text{ originated from plant dynamics representable by model } \ell \\ &\mathcal{E}_0 : y(k) \text{ did not originate from plant dynamics representable by model } \ell \end{aligned} \tag{7.25}$$

Next, we derive the recursive equations for the coefficient estimate and covariance updates that take the posterior probabilities for each model into account. For the events in (7.25), the estimate of model ℓ's coefficients can be expressed using the total probability theorem as follows:

$$\begin{aligned} \hat{\boldsymbol{\beta}}_\ell(k) &= \mathbf{E}\left[\boldsymbol{\beta}_\ell(k) \mid Y^{(k)}\right] \\ &= \mathbf{E}\left[\boldsymbol{\beta}_\ell(k) \mid \mathcal{E}_0, Y^{(k)}\right]\Pr\left\{\mathcal{E}_0 \mid Y^{(k)}\right\} + \mathbf{E}\left[\boldsymbol{\beta}_\ell(k) \mid \mathcal{E}_\ell, Y^{(k)}\right]\Pr\left\{\mathcal{E}_\ell \mid Y^{(k)}\right\} \end{aligned}$$

$$= \hat{\boldsymbol{\beta}}_\ell^{(0)}(k)\left[1-\gamma_\ell(k)\right] + \hat{\boldsymbol{\beta}}_\ell^{(\ell)}(k)\gamma_\ell(k) \tag{7.26}$$

where $\gamma_\ell(k)$ is given by (7.12) and $\hat{\boldsymbol{\beta}}_\ell^{(i)}(k)$ denotes the coefficient estimates at k conditioned on $\mathcal{E}_i$. For $\mathcal{E}_0$, the model coefficient estimates and their covariance should remain unchanged from their previous estimates since the plant dynamics represented by model ℓ did not play a role in generating the measurement. Accordingly, we have:

$$\hat{\boldsymbol{\beta}}_\ell^{(0)}(k) = \hat{\boldsymbol{\beta}}_\ell(k-1)$$
$$\mathbf{P}_\ell^{(0)}(k) = \mathbf{P}_\ell(k-1)$$

where $\mathbf{P}_\ell^{(i)}(k)$ denotes the covariance of $\hat{\boldsymbol{\beta}}_\ell^{(i)}(k)$ (i.e., the covariance conditioned on $\mathcal{E}_i$). For $\mathcal{E}_\ell$, the standard RLS equations with a forgetting factor, written specifically for model ℓ, can be used to update model ℓ's coefficients and their covariance:

$$\begin{aligned}
v_\ell(k) &= y(k) - \bar{\boldsymbol{x}}(k)\hat{\boldsymbol{\beta}}_\ell(k-1) \\
\hat{\boldsymbol{\beta}}_\ell^{(\ell)}(k) &= \hat{\boldsymbol{\beta}}_\ell(k-1) + \boldsymbol{k}_\ell(k)v_\ell(k) \\
\boldsymbol{k}_\ell(k) &= \frac{\mathbf{P}_\ell(k-1)\bar{\boldsymbol{x}}'(k)}{\bar{\boldsymbol{x}}(k)\mathbf{P}_\ell(k-1)\bar{\boldsymbol{x}}'(k) + \lambda\sigma_y^2} \\
\mathbf{P}_\ell^{(\ell)}(k) &= \frac{1}{\lambda}\left[\mathbf{I} - \boldsymbol{k}_\ell(k)\bar{\boldsymbol{x}}(k)\right]\mathbf{P}_\ell(k-1)
\end{aligned}$$

Combining the expressions for $\hat{\boldsymbol{\beta}}_\ell^{(0)}(k)$ and $\hat{\boldsymbol{\beta}}_\ell^{(\ell)}(k)$ into (7.26) yields the update equation in the theorem (7.22). Equation (7.26) shows that the final estimate is a weighted sum of two conditional estimates. In this case, the final estimate's covariance is a weighted sum of the two conditional covariances, $\mathbf{P}_\ell^{(0)}(k-1)$ and $\mathbf{P}_\ell^{(\ell)}(k-1)$, and an additional covariance term, $\tilde{\mathbf{P}}_\ell(k)$, arising from the measurement origin uncertainty. This term, given by $\tilde{\mathbf{P}}_\ell(k) = \boldsymbol{k}_\ell(k)\left[\gamma_\ell(k)v_\ell^2(k) - \gamma_\ell(k)^2 v_\ell^2(k)\right]\boldsymbol{k}_\ell(k)'$, is similar to the 'spread of the means' term in the covariance of a random variable that is the sum of two random variables. Its lengthy derivation is available in existing literature [15] and omitted here for brevity. The final covariance update equation is given by:

$$\mathbf{P}_\ell(k) = \mathbf{P}_\ell(k-1)\left[1-\gamma_\ell(k)\right] + \frac{\gamma_\ell(k)}{\lambda}\left[\mathbf{I} - \boldsymbol{k}_\ell(k)\bar{\boldsymbol{x}}(k)\right]\mathbf{P}_\ell(k-1) + \tilde{\mathbf{P}}_\ell(k)$$

In summary, by weighting the updated model coefficients (and their covariance) by the probability that they should have been updated and by using the standard RLS equations to compute the updated coefficients, the final coefficient estimates minimize the least squares criterion in (7.4) in the probabilistic sense.

□

Remark 7.5 Compared to the standard RLS algorithm, using the PRLS algorithm for updating each model's coefficients provides more flexibility for controlling the adaptation. First, the forgetting factor for each estimator can be tuned independently. More importantly, the local adaptations can be made more aggressive while maintaining more precise coefficient estimates. This is because the covariance increase from discounting old data is weighted by the (posterior) probability, $\gamma_\ell(k)$. In this way, covariance increases only occur during the appropriate periods when the local model adaptation is active. With the standard RLS algorithm, the covariance associated with all the models can increase at every sampling instant whether or not the model should have been updated.

Remark 7.6 In general, no conclusions can be drawn as to which RLS update strategy, conventional or probabilistic, will perform better in practice. The two strategies adopt different approaches in updating the models. In the PRLS algorithm, the models are updated independently/locally according to their (posterior) probabilities of being consistent with the current plant dynamics. This approach is more suitable for specific situations than a global update of all model coefficients. For instance, when a batch is operating near a cluster's center point, it is more meaningful to update the cluster's corresponding model rather than all the models during this period. Note also that due to the probabilistic nature of the update, the PRLS algorithm is still capable of meaningful adaptations when the operating conditions are somewhere between different cluster center points.

Remark 7.7 One advantage of the proposed PRLS algorithm is that it has lower computational requirements than the standard RLS algorithm (for $L > 1$). Since the standard RLS algorithm for a given output is interpretable as a Kalman filter for the following system [13]:

$$\begin{aligned}\hat{\boldsymbol{\beta}}(k) &= \hat{\boldsymbol{\beta}}(k-1)\\ y(k) &= \boldsymbol{h}(k)\hat{\boldsymbol{\beta}}(k) + v(k)\end{aligned}$$

its computational requirements are approximately proportional to $(L \times n_{\boldsymbol{\beta}_\ell})^3$ where $n_{\boldsymbol{\beta}_\ell}$ is the number of coefficients in each local model. This follows from the fact that the computational requirements of a Kalman filter are proportional to n^3 where $n = \max\{n, p\}$ with n and p denoting the number of states and outputs (respectively) for the system under consideration [15]. By comparison, since each model is updated individually in the PRLS algorithm, its computational requirements are proportional to $\sum_{\ell=1}^{L} n_{\boldsymbol{\beta}_\ell}^3 = L \times n_{\boldsymbol{\beta}_\ell}^3$. Thus, the computational requirements are lower by a factor of L^2, which may be significant for a large L.

7.4 Simulation Example

In this section, we first apply the data-based modeling methodology reviewed in the preliminaries on a nylon-6,6 batch polymerization process to identify models for the measurable process variables. Next, under open-loop conditions, we compare the performance of the RLS and PRLS algorithms. Finally, we formulate an adaptive trajectory tracking predictive controller and benchmark its tracking performance against its nonadaptive counterpart.

For our studies, we utilized the mathematical model of nylon-6,6 batch polymerization in which nylon-6,6 polymer is produced by the amidation of adipic acid and hexamethylenediamine (HMD) [16]. In this model, the reaction is carried out in a batch autoclave reactor with a steam jacket for providing the heat needed for vaporization and reaction and a valve for venting vaporized water. The reaction model, modeling assumptions (and their explanations), parameter values, and kinetic relationships are available in the provided Ref. [16] and omitted here for brevity. The final state-space model consists of nine coupled ordinary differential equations (ODEs) with the state vector comprised of the molar amounts of each functional group and evaporated HMD, the reaction medium mass, temperature, and volume, and reactor pressure. The final model takes the general form shown in (7.1).

The control objective considered in this chapter was to track a predetermined set of process variable trajectories. In existing work [16], industrially popular trajectory tracking strategies and loop-pairings for this process are evaluated (using PID controllers) in terms of robustness to common disturbances (specifically heat transfer disturbances and the initial water content in the feed) and according to their ability to produce the desired product quality at batch termination. For this chapter, we selected the strategy wherein reference trajectories for the reaction mixture temperature, T (K), and reactor pressure, P (psia), denoted by T_{ref} and P_{ref} (respectively), are tracked by manipulating the steam jacket pressure, P_j (psi), and vent rate, v (kg/h). Thus, we have: $\boldsymbol{y} = \left[T \; P\right]'$ and $\boldsymbol{u} = \left[P_j \; v\right]'$. The inputs were constrained between $\boldsymbol{u}_{\min} = \left[700 \; 0\right]'$ and $\boldsymbol{u}_{\max} = \left[1800 \; 2000\right]'$. The duration of the batch was assumed to be fixed at $t_f = 3\,\text{h}$, and it is sampled every 1 min. The current chapter assumes that the reference trajectories have been identified appropriately in some fashion, and Fig. 7.1 presents these specific trajectories. Since we are focusing on the complexities associated with the trajectory tracking problem (as opposed to meeting the desired product quality), the profiles in Fig. 7.1 were generated arbitrarily.

7.4.1 Data-Based Model Development

Empirical models for the two outputs are developed in this section from an artificially generated batch database comprised only of input and output measurements. The database was generated by simulating the fundamental model 15 times from different initial conditions. Five of the 15 batches were reserved as the validation data set. To mimic a typical industrial batch database, which consists mostly of 'successful' batches, the reference profiles in Fig. 7.1 were tracked using two PI controllers for the simulations. For the PI loop-pairing, the vent rate was used to track the reactor pressure, while the steam jacket pressure was used to track the reaction mixture temperature. Both controllers were tightly tuned for one set of initial conditions and

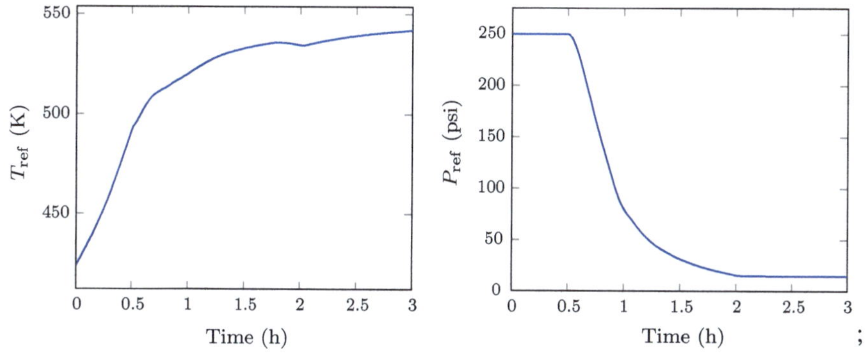

Fig. 7.1 Reference trajectories of the process outputs that are required to be tracked in the control objective

fixed for the 14 remaining batches. The tuning goal was to minimize the integral of time-weighted absolute error (ITAE) while attaining reasonably smooth input trajectories.

The pressure dynamics in the nylon-6,6 polymerization model are assumed to be significantly faster than the reaction mixture dynamics, leading to a weak correlation between the two outputs. Consequently, individual models for the outputs were identified with PCR as opposed to identifying one PLS model that predicted both outputs. The model identification procedure for a given output was as follows. For a fixed lag structure, the number of clusters was varied from $L = 1$ to $L = 20$. For each choice of L, the $\bar{\mathbf{X}}$ matrix was constructed (lagged outputs and inputs), decomposed using PCA, and clustered. Since the model weights were then known, multiple ARX models were estimated using PCR. This was repeated for all possible lag structures with a lag range of $0 - 2$ for each variable. The goodness of each fit was judged using its root-mean-squared error (RMSE) in predicting back the fit and validation batches. The RMSE was defined as follows.

$$\text{RMSE} \triangleq \sqrt{\frac{1}{B}\frac{1}{K}\sum_{b=1}^{B}\sum_{k=1}^{K}\left[\hat{y}^{(b)}(k) - y^{(b)}(k)\right]^2} \tag{7.27}$$

where b indexes the batch number, and B and K are the number of batches and time steps in each batch (respectively) (i.e., $B = 15$ and $K = 180$). The lag structure and number of clusters, L, for the two outputs that yielded the lowest RMSE values are tabulated in Table 7.1. In Table 7.1, a zero lag indicates that the variable was excluded from the model. Observe that the reactor pressure was not used in predicting the reaction mixture temperature, and its dynamics were best captured with a single linear first-order model. These results are consistent with the fundamental process model; the significantly faster pressure dynamics assumed in the state-space model leads to a decoupling of the pressure from the other states (i.e., the pressure does not

Table 7.1 Final lag structures, number of clusters, L, and RMSE values

Output	Lags				L	RMSE
	T	P	P_j	v		
T	1	0	1	1	5	1.648
P	0	1	0	1	1	0.1791

influence any of the other states and vice-versa), and the pressure ODE is simply the product of a constant gain term and the vent rate (a linear first-order model). Despite the decoupling of the outputs, the control problem cannot be decomposed into two single-input, single-output (SISO) problems because one of the inputs, the vent rate, affects both outputs. Another observation from Table 7.1 is that the lag structure for the temperature models corresponds to a first-order model between the outputs and inputs. One explanation for this behavior is the assumption of the same lag structure for all the local models. With this assumption, using all first-order models minimizes the possibility of overfitting and, in this case, yielded the lowest RMSE values.

The outputs from the data-based models, $\hat{T}$ and $\hat{P}$, are compared with those from a batch in the validation data set, denoted by T and P, in Fig. 7.2. Validation batches were chosen (one for each output) for which the data-based models displayed the poorest predictions (highest RMSE values). The predicted temperature, particularly during the finishing stages, demonstrates the need/room for improvement of the temperature model. This is addressed in the next section by incorporating online learning ability into the model using RLS and PRLS. The pressure model, on the other hand, did not require any further improvements. Note that in this chapter we focus specifically on adapting the models in an intra-batch (i.e. within-batch) fashion. Inter-batch (batch-to-batch) adaptation could also be done where all the model parameters (ARX model coefficients, lags, number of clusters, and center points) are updated using the results of the current batch. A simple way to do this is by appending the completed batch data to the original fit matrices and repeating the fit procedure.

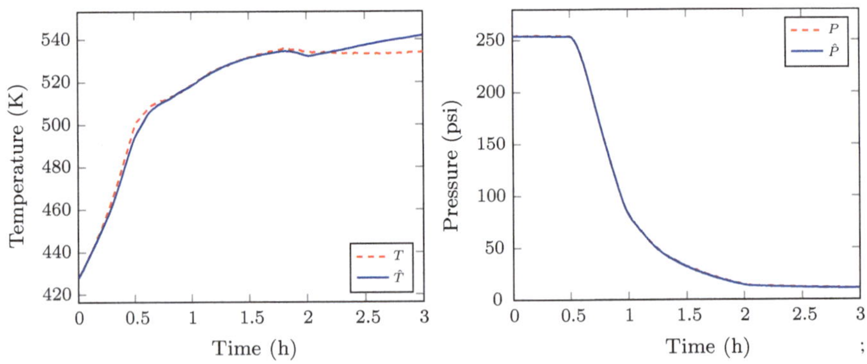

Fig. 7.2 Output from using the data-based model and the corresponding batch trajectory in the validation data set

Table 7.2 Performance summary of RLS and PRLS algorithms with different forgetting factors, λ

	RLS		PRLS	
λ	RMSE	MSV	RMSE	MSV
1	0.9074	7.861	0.8219	5.657
0.995	0.7927	12.49	0.8073	5.918
0.95	0.4980	3.349×10^3	0.6721	11.57
0.90	0.4399	1.465×10^6	0.5631	50.83

7.4.2 *Online Update of the Temperature Model Using RLS and PRLS*

In this section, we use an adaptive temperature model to predict the temperature of the 15 batches in the database (training and validation data). The adaptation performance was evaluated based on prediction accuracy and precision of the model coefficient estimates. The prediction accuracy was quantified using the RMSE metric (7.27) defined in the previous section. The precision was quantified using the covariance matrix as follows. At sampling instant k for batch b, the sum of the variances of the coefficient estimates is trace $\left\{\mathbf{P}^{(b)}(k)\right\}$. The average of this trace over the batch duration and number of batches was termed the mean sum of variances (MSV) and used as the metric to asses the estimation precision.

$$\text{MSV} = \frac{1}{B}\frac{1}{K}\sum_{b=1}^{B}\sum_{k=1}^{K} \text{trace}\left\{\mathbf{P}^{(b)}(k)\right\}$$

The results from applying the RLS and PRLS algorithms on the temperature model with various forgetting factors are summarized in Table 7.2. The forgetting factor in each PRLS estimator was kept the same for simplicity.

The improved prediction accuracy obtained by adapting the model is evident by comparing the RMSE value in Table 7.1 with those in Table 7.2. Adapting the model with $\lambda = 1$ improves the prediction accuracy by 45% (RLS) and 50% (PRLS). As the forgetting factor is decreased to make the adaptation more aggressive, there is a trade-off between the prediction accuracy and parameter precision. However, the loss in precision is significantly less sensitive for the PRLS algorithm, permitting aggressive adaptation (low values of λ) with acceptable parameter variances. Thus, the PRLS algorithm was concluded to offer a better management of the trade-off between the prediction accuracy and parameter precision. The temperature prediction error magnitude, $|\hat{T} - T|$, for the batch in Fig. 7.2 is shown in Fig. 7.3 before and after incorporating model adaptation. Forgetting factors of 0.995 and 0.90 for the RLS and PRLS estimators (respectively) are selected for a fair comparison since they resulted in comparable estimation precision. These forgetting factors are also used for adapting the model in the closed-loop simulations.

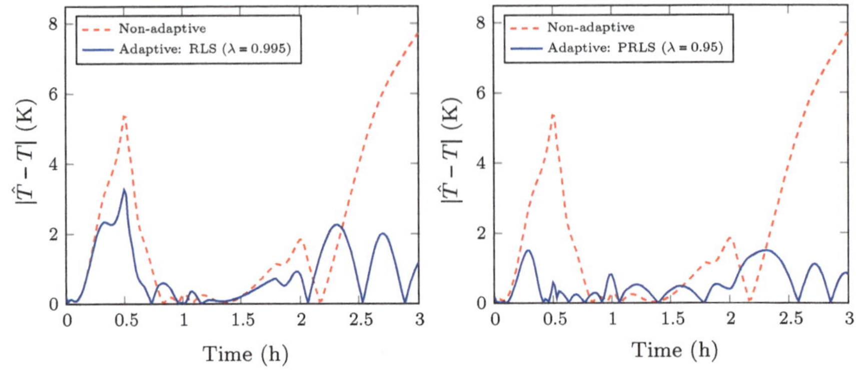

Fig. 7.3 Temperature prediction error magnitudes (for the batch in Fig. 7.2) before and after using RLS and PRLS to adapt the model

7.4.3 Closed-Loop Simulations

Closed-loop simulation results of a trajectory tracking predictive controller that uses the data-based models as its underlying models are presented in this section. The temperature model is updated at each sampling instant prior to the controller computations, making the controller adaptive. The control action at each sampling instance is computed by solving the optimization problem below.

$$\min_{\boldsymbol{u}_{\min} \le \boldsymbol{u}(k) \le \boldsymbol{u}_{\max}} \quad J = \sum_{k=1}^{P} \|\hat{\boldsymbol{y}}(k) - \boldsymbol{y}_{\text{ref}}(k)\|_{\mathbf{Q}}^2 + \|\Delta \boldsymbol{u}(k)\|_{\mathbf{R}}^2 \tag{7.28}$$

$$\text{subject to: 7.2} \tag{7.29}$$

$$\hat{\boldsymbol{y}}(0) = \boldsymbol{y}(t) \tag{7.30}$$

where the notation, $\|\cdot\|_{\mathbf{Q}}^2$, refers to the weighted norm, defined by $\|\boldsymbol{x}\|_{\mathbf{Q}}^2 = \boldsymbol{x}^{\mathrm{T}}\mathbf{Q}\boldsymbol{x}$, and $\Delta \boldsymbol{u}$ denotes a vector in which each element is the difference between successive input moves. The constraint in Eq. (7.29) signifies that the data-based models are the predictive models for the two outputs. Equation (7.30) represents the initialization of the optimization problem at the current process conditions and can be understood as the feedback mechanism to account for plant-model mismatch. The first term in the objective function penalizes discrepancies between the predicted output trajectories, $\hat{\boldsymbol{y}}$, and the output reference trajectories, $\boldsymbol{y}_{\text{ref}}$, over the prediction horizon, P, and the second term is a move suppression term that penalizes the magnitude of input changes (i.e., the control rate). The positive-definite, diagonal matrices, $\mathbf{Q}$ and $\mathbf{R}$, are used to trade-off the relative importance of the output and input performance. The controller was tuned for the nonadaptive model and left unchanged for the adaptive cases to avoid confounding the results with tuning. The tuning parameters were set as follows: $\mathbf{Q} = \text{diag}\,\{20, 50\}$, $\mathbf{R} = \text{diag}\,\{0.05, 0.05\}$, and $P = 10$.

Table 7.3 Closed-loop performance summary from using the trajectory tracking predictive controller for initial conditions within the training data range

	Adaptation algorithm	
	None RLS ($\lambda = 0.995$)	PRLS ($\lambda = 0.95$)
ITAE for T: 2.825	1.108	1.034
ITAE for P: 4.146	2.805	1.357

Table 7.4 Closed-loop performance summary from using the trajectory tracking predictive controller for initial conditions outside the training data range

	Adaptation algorithm	
	None RLS ($\lambda = 0.995$)	PRLS ($\lambda = 0.95$)
ITAE for T: 30.70	1.676	1.211
ITAE for P: 29.20	2.881	1.474

The performance of the adaptive predictive controller was compared against the nonadaptive version when encountering disturbances in the initial conditions. Note that the superior tracking ability of the nonadaptive version of the design over existing trajectory tracking approaches, namely PI control and latent variable MPC [17], has been previously established in our earlier work [18]; thus, in this chapter, we are mainly concerned with highlighting the potential gains in closed-loop performance from including model adaptations. Two initial conditions, one within and one outside the range of initial conditions in the training data, were considered for the simulations. The metric used to assess the tracking performance of the controllers was the ITAE. The closed-loop performance is summarized in Tables 7.3 and 7.4 for both initial conditions.

Focusing first on the case when the initial conditions are within the training data range incorporating standard RLS with $\lambda = 0.995$ offered improvements of 61 and 32% in temperature and pressure tracking, respectively. The temperature and pressure tracking were further improved by 7 and 50% (respectively) when the PRLS algorithm was used. These results illustrate that considerably better tracking can be achieved by using an adaptive model in the controller.

From the results in Table 7.4, model adaptation is crucial for achieving acceptable closed-loop performance for large disturbances in the initial conditions (i.e., when the initial conditions are outside the training data range). The tracking errors and input profiles for these initial conditions are shown in Fig. 7.4. There is substantial deviation from the reference trajectories during the finishing stages of the batch with the nonadaptive controller. With these initial conditions, foreign operating conditions (and therefore dynamics) were likely encountered for the temperature model that are not originally modeled. This led to poor temperature predictions in the controller calculations, and the coupled nature of the control problem led to poor tracking for both outputs. The adaptive designs, on the other hand, are able to learn the new dynamics using the plant measurements, leading to more accurate predictions and

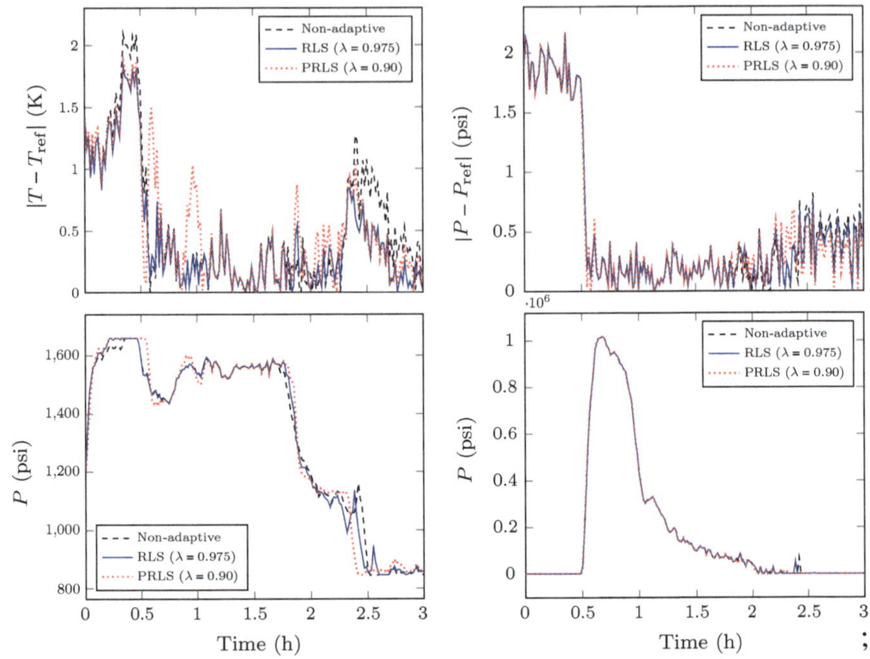

Fig. 7.4 Tracking error magnitudes and input profiles for the predictive controller before and after adaptation (for initial conditions outside the training data range)

significantly improved tracking performance. While there is no general guarantee that the PRLS algorithm will always outperform standard RLS, among the two adaptation approaches, using the PRLS algorithm produced lower ITAE values for both outputs consistently for this simulation example.

7.5 Conclusions

In this chapter, online learning ability was integrated within a previously developed data-based modeling methodology for batch processes. First, it was demonstrated how the standard RLS algorithm with a forgetting factor can be applied in a straightforward manner to provide online updates of the model parameters. Next, a probabilistic RLS (PRLS) estimator (also with a forgetting factor) was developed that updates each model individually according to its probability of being representative of the local dynamics. The advantage of the PRLS algorithm was tuning flexibility. Specifically, the forgetting factors for the individual estimators can be tuned independently and also more aggressively than in standard RLS while maintaining good precision (i.e., low parameter variances). The benefits from incorporating the two RLS algorithms in the modeling approach were demonstrated via simulations of a nylon-6,6 batch polymerization reactor. Open-loop simulations verified that the precision of the PRLS algorithm is less sensitive to the adaptation aggressiveness compared to conventional RLS. Closed-loop simulations indicated that both RLS algorithms can help improve the performance of a trajectory tracking predictive controller, particularly for large disturbances in the initial conditions.

References

1. Aumi, S., Mhaskar, P.: Integrating data-based modeling and nonlinear control tools for batch process control. AIChE J. (2011). https://doi.org/10.1002/aic.12720 (in press)
2. Kiparissides, C., Shah, S.: Self-tuning and stable adaptive control of a batch polymerization reactor. Automatica **19**(3), 225–235 (1983)
3. Welz, C., Srinivasan, B., Marchetti, A., Bonvin, D., Ricker, N.L.: Evaluation of input parameterization for batch process optimization. AIChE J. **52**(9), 3155–3163 (2006)
4. Ruppen, D., Bonvin, D., Rippin, D.: Implementation of adaptive optimal operation for a semi-batch reaction system. Comput. Chem. Eng. **22**, 185–199 (1998)
5. Srinivasan, B., Bonvin, D.: Real-time optimization of batch processes by tracking the necessary conditions of optimality. Ind. Eng. Chem. Res. **46**, 492–504 (2007)
6. Kadam, J.V., Marquardt, W., Srinivasan, B., Bonvin, D.: Optimal grade transition in industrial polymerization processes via NCO tracking. AIChE J. **53**, 627–639 (2007)
7. Ljung, L.: System Identification: Theory for the User, 2nd edn. Prentice Hall PTR, Upper Saddle River (1998)
8. Bezdek, J.: A convergence theorem for the fuzzy isodata clustering algorithms. IEEE Trans. Pattern Anal. Mach. Intell. **2**(1), 1–8 (1980)
9. Xie, X.L., Beni, G.: A validity measure for fuzzy clustering. IEEE Trans. Pattern Anal. Mach. Intell. **13**, 841–847 (1991)
10. Huang, B., Ding, X.S., Qin, S.J.: Closed-loop subspace identification: an orthogonal projection approach. J. Process Control **15**(1), 53–66 (2005)
11. Wang, J., Chen, T., Huang, B.: Closed-loop identification via output fast sampling. J. Process Control **14**(5), 555–570 (2004)
12. Geladi, P., Kowalski, B.: Partial least-squares regression: a tutorial. Anal. Chim. Acta **185**, 1–17 (1986)
13. Aström, K., Wittenmark, B.: Adaptive Control, 2nd edn. Addison-Wesley Longman, Boston (1994)
14. Fortescue, T., Kershenbaum, L., Ydstie, B.: Implementation of self-tuning regulators with variable forgetting factors. Automatica **17**(6), 831–835 (1981)
15. Bar-Shalom, Y., Li, X.: Multitarget-Multisensor Tracking: Principles and Techniques, 3rd edn. University of Connecticut, USA (1995)
16. Russell, S.A., Robertson, D.G., Lee, J.H., Ogunnaike, B.A.: Control of product quality for batch nylon-6,6 autoclaves. Chem. Eng. Sci. **53**(21), 3685–3702 (1998)
17. Flores-Cerrillo, J., MacGregor, J.F.: Latent variable MPC for trajectory tracking in batch processes. J. Process Control **15**, 651–663 (2005)
18. Aumi, S., Corbett, B., Mhaskar, P.: Data-based modeling and control of nylon-6,6 batch polymerization. In: Proceedings of the American Control Conference (ACC), pp. 2540–2545 (2011)

Chapter 8
Data-Driven Model Predictive Quality Control of Batch Processes

8.1 Introduction

In this part of the book, we thus far presented modeling approaches to achieve good trajectory tracking for batch processes. While trajectory tracking controllers can reject disturbances online, even with perfect tracking, there is no guarantee that the desired quality will be met if there is significant variation in the initial conditions from batch to batch. This is because disturbances encountered during the new batch could alter the relationship between the product quality and the trajectories of the process variables. Thus, implementing the same reference trajectories batch-to-batch is not guaranteed to consistently produce on-spec product.

Inferential quality control refers to the control approach aimed at controlling quality directly, and is most commonly achieved through multivariate statistical process control (SPC) approaches, particularly those utilizing latent variable tools, such as principal component analysis (PCA) or partial least squares (PLS) regression [1]. For batch processes, the model development for the majority of SPC applications begins with the so-called batch-wise unfolding of multi-way batch data [2, 3]. The unfolded data is regressed (commonly via PLS regression) onto a matrix of final quality measurements to obtain an inferential PLS quality model [4] that is usable for predicting the final quality prior to batch completion. For batches with multiple phases or stages with distinct dynamics, multiple phase-specific (and transition) models can also be constructed [5, 6]. During the batch evolution, the final quality can be predicted (at every sampling instant or predetermined decision points) and if the prediction exceeds the control limits, appropriate remedial action can be taken to correct the batch. The nature of the corrective action may be heuristics or knowledge based or more systematic wherein the quality model is inverted (one way or another) to directly compute the future input trajectories that recover the batch. The latter approach has been classified as a mid-course correction (MCC) control strategy [7, 8]. Since it requires model inversion, the effectiveness of a MCC approach is particularly dependent on the underlying quality model and in general, demands richer

P. Mhaskar et al., *Modeling and Control of Batch Processes*,
Advances in Industrial Control, https://doi.org/10.1007/978-3-030-04140-3_8

training data that spans a wider operating range and exhibits more input variation compared to modeling for SPC [9].

An important issue that arises in SPC and MCC approaches is that future online measurements that are required to predict the quality are incomplete. More specifically, the data arrangement in the model building process calls for the entire batch trajectory to predict the quality of the batch. However, during a batch, measurements are only available up to the current sampling instant, and the future data is required to predict the final quality. In the literature, this has so far been treated as a 'missing data' problem, with the choice of the data completion technique playing a key role in the overall performance of the control design. Prediction error in the future data is propagated to the quality prediction error, and both of these errors add uncertainty to any control action computed using the model. This problem is particularly prevalent during the early stages of the batch when most of the information is unknown. In fact, with poor prediction of the future batch behavior, inputs determined from using the model can drive the batch to a point from where good quality product cannot be produced. This characteristic is typical for methods that lack a causal relationship between the inputs and the process trajectory, and in turn, the quality, which leads to the treatment of the future trajectory as a 'missing data' problem [7].

A variety of ad hoc approaches exist to handle this 'missing data' problem. Many methods utilize missing data algorithms available for latent variable methods. These missing data algorithms work on the assumption that the correlation structure between the collected measurements and future measurements for the new batch is the same as in the training data. Another approach has been to build a finite set of quality models at predetermined decision points (possibly at every sampling instant), and in building each model, rather than using the entire batch trajectory, data only up to the decision point is used [9, 10]. This idea of an evolving model has also been modified for improving the quality prediction in multi-stage batches through consideration of critical-to-quality time periods at specific phases of the batch [11, 12]. One issue with these multi-model based approaches, however, is that quality models developed at early time points may be highly inaccurate since they will not capture the effects of large periods of the batch duration toward the batch quality. While these missing data approaches are useful for predicting the quality and monitoring applications, when the inferential model is used in a control design, the need to consider the nonlinear casual relationship between the future input–output behavior is obvious. The quality control problem, therefore, stands to gain from the use of a causal, nonlinear model that does not treat the future trajectory as a missing data problem and instead recognizes it as the problem of choosing the remaining input trajectory, which determines the rest of the measured variables trajectory and in turn the final quality.

Motivated by the above considerations, in this chapter we develop a within-batch quality control strategy for batch processes that unites a single PLS inferential quality model with a previously developed nonlinear, data-driven modeling approach. By properly representing the future behavior using a causal model, control inputs can be chosen that result in improved quality control. The rest of this manuscript is organized as follows: first, we describe the class of processes being considered and review the key concepts, namely the different modeling approaches used in the control design. Next, we present the details of a predictive controller that is designed to drive a batch process to a desired specified product quality by batch termination. The efficacy of the control design is then demonstrated via simulations of a nylon-6,6 batch polymerization system. Finally, we summarize our results.

8.2 Preliminaries

In this section, we first describe the class of batch processes considered. Next, we discuss how an inferential quality model can be identified from existing batch data through multi-way analysis. This is followed by an overview of partial least squares (PLS) regression, a latent variable technique used to compute the inferential quality model parameters. Finally, we review a previously developed data-based modeling approach [13, 14] that serves as the basis for predicting the future state trajectory (for a candidate input profile) when using the inferential quality model for quality control.

8.2.1 Process Description

We consider batch processes subject to input constraints and limited measurements described by:

$$\begin{aligned}
\dot{\boldsymbol{x}} &= \boldsymbol{f}(\boldsymbol{x}, \boldsymbol{u}) \\
\boldsymbol{y} &= \boldsymbol{g}(\boldsymbol{x}, \boldsymbol{u}) + \boldsymbol{v} \\
\boldsymbol{q} &= \boldsymbol{h}(\boldsymbol{x}, \boldsymbol{u}) \\
t &\in [t_0, t_{\mathrm{f}}], \boldsymbol{u}(\cdot) \in \mathcal{U}, \boldsymbol{x}(t_0) = \boldsymbol{x}_0
\end{aligned} \tag{8.1}$$

Note that the above model is used only to illustrate the class of processes that the proposed modeling approach is expected to work, and the proposed approach does not require a model of the form of Eq. (8.1) to be available. The times, $t = t_0$ and $t = t_{\mathrm{f}}$, are the initial and termination times, respectively. The vectors, $\boldsymbol{x} \in \mathbb{R}^n$, $\boldsymbol{y} \in \mathbb{R}^p$, and $\boldsymbol{q} \in \mathbb{R}^q$ denote the state variables, measurable process variables, and quality variables, respectively. Measurements of $\boldsymbol{y}$ are assumed to be available at every sampling instant, whereas the elements of $\boldsymbol{q}$ are only measured once following batch completion. The vector $\boldsymbol{v} \in \mathbb{R}^p$ represents zero-mean, normally distributed measurement noise. The vector $\boldsymbol{u} \in \mathbb{R}^m$ consists of constrained manipulated inputs, taking values in a nonempty, convex set, $\mathcal{U} \triangleq \{\boldsymbol{u} \mid \boldsymbol{u}_{\min} \le \boldsymbol{u} \le \boldsymbol{u}_{\max}\} \subset \mathbb{R}^m$ where $\boldsymbol{u}_{\min}$ and $\boldsymbol{u}_{\max}$ define the minimum and maximum (respectively) allowable input values.

8.2.2 Inferential Quality Model

To understand how to build a quality model that can be used during batch evolution to predict the quality at batch termination, we first describe the nature of data available in a batch database. Consider a typical batch run in which $j = 1, 2, \ldots, J = m + p$ variables are measured at $k = 1, 2, \ldots, K$ sampling instants. For $b = 1, 2, \ldots B$ batches, this data can be organized into a three-dimensional array, $\mathbf{X}$ $(B \times J \times K)$, as shown in Fig. 8.2a.

Each vertical slice in this array, which has dimensions $B \times J$, represents the values of all the measurable variables for all batches at a common sampling instant. In addition to $\mathbf{X}$, measurements of $q = 1, 2, \ldots Q$ quality variables taken post-batch can be summarized in a $B \times Q$ quality matrix, $\mathbf{Q}$, as shown in Fig. 8.2b. Finally, information about the initial conditions for each batch is also typically available (i.e., feed-stock properties, measured raw material properties and compositions, and charges of each ingredient), and this can be summarized in a $B \times M$ matrix, $\mathbf{Z}_0$, (see Fig. 8.2c) where M is the number of known variables related to the initial conditions.

To identify an inferential quality model that can be used to predict the batch quality, the three-dimensional array, $\mathbf{X}$, is first transformed into a two-dimensional $B \times JK$ matrix by unfolding it 'batch-wise' such that each of its vertical slices is arranged side-by-side [2, 3] (note that the dynamic model that we use to predict the batch process dynamics [13] does not require this kind of unfolding). Next, the initial conditions matrix, $\mathbf{Z}_0$, is concatenated to the unfolded matrix, forming a regressor matrix, $[\mathbf{Z}_0\ \mathbf{X}]$, as shown below in Fig. 8.3.

The matrix in Fig. 8.3 can be regressed onto the quality data matrix (Fig. 8.2b) using linear regression, yielding a model that relates the initial and process conditions to the quality characteristics as shown below:

$$\hat{\mathbf{Q}} = [\mathbf{Z}_0\ \mathbf{X}]\, \boldsymbol{\Lambda} \tag{8.2}$$

where $\hat{\mathbf{Q}}$ is the predicted quality and $\boldsymbol{\Lambda}$ is a $(M + JK) \times Q$ matrix of the quality model coefficients. Due to the high dimensionality/multivariate nature of the regressor matrix and the likely presence of correlations among its variables, a latent variable regression technique, such as partial least squares (PLS) regression or principal component regression (PCR), is necessary for appropriately estimating $\boldsymbol{\Lambda}$.

Remark 8.1 An assumption made during the batch-wise unfolding scheme is that all batches are of equal length and the trajectories are synchronized (a requirement relaxed in recent results [13], albeit for the purpose of dynamic modeling, not quality models). In practice, this assumption may not hold for raw batch data. Consequently, several methods have been proposed for addressing unequal batch lengths and to synchronize trajectories. The most common method involves choosing a monotonic indicator variable common to all batches, such as conversion, and resampling the variables with respect to this variable instead of time [15]. Additional methods for aligning batch trajectories include dynamic time-warping [16] and curve registration [17].

8.2.3 Partial Least Squares (PLS) Regression

When variables in the regressor matrix are (auto/cross) correlated, the use of ordinary least squares (OLS) regression leads to parameter estimates with large variances (due to covariance of the regressor matrix being ill-conditioned). Correlations can result from data being collected under closed-loop conditions [18] and/or from lagging the variables. One way to handle the numerical issues arising from correlations is through partial least squares (PLS) regression. Geometrically, in PLS regression, the variables in the regressor and response matrices, $\left[\mathbf{Z}_0\ \mathbf{X}\right]$ and $\mathbf{Q}$, are projected onto corresponding orthogonal subspaces of A-pairs of latent variables. Each pair of latent variables accounts for a certain percentage of the variance in the original matrices. Mathematically, PLS regression consists of decomposing $\left[\mathbf{Z}_0\ \mathbf{X}\right]$ and $\mathbf{Q}$ as the sum of the outer products of a score and loading vector:

$$\left[\mathbf{Z}_0\ \mathbf{X}\right] = \sum_{a=1}^{A} \boldsymbol{t}_a \boldsymbol{p}_a' + \mathbf{E} \tag{8.3}$$

$$\mathbf{Q} = \sum_{a=1}^{A} \boldsymbol{d}_a \boldsymbol{r}_a' + \mathbf{F} \tag{8.4}$$

where $\boldsymbol{t}_a$ and $\boldsymbol{d}_a$ are the input and output scores representing the projections of the variables in $\left[\mathbf{Z}_0\ \mathbf{X}\right]$ and $\mathbf{Q}$ on the subspaces, $\boldsymbol{p}_a$ and $\boldsymbol{r}_a$ define the orientation of the corresponding subspaces, and $\mathbf{E}$ and $\mathbf{F}$ denote residual matrices. Because it is desired to obtain an useful relationship between the original data matrices, the two matrices are linked by an inner relation between their scores of the form:

$$\boldsymbol{d}_a = \boldsymbol{b}_a \boldsymbol{t}_a + \boldsymbol{e}_a \quad \forall a \in [1, A] \tag{8.5}$$

where $\boldsymbol{b}_a$ are the coefficients of the inner relationship and $\boldsymbol{e}_a$ are the residuals. In PLS algorithms, such as Nonlinear Iterative Partial Least Squares (NIPALS) [1], the subspace orientation and scores for both matrices are determined simultaneously to maximize the correlation between $\left[\mathbf{Z}_0\ \mathbf{X}\right]$ and $\mathbf{Q}$ and therefore obtain the optimal fit for the inner relationship. The final result from PLS regression is a linear model between $\left[\mathbf{Z}_0\ \mathbf{X}\right]$ and $\mathbf{Q}$ where the coefficients are functions of the scores and loadings from the matrix decompositions.

Remark 8.2 In the PLS model, trajectories of the process variables over the entire batch duration and initial conditions are projected onto a latent variable subspace, and the values of the latent variables in this space are related to the scores of the final quality through the inner relationship. The projection essentially represents an estimation of the batch states at the end of the batch while the inner relationship is a 'measurement' function relating the states at batch termination to the final quality. For a *new* batch, at sampling instant k, process variable trajectories are only available only up to k. More specifically, the process outputs are available up to sampling instant k and the inputs are available up to $k-1$. As a result, the vector required to make the state estimation at batch termination is incomplete. There are ways to eliminate this problem in monitoring applications (e.g., by using multiple models [9], lookup-tables [19], or a different unfolding scheme [2]); however, when using the model for *control*, the prediction of the future behavior for a given input is a necessity. Rather than eliminating the need for future data, we recognize the causal nature of the inputs in determining the future trajectory and in turn the quality.

Remark 8.3 In conventional PLS modeling, a common preprocessing step is to normalize the regressor and response matrices to zero-mean and unit variance. Scaling to unit variance gives each variable equal importance during model identification; however, in many batch systems, there are specific periods during the batch that play a more critical role in determining the final quality than others. A simple way to account for these quality-critical periods within the PLS regression framework is to multiply the appropriate columns in the regressor matrix by weighting factors that make them more influential during the computation of the model parameters (and therefore during quality prediction). More formalized approaches for considering time-specific effects are also available [10, 11].

8.2.4 Multi-model Data-Driven Modeling for Batch Systems

In this section, we review a multi-model, data-driven modeling approach that is used for predicting the future output behavior when using the inferential quality model [13, 14]. Mathematically, the model for the process outputs takes the form of a weighted combination of L linear dynamic models as shown below in Eq. (8.7).

$$\hat{\boldsymbol{y}}(k) = \sum_{\ell=1}^{L} w_\ell(k)\hat{\boldsymbol{\beta}}_\ell \bar{\boldsymbol{x}}(k) \tag{8.6}$$

$$= \sum_{\ell=1}^{L} w_\ell(k)\hat{\boldsymbol{\beta}}_\ell \left[\boldsymbol{y}'(k-1) \cdots \boldsymbol{y}'(k-n_y) \cdots \boldsymbol{u}'(k-1) \cdots \boldsymbol{u}'(k-n_u)\right]' \tag{8.7}$$

where $w_\ell(k)$ is model ℓ's weight at sampling instant, k, $\hat{\boldsymbol{\beta}}_\ell$ defines the ℓth local model, and $\bar{\boldsymbol{x}}(k)$ is a vector of lagged inputs and outputs. The scalars, n_y and n_u, denote the number of lags in the outputs and inputs (respectively). For notational simplicity, we have assumed the same number of lags, n_y and n_u, for each output and input variable (respectively) and the same lag structure for all L models. These assumptions can be readily relaxed.

Using the following definitions,

$$\hat{\boldsymbol{\beta}} \triangleq \left[\hat{\boldsymbol{\beta}}_1 \cdots \hat{\boldsymbol{\beta}}_\ell \cdots \hat{\boldsymbol{\beta}}_L\right]$$
$$\boldsymbol{h}(k) \triangleq \left[w_1(k)\bar{\boldsymbol{x}}'(k) \cdots w_\ell(k)\bar{\boldsymbol{x}}'(k) \cdots w_L(k)\bar{\boldsymbol{x}}'(k)\right]'$$

Equation (8.7) can be rewritten in the vector form:

$$\hat{y}(k) = \hat{\boldsymbol{\beta}}\boldsymbol{h}(k) \tag{8.8}$$

The model identification procedure consists of an initial clustering step followed by solving a linear regression problem. In the first step, for a given lag structure, a matrix, $\bar{\mathbf{X}}$, corresponding to $\bar{\boldsymbol{x}}(k)$, is generated by sorting the plant data sample-wise (not time or critical variable-wise, as is done in existing batch modeling approaches) and then, $\bar{\mathbf{X}}$ (or its equivalent latent variable space [13]) is clustered into L clusters using an appropriate clustering technique. Thus, observations of lagged inputs and outputs are clustered together, and, to illustrate our results, in this chapter we use fuzzy c-means clustering. In fuzzy c-means clustering, points that are mathematically 'similar' according to the Euclidean 2-norm are clustered into overlapping spherical clusters with corresponding center points [20]. Each cluster represents a region in the $\bar{\mathbf{X}}$ space where an associated model has the highest degree of validity, and in this way, the cluster's center point represents the linearization point of its associated model. Using the results of the clustering step (the cluster center points), the weights, $w_\ell(\cdot)$, for the training data can be computed prior to the model coefficients (to be discussed shortly). Consequently, the $\boldsymbol{h}(k)$ vector in Eq. (8.8) is completely specified for the training data. Thus, a regressor matrix corresponding to $\boldsymbol{h}(k)$ can be constructed, and the local linear models, $\hat{\boldsymbol{\beta}}$, are computable using linear regression.

Intuitively, from Eq. (8.1), the model weights, $w_\ell(\cdot)$, should depend on the current values of the states and inputs since they define the system dynamics through $\boldsymbol{f}(\cdot)$. In other words, the local models should be weighted according to the current process conditions. In the absence of state measurements, the vector of lagged outputs and inputs, $\bar{\boldsymbol{x}}(k)$, can be used to infer the current process conditions, and each model's weight can be assigned based on the proximity of the operating conditions to its center point. For instance, denoting model ℓ's center point as $\boldsymbol{c}_\ell$, its weight should be inversely proportional to the squared distance between $\bar{\boldsymbol{x}}(k)$ and $\boldsymbol{c}_\ell$:

$$w_\ell(k) \propto \|\bar{\boldsymbol{x}}(k) - \boldsymbol{c}_\ell\|^{-2}$$

Normalizing this expression over all clusters yields the following normalized weighting function:

$$w_\ell(k) = \frac{\|\bar{\boldsymbol{x}}(k) - \boldsymbol{c}_\ell\|^{-2}}{\sum_{i=1}^{L} \|\bar{\boldsymbol{x}}(k) - \boldsymbol{c}_i\|^{-2}}$$

The number of clusters is an important parameter in this approach. Well-defined criteria (based on the cluster geometry) to iteratively refine the number of clusters are available [21]. Additionally, to evaluate the goodness of the final fuzzy partitions, many validation measures have also been proposed with the most popular being the Xie-Beni index [22], which is a ratio of the total within-cluster variance to the separation of the cluster centers (and therefore should be minimal for the best partition). In this chapter, we picked the number of clusters (iteratively) based on how well an independent validation data set was predicted. Thus, there was a balancing of the number of clusters and prediction error.

8.3 Model Predictive Quality Control

In this section, the data-driven modeling approach reviewed in the previous subsection is used in conjunction with an inferential quality model in a MPC design. The quality model captures the (time-cumulative) effects of the entire batch trajectory on the final quality while the multiple linear models for the measurable process variables take the causality and nonlinear relationship between the inputs and outputs into account. The benefit from this approach is the ability to account for the direct connection between the control action and the quality, something that is both expected and desired.

Given a batch database, consider the case where the quality and process variable models have been identified. The following MPC optimization problem is solved to compute the control action with the objective of achieving a desired product quality, $\boldsymbol{q}_{\text{des}}$.

$$\min_{\boldsymbol{u}(k)\in\mathcal{U}} \left(\hat{\boldsymbol{q}} - \boldsymbol{q}_{\text{des}}\right)' \boldsymbol{\Psi} \left(\hat{\boldsymbol{q}} - \boldsymbol{q}_{\text{des}}\right) + \sum_{i=k}^{K} \Delta\boldsymbol{u}'(i)\boldsymbol{\Phi}\Delta\boldsymbol{u}(i) \tag{8.9}$$

$$\text{s.t.: } \hat{\boldsymbol{y}}(k) = \boldsymbol{y}(t) \tag{8.10}$$

$$\hat{\boldsymbol{y}}(k) = \sum_{\ell=1}^{L} w_{\ell}(k)\hat{\boldsymbol{\beta}}_{\ell}\bar{\boldsymbol{x}}(k) \tag{8.11}$$

$$\boldsymbol{x}'_{\text{future}} = \left[\boldsymbol{u}'(k)\ \hat{\boldsymbol{y}}'(k+1)\ \boldsymbol{u}'(k+1)\ \cdots\ \boldsymbol{u}'(K-1)\ \hat{\boldsymbol{y}}'(K)\right]' \tag{8.12}$$

$$\hat{\boldsymbol{q}} = \left[\boldsymbol{x}_{\text{past}}\ \boldsymbol{x}_{\text{future}}\right]\boldsymbol{\Lambda} \tag{8.13}$$

In this optimization problem, the objective function consists of a term for minimizing the discrepancy between the target product quality and the predicted quality, $\hat{\boldsymbol{q}}$, and a move suppression factor. Each term's relative importance is traded-off using the positive-definite weighting matrices, $\boldsymbol{\Psi}$ and $\boldsymbol{\Phi}$. Equation (8.10) is the MPC initialization at the current plant conditions, and Eq. (8.11) represents the prediction of the future process variables using the data-driven model (given the current input trajectory in the optimizer). The predicted process outputs and optimizer inputs are stored appropriately in the row vector, $\boldsymbol{x}_{\text{future}}$, through Eq. (8.12). This vector is concatenated with a vector of previous plant outputs and implemented inputs, $\boldsymbol{x}_{\text{past}}$. Note that

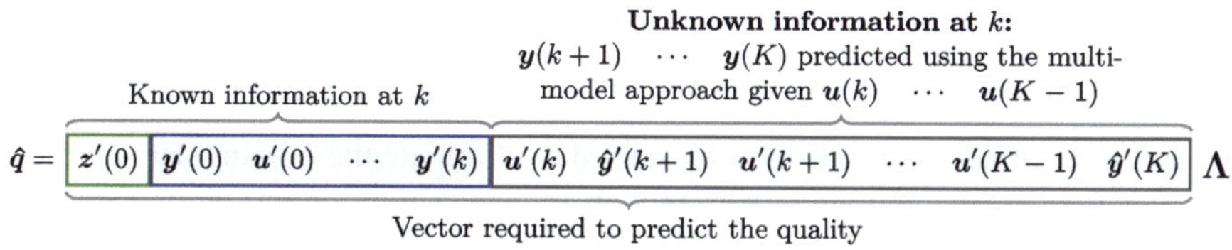

Fig. 8.1 Schematic of how the multi-model approach is used with the inferential quality model to predict the future (unknown) output trajectories

the vector, $\boldsymbol{x}_{\text{past}}$, is known prior to solving the MPC optimization problem; specifically, $\boldsymbol{x}'_{\text{past}} = \left[z'(0)\ \boldsymbol{y}'(0)\ \boldsymbol{u}'(0) \cdots \boldsymbol{y}'(k)\right]'$ where $z'(0)$ denotes all the information known prior to starting the batch (i.e., the initial conditions). The concatenated vector, $\left[\boldsymbol{x}_{\text{past}}\ \boldsymbol{x}_{\text{future}}\right]$, is used to predict the quality through Eq. (8.13).

The way the multi-model approach is integrated with the inferential quality model is illustrated in Fig. 8.1. At sampling instance k, the vector required to predict the quality is incomplete; the inputs and outputs beyond k are unknown. However, when the model is embedded in the MPC optimization problem (wherein the prediction horizon extends to the end of the batch), a candidate input trajectory exists in the optimizer that can be used to predict the future outputs up to batch termination. In our work, we use the multi-model approach to predict these future outputs. For a candidate input trajectory, the final product quality can be thus predicted, allowing direct control to a desired quality.

Remark 8.4 A distinguishing feature of this MPC design includes the use of a causal (and effectively nonlinear) model for predicting the future observations expected if a particular candidate input trajectory is implemented. The majority of control designs that have utilized a multivariate, PLS-based inferential quality model instead treat this as a missing data problem. Based on the data collected up to the current sampling instant, these algorithms essentially invert the *linear*, PLS model such that the future behavior maintains the same correlation structure as previous batches. This leads to an inherent mismatch in the sense that the predicted future behavior is based on past data (that typically uses existing PI controllers) which in turn is used to compute the current control action via a different control algorithm than in the data set. In contrast, the proposed approach recognizes that the problem is not that of missing data, because the future trajectories depend on state and input trajectories up to the current point as well as future input moves. The only 'missing' part therefore is the part that needs to be computed by the controller—the set of future control moves. By utilizing an appropriate model (which captures the process nonlinearity) that links the future inputs to the future process outputs and in turn to the quality, the controller then computes the set of input moves that would yield the desired quality. The problem of unequal batch lengths can also be readily handled by the proposed quality prediction approach by virtue of using a dedicated (nonlinear) model for predicting the future batch behavior. In particular, all the batch data can be used to build the dynamic model, while the quality model can be built using a common time from the end of batch for all batches (i.e., using the batch time of the shortest batch, if not significantly shorter than the other batches).

Remark 8.5 Other control designs have tried to eliminate the missing data problem completely through evolving quality models (at each sampling instant or a selected number of predetermined decision points), which utilize measurements only up to a given time. These models are designed to forecast the final quality without the future batch trajectories and inherently rely on the assumption that the same control action is implemented for the rest of the batch. Therefore, while such methods may be good to 'predict' the quality under an existing controller, they are not well suited for use in a control design aimed at computing the control action to yield the desired quality.

Remark 8.6 The prediction horizon for the MPC optimization problem above must extend to the end of the batch; thus, the prediction horizon, $P = K - k$, shrinks at every sampling instant. During the early stages of the batch when k is low, the MPC optimization problem may be too computationally demanding for real-time implementation. Under such circumstances, the optimization problem can be used to update the reference trajectories for local controllers rather than directly computing the inputs. Specifically, while the optimization problem is being solved, trajectory tracking controllers can be used to track the nominal reference trajectories, and upon completion of the optimization problem, the trajectories of the measurable output variables from the solution can be specified as the updated nominal reference trajectories in the local trajectory tracking controllers.

8.4 Simulation Results

In this section, we demonstrate the efficacy of the proposed MPC design through closed-loop simulations of a nylon-6,6 batch polymerization process. First, we give an overview of the process and state the control objective. Next, we develop data-driven models for the measurable process variables and product quality from a database generated using a rigorous first-principles model (also used as a test bed for implementation of the control design). Finally, using these models, we implement the proposed MPC design and compare its performance against trajectory tracking control.

8.4.1 Process Overview and Control Objective

For this chapter, we focus on nylon-6,6 production by the amidation of adipic acid and hexamethylenediamine (HMD) in a batch reactor. The reactor is initially charged with molten adipic acid and HMD (from an evaporator) in approximately stoichiometric (1:1) proportions. The polymerization reaction (a polyamidation) is summarized as follows: Carboxylic end groups (C) on adipic acid or the polymer chain react (reversibly) with amine end groups (A) on HMD or the polymer chain, producing a polymer chain link (L) and a water molecule (W):

$$\text{Polyamidation:} \quad \mathrm{C + A} \leftrightarrows \mathrm{L + W} \tag{8.14}$$

This polymerization is typically carried out in an autoclave reactor equipped with a steam jacket for providing the heat needed for vaporization (and reaction) and a valve for venting vaporized water. The polymerization occurs in three phases: (1) initial heating phase, (2) boiling phase, and (3) finishing phase. During the initial heating phase, the vent valve is closed to prevent the loss of volatile HMD and heat is supplied through the steam jacket, driving the polymerization reaction. After a certain extent of reaction, the valve is opened, initiating the boiling phase. During this phase, excess water is removed, which is important for achieving high molecular weight of the final polymer. After venting water for an appropriate amount of time, the vent is closed, and the finishing phase begins during which the final quality characteristics of the polymer are developed.

For our studies, we utilized a mathematical model of this process [23] that takes the general form shown in Eq. (8.1). The reaction model, modeling assumptions, equations, and parameter values [23] are omitted here for brevity. The state vector is comprised of the molar amounts of each functional group and evaporated HMD, and the reaction medium mass and temperature. The states were assumed to be measured only once at the initial time, but note that many of the states were trivially zero due to the absence of any reaction. The manipulated inputs were taken to be the steam jacket pressure, P_j (psi), and vent rate, v (g/h): $\boldsymbol{u} = \left[P_j \ v\right]'$, and the constraints were defined as follows: $\boldsymbol{u}_{\min} = \left[700 \ 0\right]'$ and $\boldsymbol{u}_{\max} = \left[1.8 \times 10^3 \ 2.5 \times 10^6\right]'$. All batches were assumed to be 3 h, eliminating the requirement for any trajectory synchronization, with a sampling period of 1 min.

The measurable process variables were taken to be the reaction mixture temperature, T (K), volume, V (L), and the viscosity, η (cP): $\boldsymbol{y} = \left[T \ V \ \eta\right]'$. Note that in practice, while the viscosity may not be directly measurable at the sampling period of 1 min, stirrer torque measurements are typically available in real time at every sampling instant. The stirrer torque is strongly correlated with the solution viscosity with a more viscous polymer resulting in higher torque (for a fixed RPM stirrer motor). Thus, the torque measurements provide important information about the viscosity evolution.

The product quality of nylon-6,6 polymer is defined by the number of average molecular weight, MW, and the residual amide concentration, $R_{\mathrm{NH_2}}$ (mol/g): $\boldsymbol{q} = \left[\mathrm{MW} \ R_{\mathrm{NH_2}}\right]'$. Both qualities are related to the state variables through highly nonlinear relationships [23]. The control objective considered in this chapter was to achieve end-point qualities of $\boldsymbol{q}_{\mathrm{des}} = \left[5569 \ 136\right]'$.

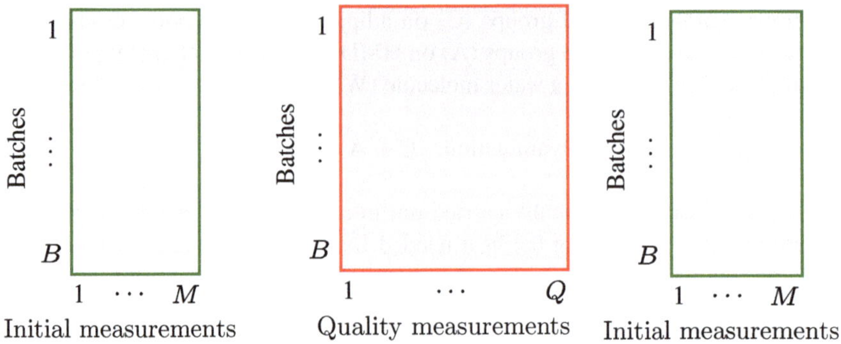

(a) Process variable trajectories, X (b) Quality data, Q (c) Initial conditions, Z_0

Fig. 8.2 Nature of data in a typical batch database

8.4.2 Inferential Quality and Process Variable Models

To develop data-driven models for the quality and process variables, an artificial batch database of the form in Fig. 8.2 was first generated. To this end, the deterministic model was simulated 80 times from different initial conditions (30 batches were reserved as the validation data set). In generating the database, a set of reference T and V profiles presented in Fig. 8.6a were tracked using P_j and v (respectively) via two tightly tuned PI controllers. In addition to these closed-loop trajectories, the database was supplemented with four open-loop identification batches. For these batches, low amplitude, pseudo-random-binary-sequence (PRBS) signals were added on top of the nominal input trajectories. In Fig. 8.6b, the input trajectories for one of these identification batches are shown together with the nominal trajectories (that correspond to the T and V trajectories in Fig. 8.6a).

The final database consisted of measurements of the states at the initial time, T, V, and η at every sampling instant, and the qualities at batch termination. Prior to developing the models, the η measurements were replaced by $\ln(\eta)$, in recognition of the fact that viscosity typically satisfies a power law correlation. Using this database, an inferential quality model was developed using PLS regression. As discussed in Remark 8.3, quality-critical periods during the batch can be given more weight prior to computing the model parameters. For the nylon-6,6 process, the initial conditions and process behavior during the boiling and finishing phases are more influential to the final quality compared to the heating phase; consequently, columns corresponding to these conditions in the unfolded regressor matrix (see Fig. 8.3) were given 6, 2, and 4 times more weight than the heating phase (respectively).

The motivation behind placing the lowest weight on the heating phase was because it corresponded to a limited extent of the polymerization compared to the other phases. The weights for the other phases and initial conditions were found iteratively and chosen to minimize the root mean squared error (RMSE) in the predicted qualities

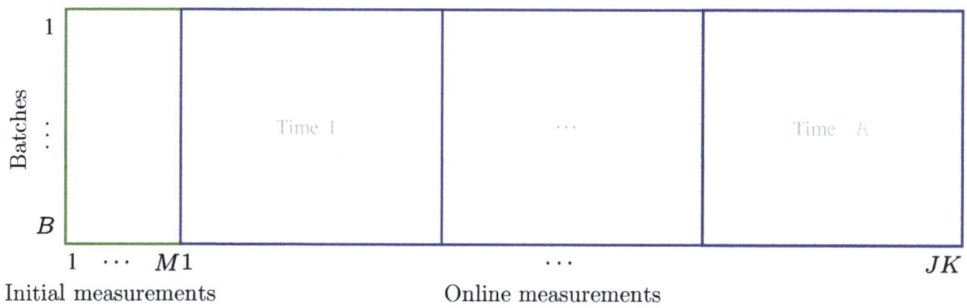

Fig. 8.3 Rearrangement of the batch data in Fig. 8.2 to form the regressor matrix for identifying the quality model in Eq. (8.2)

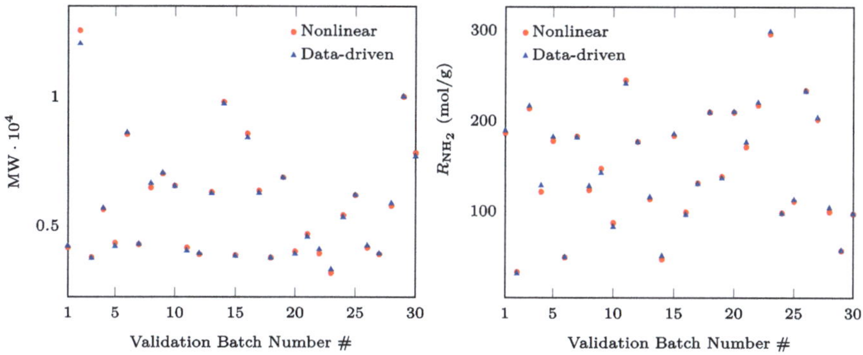

Fig. 8.4 Comparison of the predicted quality by the inferential PLS-based data-driven model with the nonlinear model

of the validation batches. Thus, these weights were essentially tuning parameters in the model. High weights were placed on the initial conditions to compensate for the fact that they constituted a very small portion of the regressor matrix compared to the other phases. In Fig. 8.4, the qualities predicted by the PLS model for the 30 validation batches are displayed along with the database qualities. The number of principal components in the PLS model, 24, was selected to minimize the RMSE in the predicted qualities of the validation batches. From the discussion in Remark 8.2, this meant 24 latent variables were required to estimate the states at batch termination. Note that the total number of columns in the regressor matrix was over 900; thus, 24 latent variables still represent a significant reduction in the dimensionality of the process. Additionally, the training data contained four identification batches that expanded the range of typical operating conditions, calling for additional latent variables. Overall, the inferential quality model predicted the final qualities with relatively good accuracy.

Next, multiple local linear models of the form described in Sect. 8.2.4 were developed for T, V, and $\ln(\eta)$ since these variables must be predicted (for a given input trajectory) as part of the quality prediction. To cluster the database and compute the

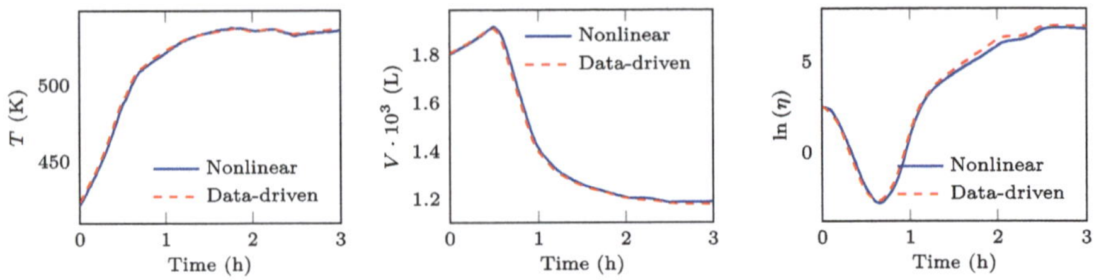

Fig. 8.5 Comparison of the predicted (measurable process) variables by the data-driven model with the nonlinear model for a random validation batch

resulting local linear models, the following parameters were required to be specified: the lag structure (n_y and n_u) and the number of clusters, L. The model identification procedure was as follows. For a given lag structure (which sets the dimensions of $\bar{\mathbf{X}}$), the $\bar{\mathbf{X}}$ matrix was constructed, decomposed using PCA, and clustered for a range of clusters, $L = \{1, 2, \ldots, 20\}$. For each cluster choice and a lag structure, a PLS model was identified. This was repeated for all possible lag structures with a lag range of 0–2 for each variable. The goodness of each fit (and the number of latent variables to retain in the PLS regression) was judged using its RMSE in predicting the validation batches. The number of input and output lags were found to be 1 (i.e., $n_y = n_u = 1$) and the number of clusters, L, was 7. One explanation for requiring only one lag is the assumption of the same lag structure for all the local models (note that this assumption can be readily relaxed if needed). With this assumption, using all first-order models minimized the possibility of overfitting, and in this case, yielded the lowest RMSE values. In Fig. 8.5, we compare the output of the nonlinear model with the model output from the data-driven model for a set of initial conditions in the validation data set. Overall, the multi-model approach captured the major nonlinearities and provided relatively reliable predictions.

8.4.3 Closed-Loop Simulation Results

Using the models developed in the previous section, in this section, we present the results from implementing the predictive controller and compare its control performance against trajectory tracking via PI controllers. For these simulations, we considered 21 new initial conditions that were not in the training or validation data sets. The reference trajectories for the PI-based trajectory tracking simulations were those presented earlier in Fig. 8.6a, and the loop-pairings and tunings were kept consistent with the database generation procedure. In solving the MPC optimization problem, the initial guess for the input trajectories was set to the nominal trajectories at $t = 0$ and the tail of the solution at the previous sampling instant for all subsequent sampling instants. The computation time (as reported by the Matlab functions, `tic` and `toc`) for the predictive controller at $t = 0$ was 1.5 s on an Intel Quad Core machine using GAMS with IPOPT as the optimization software. The computation times for

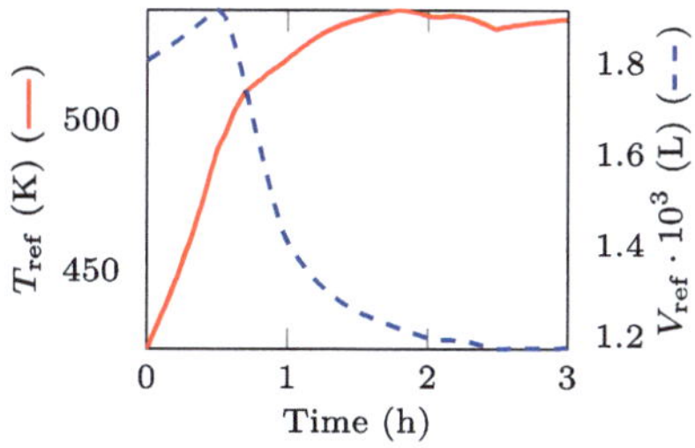

(a) References trajectories

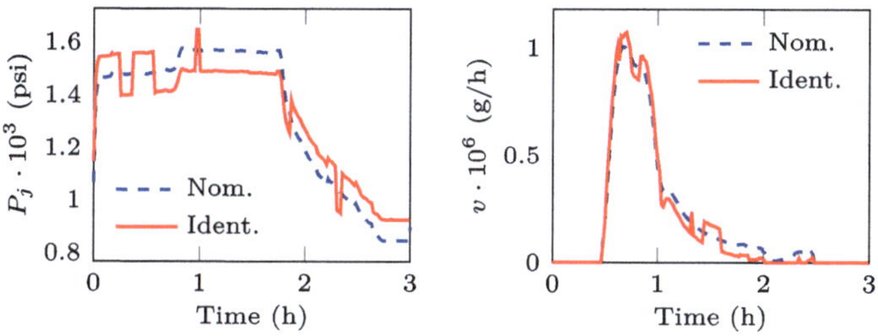

(b) Representative input trajectories for the identification (ident.) batches along with the nominal (nom.) input trajectories

Fig. 8.6 Trajectories used during database generation

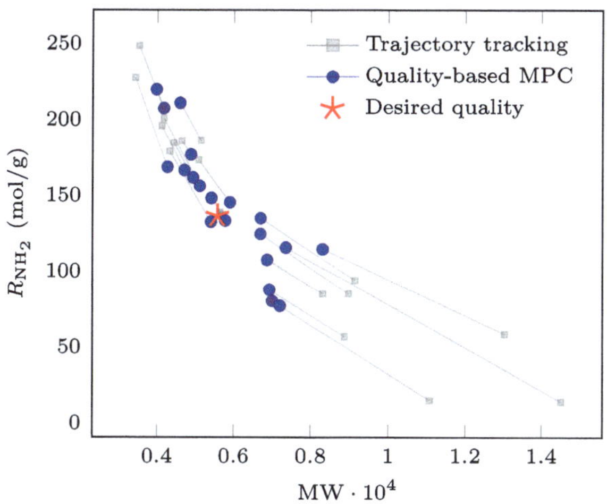

Fig. 8.7 Comparison of the final qualities obtained from trajectory tracking and quality-based MPC design for 21 new initial conditions

all successive sampling times were lower due in part to the shrinking horizon nature of the optimization problem.

In Fig. 8.7, the final qualities yielded from trajectory tracking are compared with those from the proposed MPC design. On average, there was a significant improvement in meeting the specified quality. With trajectory tracking, the standard deviations from the target quality were 3392 and 69.69 mol/g for MW and R_{NH_2},

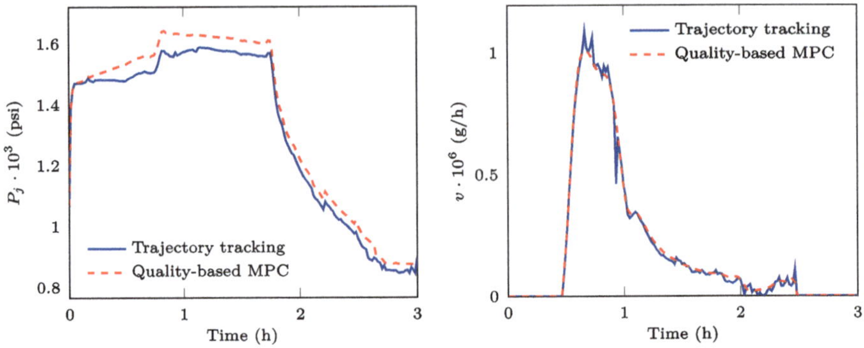

Fig. 8.8 Inputs prescribed by the predictive and trajectory tracking controllers for one of the batches in the closed-loop simulations

respectively. These values were reduced to 1240 and 40.71 mol/g by the predictive controller.

Input trajectories (from both controllers) for one of the batches are shown in Fig. 8.8. For these inputs, trajectory tracking yielded qualities of 6434 and 110 mol/g for MW and R_{NH_2} (respectively) while MPC yielded qualities of 5559 and 131 mol/g. Recall that the desired values for the final MW and R_{NH_2} were 5569 and 136 mol/g. Comparing the input trajectories in Fig. 8.8, we observe that the MPC-prescribed inputs followed the same general trends as with trajectory tracking but with sufficient refinements to significantly improve upon the quality.

8.5 Conclusions

In this chapter, we proposed a predictive control design for batch systems designed to drive the batch to a specified quality by batch termination. The MPC design utilized two types of models: a linear, PLS model which relates the process conditions over the entire batch duration to the final product quality and weighted local linear models that, for a candidate input trajectory, predict the future process conditions up to batch termination. Accounting for the causality and nonlinear relationships between the future input and process measurements, trajectory through the multiple local linear models led to more effective control action. The proposed control design was demonstrated via simulations of a highly nonlinear nylon-6,6 batch polymerization system, and it significantly improved on the final quality variance over that obtained using trajectory tracking control.

References

1. Geladi, P., Kowalski, B.: Partial least-squares regression: a tutorial. Anal. Chim. Acta **185**, 1–17 (1986)
2. Wold, S., Geladi, P., Esbensen, K., Ahman, J.: Multi-way principal components-and PLS-analysis. J. Chemom. **1**(1), 41–56 (1987)
3. Nomikos, P., MacGregor, J.F.: Monitoring batch processes using multiway principal component analysis. AIChE J. **40**(8), 1361–1375 (1994)
4. Flores-Cerrillo, J., MacGregor, J.F.: Within-batch and batch-to-batch inferential-adaptive control of semibatch reactors: a partial least squares approach. Ind. Eng. Chem. Res. **42**(14), 3334–3345 (2003)
5. Undey, C., Cinar, A.: Statistical monitoring of multiphase, multistage batch processes. IEEE Control Syst. Mag. **22**, 40–52 (2002)
6. Ng, Y.S., Srinivasan, R.: An adjoined multi-model approach for monitoring batch and transient operations. Comput. Chem. Eng. **33**(4), 887–902 (2009)
7. Flores-Cerrillo, J., MacGregor, J.F.: Control of particle size distributions in emulsion semibatch polymerization using mid-course correction policies. Ind. Eng. Chem. Res. **41**(7), 1805–1814 (2002)
8. Yabuki, Y., MacGregor, J.F.: Product quality control in semibatch reactors using midcourse correction policies. Ind. Eng. Chem. Res. **36**(4), 1268–1275 (1997)
9. Wang, D., Srinivasan, R.: Multi-model based real-time final product quality control strategy for batch processes. Comput. Chem. Eng. **33**(5), 992–1003 (2009)
10. Gunther, J.C., Conner, J.S., Seborg, D.E.: Process monitoring and quality variable prediction utilizing PLS in industrial fed-batch cell culture. J. Process Control **19**(5), 914–921 (2009)
11. Lu, N., Gao, F.: Stage-based process analysis and quality prediction for batch processes. Ind. Eng. Chem. Res. **44**(10), 3547–3555 (2005)
12. Zhao, C., Wang, F., Mao, Z., Lu, N., Jia, M.: Quality prediction based on phase-specific average trajectory for batch processes. AIChE J. **54**(3), 693–705 (2008)
13. Aumi, S., Mhaskar, P.: Integrating data-based modelling and nonlinear control tools for batch process control. AIChE J. **58**, 2105–2119 (2012)
14. Aumi, S., Corbett, B., Mhaskar, P., Clarke-Pringle, T.: Data-based modelling and control of nylon-6, 6 batch polymerization. IEEE Trans. Control Syst. Technol. (2011) (in press)
15. Kourti, T., Lee, J., Macgregor, J.F.: Experiences with industrial applications of projection methods for multivariate statistical process control. Comput. Chem. Eng. **20**, 745–750 (1996)
16. Kassidas, A., MacGregor, J.F., Taylor, P.A.: Synchronization of batch trajectories using dynamic time warping. AIChE J. **44**(4), 864–875 (1998)
17. Undey, C., Williams, B.A., Cinar, A.: Monitoring of batch pharmaceutical fermentations: data synchronization, landmark alignment, and real-time monitoring. In: Proceedings of the IFAC World Congress on Automatic Control, vol. 15. Barcelona (2002)
18. Huang, B., Ding, X.S., Qin, S.J.: Closed-loop subspace identification: an orthogonal projection approach. J. Process Control **15**(1), 53–66 (2005)
19. Patel, S., Yelchuru, R., Ryaliand, S., Gudi, R.D.: Discriminatory learning based performance monitoring of batch processes. In: Proceedings of the American Control Conference, pp. 2552–2557 (2011)
20. Bezdek, J.: A convergence theorem for the fuzzy isodata clustering algorithms. IEEE Trans. Pattern Anal. Mach. Intell. **2**(1), 1–8 (1980)
21. Frigui, H., Krishnapuram, R.: Clustering by competitive agglomeration. Pattern Recognit. **30**(7), 1109–1119 (1997)
22. Xie, X.L., Beni, G.: A validity measure for fuzzy clustering. IEEE Trans. Pattern Anal. Mach. Intell. **13**, 841–847 (1991)
23. Russell, S.A., Robertson, D.G., Lee, J.H., Ogunnaike, B.A.: Control of product quality for batch nylon-6,6 autoclaves. Chem. Eng. Sci. **53**(21), 3685–3702 (1998)

Chapter 9
Model Predictive Quality Control of Polymethyl Methacrylate

9.1 Introduction

Polymethyl methacrylate (PMMA) is a staple of the polymer industry. Originally marketed as Plexiglass, PMMA now has applications in many products ranging from its traditional application as a glass substitute to medical implants. The desirable properties of PMMA are known to be strongly dependent on the molecular weight distribution, often characterized through the number and weight average molecular weight. The primary control objective in the PMMA process, therefore, is to reach a specified weight average and number average molecular weight, resulting in high-quality product, motivating their production in a batch fashion. Polymerization of PMMA in this fashion is representative of a class of polymerization reactions that is common in industry (batch production when high-quality products are required).

PMMA production is carried out in a typical batch characterized by finite duration, nonlinear dynamics over a wide operating range and the absence of equilibrium points. The absence of a nominal equilibrium point precludes the direct application of controllers designed for continuous processes. Furthermore, the molecular weight distribution (or the number and weight average molecular weight) is typically not measured online during the batch process, but only made off-line after batch completion, making the direct control of the quality variables infeasible.

In this chapter, we present a within-batch molecular weight distribution modeling and control strategy for the PMMA process. The approach unites a single PLS inferential model with a nonlinear, data-driven dynamic modeling approach, while accounting for the variation in information during the batch and also ensuring that the process stays within the range of model validity. The rest of the manuscript is organized as follows: In Sect. 9.2, we first present the process description, a first-principles model used as a 'test bed' for the implementation of the proposed approach and discuss the data set assumed to be available and define the control objectives. In Sect. 9.3, the data is utilized to develop a data-driven dynamic model for the PMMA process and the 'goodness' of the model demonstrated. Then in Sect. 9.4, an integrated quality model integrating the dynamic model of Sect. 9.3 with an inferential

P. Mhaskar et al., *Modeling and Control of Batch Processes*,
Advances in Industrial Control, https://doi.org/10.1007/978-3-030-04140-3_9

Table 9.1 Reaction mechanism for MMA polymerization

Initiation	$I \rightarrow 2\phi$ $\phi + M \rightarrow R_1$
Propagation	$R_i + M \rightarrow R_{i+1}$
Termination by combination	$R_i + R_j \rightarrow R_{i+j}$
Termination by disproportionation	$R_i + R_j \rightarrow P_i + P_j$
Chain transfer to monomer	$R_i + M \rightarrow P_i + R_1$
Chain transfer to solvent	$R_i + S \rightarrow P_i + S$

quality model is developed and shown to be able to predict quality reliably for a set of candidate control input moves. In Sect. 9.5, the data-driven model is utilized within a model predictive control design that exploits the constraint handling ability of MPC to not only ensure input constraint and acceptable minimum conversion satisfaction, but also to ensure that the candidate control moves keep the process within the region where the model remains 'good.' The results of the control design are shown in Sect. 9.5.2, and finally in Sect. 9.6, conclusions are presented.

9.2 Process Overview and Control Objective

In this section, we first describe the process and a detailed first-principles model that will be used as a test bed for the implementation, followed by the control objectives.

9.2.1 Process Description

The process is carried out in a reactor with a heating/cooling jacket. The deterministic model used in this chapter is adapted [1] while making suitable alterations [2, 3]. Therefore, the complete model is presented in Sect. 9.7 to allow reproducibility. The underlying kinetic mechanism for the free radical polymerization of PMMA is shown in Table 9.1 where I is the initiator, M is the monomer, R_i is a live polymer with i monomer units, P_i is a dead polymer with i units, and S is the solvent [3].

The key states in the deterministic model are the three moments of the living and dead polymer distributions. These six states can be related to end product quality. In particular, they can be used to calculate (through expressions presented in the appendix) the weight and number average molecular weight as well as the conversion of monomer. The weight and number average molecular weight are particularly important properties for end use.

Table 9.2 Initial conditions for batches

Variable	Nominal	Standard deviation
Initiator Conc. [kg/m^3]	2.06×10^{-2}	2×10^{-3}
Monomer Conc. [kg/m^3]	4.57	1.5
Reactor temperature [°C]	61.0	5.0

9.2.2 Control Objectives

For the PMMA process, the objective is, as for most batch processes, to reach some target qualities. The key variables that affect the final quality for the PMMA process are the initial ingredient recipe and the jacket temperature trajectory. In this chapter, we focus primarily on the problem of determining the jacket temperature trajectory that in turn affects the reactor temperature trajectory to yield the desired weight and number average molecular weight at the end of the batch.

9.2.3 Database

The proposed dynamic model building approach relies on the availability of past data to build the dynamic (and quality) models. We use the model described above to build this data of past batches. The duration of a single batch was taken to be four hours with a sampling time of one minute. Each batch was started from a new initial condition to reflect normal variation in feedstock. The nominal value and standard deviation for each variable are shown in Table 9.2. In the past databases, a proportional–integral (PI) controller was used to track a nominal reactor temperature profile. This profile was based on a temperature trajectory presented in [3] (figure 9). The PI controller was tightly tuned to minimize the integral of absolute error (IAE). Furthermore, measurements recorded in the database were corrupted with random noise to replicate sensor error.

The database consisted of initial conditions (states), measurements at each sampling instance, manipulated variable moves, and quality measurements at the completion of the batch. The initial conditions were defined by monomer and initiator concentration as well as initial temperature. The measurable outputs were taken to be temperature, density, and viscosity. The recorded qualities were monomer conversion, number average molecular weight, and weight average molecular weight.

In addition to the batches tracking the nominal profile, five additional identification batches were generated. The purpose of these batches was to enrich the database for later model fitting. Of these batches, two were operated at temperatures shifted up and down from the nominal operating temperatures by $\pm 1\,°C$. The remaining three were operated with pseudorandom binary sequencing signals added on top of the nominal temperature trajectory (average switching time of six sampling instances

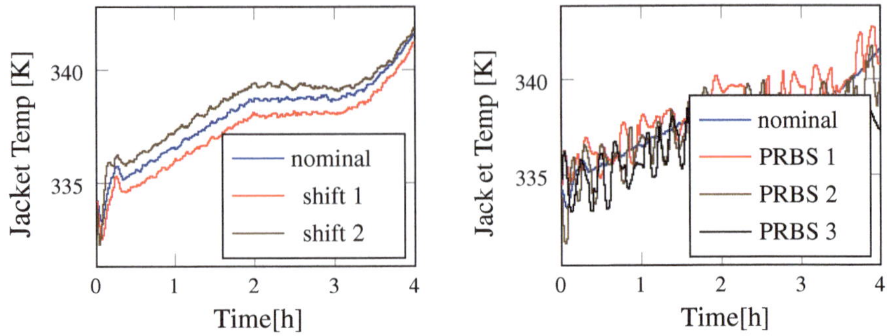

Fig. 9.1 Inputs used in identification batches

with a variance of ±0.5 °C). The input trajectories for these batches are shown in Fig. 9.1.

Remark 9.1 Note that the baseline for comparison could very well have been another data-based/first-principles-based approach for trajectory tracking. The focus in the present manuscript is not to compare PI control with advanced controllers but instead a trajectory tracking controller with a dedicated 'quality' controller. As can be seen, the PI is able to track the trajectories reasonably well, so any improvement in closed-loop performance (for quality control) is expected to be attributed to determining (online) improved control moves that better reach the target quality.

Remark 9.2 Note also that while we have available a first-principles model that captures well the inherent process dynamics, the model is not very amenable for online optimization (and control calculation). The test bed does serve well, however, as a 'replica' of the process and to test the performance of the proposed modeling and control method, which in a real application would of course be based off real process data.

9.3 Dynamic Model Development

In this section, we apply a multi-model, data-driven modeling approach to predict the future output behavior (temperature, $\ln \eta$, and ρ, and subsequently the final quality [4]) for a candidate manipulated input trajectory. Mathematically, the model for the process outputs takes the form of a weighted combination of L linear discrete time input–output models as shown below in Eq. (9.2).

$$\hat{\boldsymbol{y}}(k+1) = \sum_{\ell=1}^{L} w_\ell(k) \hat{\boldsymbol{\beta}}_\ell \boldsymbol{y}(k) \tag{9.1}$$

$$= \sum_{\ell=1}^{L} w_\ell(k) \hat{\boldsymbol{\beta}}_\ell [T(k) \quad \ln(\eta(k)) \quad \rho_{mix}(k) \quad T_j(k)]' \tag{9.2}$$

where $\mathbf{y}(k) \in \mathbb{R}^4$ is a vector of process measurements at time k. $w_\ell(k)$ is model ℓ's weight at sampling instant, k, and $\hat{\boldsymbol{\beta}}_\ell$ is a 4×4 matrix and defines the ℓth local model. Using the following definitions,

$$\hat{\boldsymbol{\beta}} \triangleq \left[\hat{\boldsymbol{\beta}}_1 \cdots \hat{\boldsymbol{\beta}}_\ell \cdots \hat{\boldsymbol{\beta}}_L\right]' \tag{9.3}$$

$$\boldsymbol{h}(k) \triangleq \left[w_1(k)\bar{\mathbf{y}}(k) \cdots w_\ell(k)\bar{\mathbf{y}}(k) \cdots w_L(k)\bar{\mathbf{y}}(k)\right]' \tag{9.4}$$

Equation (9.2) can be rewritten in the form:

$$\hat{\mathbf{y}}(k+1) = \hat{\boldsymbol{\beta}}'\boldsymbol{h}(k) \tag{9.5}$$

The model identification procedure consists of an initial clustering step (in essence determining how many local linear models need to be used, albeit not using the measurements directly, but first performing PCA to retain only independent measurements) followed by solving a linear regression problem. In the first step, a matrix $\bar{\mathbf{Y}} = \left[\, T \; \ln(\eta) \; \rho_{mix} \; T_j \,\right]$ is generated by sorting the plant data sample-wise, and then, $\bar{\mathbf{Y}}$ (or its equivalent latent variable space-see [4] for details) is clustered into L clusters using fuzzy c-means clustering (note that $\bar{\mathbf{y}}(k)$ in Eq. (9.4) represents the kth row of $\bar{\mathbf{Y}}$). In fuzzy c-means clustering, points that are mathematically 'similar' according to the Euclidean 2-norm are clustered into overlapping spherical clusters with corresponding center points [5, 6]. Each cluster represents a region in the $\bar{\mathbf{Y}}$ space where an associated model has the highest degree of validity, and in this way, the cluster's center point represents the linearization point of its associated model. Using the results of the clustering step (the cluster center points), the weights, $w_\ell(\cdot)$, for the training data can be computed prior to the model coefficients (to be discussed shortly). Consequently, the $\boldsymbol{h}(k)$ vector in Eq. (9.5) is completely specified for the training data. Thus, a regressor matrix corresponding to $\boldsymbol{h}(k)$ can be constructed, and the local linear models, $\hat{\boldsymbol{\beta}}$, are computable using linear regression.

Intuitively, the model weights, $w_\ell(\cdot)$, should depend on the current values of the states and inputs. In other words, the local models should be weighted according to the current process conditions. Because full state measurement is not available for the PMMA system, the vector of outputs and inputs, $\bar{\boldsymbol{Y}}(k)$, can be used to infer the current process conditions, and each model's weight can be assigned based on the proximity of the operating conditions to its center point. For instance, denoting model ℓ's center point as $\boldsymbol{c}_\ell$, its weight should be inversely proportional to the squared distance between $\bar{\boldsymbol{Y}}(k)$ and $\boldsymbol{c}_\ell$:

$$w_\ell(k) \propto \|\bar{\mathbf{y}}(k) - \boldsymbol{c}_\ell\|^{-2}$$

Normalizing this expression over all clusters yields the following normalized weighting function:

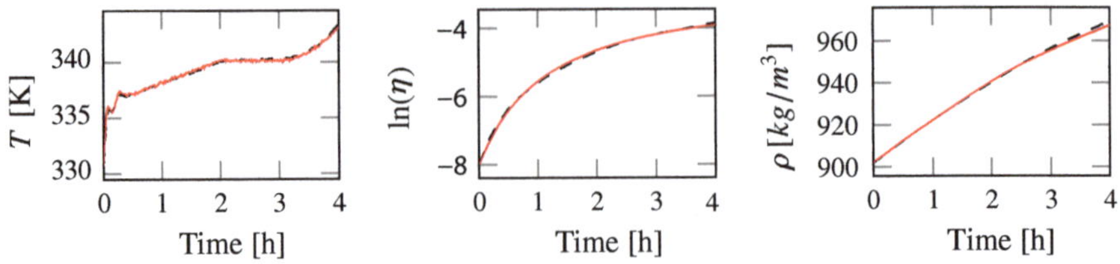

Fig. 9.2 Output trajectories predicted by the multi-model dynamic model (dashed lines) compared to actual measured output trajectories (solid lines)

$$w_\ell(k) = \frac{\|\bar{\mathbf{y}}(k) - \mathbf{c}_\ell\|^{-2}}{\sum_{i=1}^{L} \|\bar{\mathbf{y}}(k) - \mathbf{c}_i\|^{-2}}$$

Note that the choice of the number of clusters is made using cross-validation to keep the number of clusters that enable a good fit and prediction while avoiding overfitting. To do this, the second part of the training data set with 15 batches was used and the optimal number of clusters was found by evaluating several possibilities. In particular, models were fit for each combination of 1 to 8 clusters and 4 to 25 principal components. Models were evaluated in terms of the root-mean-squared error (RMSE) of prediction of the batch trajectories. It was found that the best number of clusters for this system was 6 clusters. For this number of clusters, the optimal number of principal components was found to be 25. Using these modeling parameters, the root-mean-squared error was found to be 0.22, 0.415, and 3.06 in predictions of temperature, $\ln \eta$, and ρ, respectively. Figure 9.2 shows predicted trajectories for a new batch and shows the utility of the model for feedback control (note that the predicted trajectory is entirely 'open loop', i.e., shares only the initial condition with the actual data).

9.4 Quality Model Development

In this section, the development of a model to predict end-point quality from batch trajectories is described. First, the model fitting procedure is described followed by the validation results. To obtain an inferential quality model for the PMMA process, the initial conditions and online measurements from each batch in the training data were regressed onto the final qualities. To this end, a regressor matrix, X, was formed such that each row of X contained the initial measurements and complete trajectories for a given batch. This was accomplished by first unfolding the measurement trajectories in a batch-wise fashion. To understand this unfolding, first consider a three-dimensional matrix of online measurements which split into two parts with dimensions $B \times J \times K$ where B denotes batch index ($B = 30$ for the first part to be used to compute model parameters, $B = 15$ for the second part to be used to pick the best number of principle components), J denotes measurement index ($J = 4$), and K denotes sampling time ($K = 241$). In this unfolding scheme, each row of the regressor matrix contains the complete trajectories for a single batch. Thus, the unfolded process measurements would form a two-dimensional $B \times JK$ matrix. To form the final regressor matrix, initial measurements were added to the unfolded process measurements. In the case with I initial measurements ($I = 3$), the final matrix would have dimensions $B \times (JK + I)$ (i.e., 30×967) [7]. The regressand matrix, Y, contains the quality measurements from each of the batches in the regressor matrix. In the case with Q final product qualities (for the PMMA system $Q = 3$), this would form a $B \times Q$ matrix.

Remark 9.3 Note that in the case with full state measurement, the end qualities could be calculated directly from the final state measurements in the batch. This is a result of the fact that by definition, the state of the batch describes all properties of the batch. However, continuous full state measurement is rare in batch systems. The inherent assumption made by the approach followed here (and in a sense in all inferential quality approaches) is that the final state of the system can be *estimated* from the complete output trajectory of the batch, which in turn can be used to estimate the quality.

Having formed the regressand and the regressor matrix, regression can be carried out to calculate model coefficients. Note that the columns of the regressor matrix described previously will inherently have a high degree of correlation. This is a result of the fact that the regressor contains measurements of the same variable taken one sampling instant apart. This correlation results in a near-singular matrix and will cause ordinary least squares regression to fail. To address this issue, partial least squares (PLS) regression is applied. In the next section, we present a brief overview of partial least squares regression for completeness.

9.4.1 PLS Overview

When variables in the regressor matrix are (auto/cross)-correlated, the use of ordinary least squares (OLS) regression leads to parameter estimates with large variances (due to the covariance of the regressor matrix being ill-conditioned). One way to handle this numerical issue is through partial least squares (PLS) regression. The reader is referred to Sect. 2.2.1, for the specific on the PLS modeling.

Remark 9.4 Traditionally in PLS, all variables are scaled to have unit variance. This permits each variable to have equal weight in model. However, in our application to the quality model, this results in each measurement at each sampling instant having the same weight as the initial conditions. It is clear that the initial conditions measurement (together with the input moves) should influence the final quality more than any single measurement at a given sampling instance, primarily because of there being full state measurement at the beginning for this process (some of the intermediate concentrations are known to be zero). Therefore, a technique called

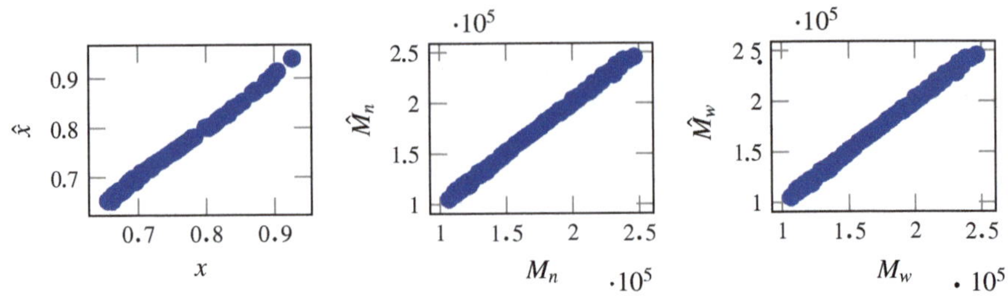

Fig. 9.3 Comparison of qualities predicted using the inferential quality model (using a completed batch trajectory) and the resulting actual quality (where x, M_n, and M_w are conversion, number and weight average molecular weight, respectively, $\hat{x}$, $\hat{M}_n$, and $\hat{M}_w$ are the predicted values of those same variables)

block scaling was used, where the variance of a given trajectory variable at a given sampling instance is reduced to $1/\sqrt{K}$, effectively increasing the influence of the initial measurement.

Remark 9.5 In PLS modeling, the key model parameter is the number of principal components used. In this chapter, the number of principal components was determined using a validation data set. To do this, a model was fit for each possible number of principal components (limited by the dimension of training data set). Then, each model was used to predict a separate validation data set. The number of principal components selected was the one that provided the smallest root-mean-squared error in predicting the validation data.

Figure 9.3 demonstrates the performance of the quality model for the PMMA system. The predictions shown in this figure are made from completed batch trajectories of new batches operated under PI trajectory tracking control. Note that the model provides reasonable prediction of both number average and weight average molecular weight and is therefore amenable for the purpose of feedback control.

9.5 Model Predictive Quality Control

In this section, the data-driven modeling approach reviewed in Sect. 9.3 is used in conjunction with the inferential quality model described in Sect. 9.4 in a quality-based MPC design, and subsequently implemented on the PMMA process.

9.5.1 MPQC Formulation

Note that the quality model captures the (time-cumulative) effects of the entire batch trajectory on the final quality, while the multiple linear models for the measurable process variables take the causality and nonlinear relationship between the inputs and outputs into account. The benefit from this approach is a more causal prediction of the final quality, which, in turn, leads to more effective control action.

Having identified the dynamic and quality model for the PMMA process, to achieve a desired quality $\boldsymbol{q}_{\text{des}}$, the entire control input trajectory from the current time to the end of the batch is computed by solving the following optimization problem (resulting in a shrinking horizon MPC implementation):

$$\min_{\boldsymbol{u}(k)\in\mathcal{U}} \left(\hat{\boldsymbol{q}} - \boldsymbol{q}_{\text{des}}\right)' \boldsymbol{\Psi} \left(\hat{\boldsymbol{q}} - \boldsymbol{q}_{\text{des}}\right) + \sum_{i=k}^{K} \Delta \boldsymbol{T}_j{}'(i) \boldsymbol{\Phi} \Delta \boldsymbol{T}_j(i) \tag{9.6}$$

$$\text{s.t.: } \hat{\boldsymbol{y}}(k) = \boldsymbol{y}(t) \tag{9.7}$$

$$\hat{\boldsymbol{y}}(k+1) = \sum_{\ell=1}^{L} w_\ell(k) \hat{\boldsymbol{\beta}}_\ell \hat{\boldsymbol{y}}(k) \tag{9.8}$$

$$\boldsymbol{x}'_{\text{future}} = [\boldsymbol{u}'(k) \quad \boldsymbol{y}'(k+1) \quad \boldsymbol{u}'(k+1) \cdots \\ \boldsymbol{y}'(k+2) \quad \boldsymbol{u}'(k+2) \cdots \boldsymbol{y}'(K)]' \tag{9.9}$$

$$SPE < SPE_{lim} \tag{9.10}$$

$$T^2 < T^2_{lim} \tag{9.11}$$

$$\hat{\boldsymbol{q}} = \left[\boldsymbol{x}_{\text{past}}\ \boldsymbol{x}_{\text{future}}\right] \boldsymbol{\Lambda} \tag{9.12}$$

where $\boldsymbol{\Psi}$ and $\boldsymbol{\Phi}$ are positive-definite weighting matrices imposing penalties on end-point quality deviation and rate of change of control moves. In this chapter, $\boldsymbol{\Psi}$ was an identity matrix and $\boldsymbol{\Phi}$ was a diagonal matrix with diagonal elements equal to 1.0×10^{-6}. Equation (9.7) is the MPC initialization at the current plant conditions, and Eq. (9.8) represents the prediction of the future process variables using the data-driven model (given the current input trajectory in the optimizer). The predicted process outputs and optimizer inputs are stored appropriately in the $\boldsymbol{x}_{\text{future}}$ vector through Eq. (9.9). This vector is concatenated with a vector of previous plant outputs and implemented inputs, $\boldsymbol{x}_{\text{past}}$. Note that the vector, $\boldsymbol{x}_{\text{past}}$, is known prior to solving the MPC optimization problem; specifically, $\boldsymbol{x}'_{\text{past}} = [\boldsymbol{z}'(0) \quad \boldsymbol{y}'(0) \quad \boldsymbol{u}'(0) \quad \boldsymbol{y}'(1) \quad \boldsymbol{u}'(1) \quad \cdots \quad \boldsymbol{y}'(k)]$ where $\boldsymbol{z}'(0)$ denotes all the information known prior to starting the batch (i.e., the initial conditions). The concatenated vector, $\left[\boldsymbol{x}_{\text{past}}\ \boldsymbol{x}_{\text{future}}\right]$, is used to predict the quality through Eq. (9.12). Note that the prediction horizon for the MPC optimization problem must extend to the end of the batch; thus, the prediction horizon, $P = K - k$, shrinks at every sampling instant.

In addition to process constraints, Eqs. (9.10) and (9.11) enforce model validity for the quality model, by enforcing constraints on the squared prediction error (SPE) and T^2, defined as:

$$SPE = (\boldsymbol{x} - \hat{\boldsymbol{x}})' (\boldsymbol{x} - \hat{\boldsymbol{x}}) \tag{9.13}$$

$$T^2 = \sum_{a=1}^{A} \left(\frac{\boldsymbol{t}_a}{\boldsymbol{s}_a} \right)^2 \tag{9.14}$$

where a represents principal component (out of A total principal components). $\boldsymbol{t}_a$ is the score associated with the ath principal component. $\boldsymbol{s}_a$ is the standard deviation in the scores of the ath component. $\hat{\boldsymbol{x}}$ is calculated by:

$$\hat{\boldsymbol{x}}' = \boldsymbol{t}\boldsymbol{p} \tag{9.15}$$

where $\boldsymbol{p}$ is the loading matrix (which defines the orientation of the latent variable space).

Enforcing constraints on SPE and T^2 ensures that the model is applied for trajectories that make the process operate in regions 'dynamically similar' and also having the same 'relationship' between process trajectories and end-point quality variables as that found in the training data. In geometric terms, SPE can be thought of as the distance of the latent variable model plane, while T^2 can be thought of as the distance along the plane from the origin of the latent variable space.

9.5.2 Data-Driven Model-Based Quality Control Results

The model predictive control scheme was implemented for the PMMA system, and results were summarized for twenty batches simulated from new initial conditions. The set-point for the quality-based control was a weight average molecular weight of 85,000 and a number average molecular weight of 150,000. The penalty on the control moves was set to 1×10^{-6}. The T squared and SPE limits were calculated from the 95 and 99.99% confidence intervals, respectively. The nonlinear quadratic program solver CONOPT was used to solve the optimization problem at each sampling instant. At the first sampling instance of each batch, the optimizer was initialized with the nominal input trajectory. At each subsequent sampling instance, the tail of the previous solution was used. In all cases, a feasible local minimum was found.

Note that the use of the multiple local model has significant computational advantages compared to using a nonlinear first-principles model for optimization, making the approach feasible for online implementation. In particular, the median time for the first control move calculation was 74 s on a dual-core Intel processor (note that the average time was 93 s but is skewed by a few outliers). The average calculation times for each subsequent sampling instances was 8.71 s which is well below the 60 s sampling time.

The results of this closed-loop simulation were evaluated both in absolute terms and also by comparison to the PI trajectory tracking performance. Figure 9.4 shows the relative error for both the PI trajectory tracking and the proposed MPQC. The average relative errors for the MPQC are 10.3 and 7.4% for number average and

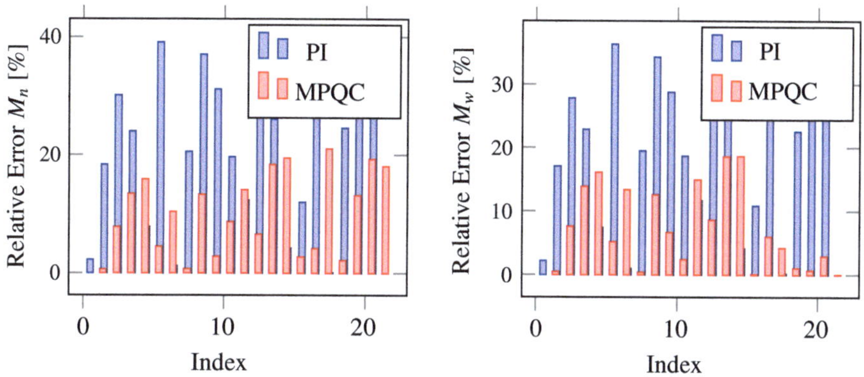

Fig. 9.4 Relative error in the molecular weight distribution qualities compared between the proposed model predictive quality control scheme and trajectory tracking PI controller

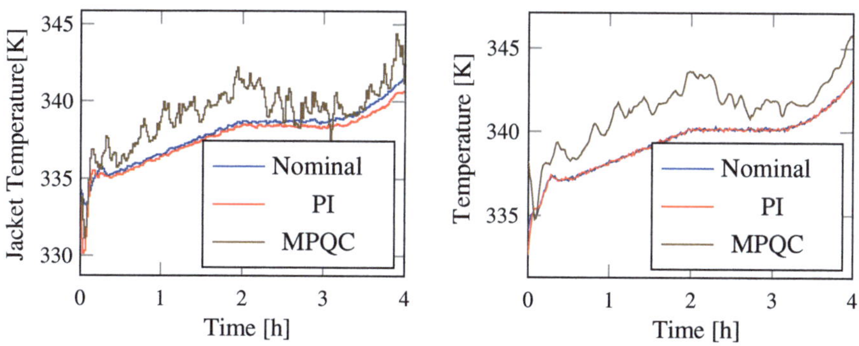

Fig. 9.5 Input moves and resulting reactor temperature compared for a typical batch under MPQC and PI control

weight average molecular weight, respectively. This is compared to 20.4 and 19.0% for the PI trajectory tracking. In 20 out of 21 cases, the MPQC produces a product closer to spec then the trajectory tracking scheme for both number and weight average molecular weight. Figure 9.5 shows a representative MPQC input trajectory. Note that the trajectory is similar to the nominal input and the PI trajectory but deviates slightly (and sufficiently, at critical times) to provide improved quality. In particular, for this case, the relative error was reduced from 30.2% in number average molecular weight to 13.4% and from 27.9% in weight average molecular weight to 13.9%.

Remark 9.6 Figure 9.6 shows the T squared and SPE values for a typical batch over the progression of the batch. Note that both T squared and SPE are moved to their limits by the optimizer. While this behavior is undesirable from a model validity perspective, it is expected. This is because in order to improve the quality from that produced by PI control, the optimizer must implement a different input–output correlation than was observed in the training data, to prevent the propagation of the

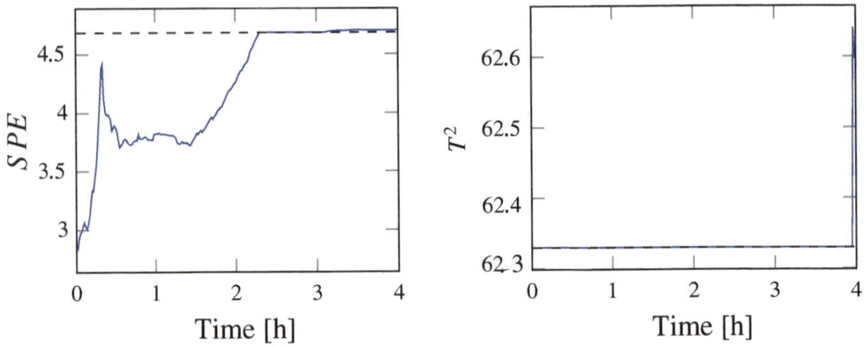

Fig. 9.6 SPE and T^2 for a typical batch (solid line), SPE and T^2 constraint (dashed line)

discrepancy in the initial condition to the final product quality. Fortunately, because of the batch-to-batch updating, each batch that is operated this way makes the models more robust and thus more accurate for future deviations in input–output correlation.

9.6 Conclusions

In this chapter, we considered the problem of quality control for a PMMA process. Our primary objective in this chapter was to develop a control strategy in which a detailed first-principles model of the process was not necessary. To do this, a first-principles model of the PMMA process was used to generate a database of batch trajectories. Then, using this data, a dynamic multiple model-based approach was used to capture the process dynamics from the data. In addition, a quality model was identified from the same data. These two models, the quality and multi-model dynamic models, were then integrated together to yield an overall model for the system. The above-described data-based models were then implemented in a model predictive controller with the objective of achieving the desired product quality while satisfying input constraints and a lower bound on the conversion, as well as additional constraints that enforce the validity of the data-driven models for the range of chosen input moves. Simulation results demonstrated the superior performance (10.3 and 7.4% relative error in number average and weight average molecular weight compared to 20.4 and 19.0% obtained by PI control) of the proposed modeling and control method over traditional trajectory tracking approaches.

9.7 First-Principles Process Model

Sym	Variable	Units
$\xi_0 - \xi_2$	0-2 live moment	kmol/m^3
$\mu_0 - \mu_2$	0-2 dead moment	kmol/m^3
$C_{i,s,m}$	Concentration *	kmol/m^3
R_g	Gas constant	kJ/kmol K
$MW_{i,s,m}$	Molecular weight *	kg/kmol
ρ_m	Density of monomer	kg/m^3
k_h	Huggins constant	-
U_0	Initial heat transfer coef.	kJ/s m^2 K
M_n	Number average molecular weight	g/mol
M_w	Weight average molecular weight	g/mol
*Subscripts i, s, m represent initiator, solvent, monomer		

$$\frac{dT}{dt} = -H_{rxn}\frac{R_p}{C_p\rho_{mix}} - UA\left(T - T_j\right)\frac{1}{C_pV\rho_{mix}} \tag{9.16}$$

$$\frac{d\xi_0}{dt} = 2fk_dC_i - k_t\xi_0^2 \tag{9.17}$$

$$\frac{d\xi_1}{dt} = 2fk_dC_i + k_pC_m\xi_0 + \ldots\left(k_{fm}C_m + k_{fs}C_s\right)(\xi_0 - \xi_1) - k_t\xi_0\xi_1 \tag{9.18}$$

$$\frac{d\xi_2}{dt} = 2fk_dC_i + (2\xi_1 + \xi_0)k_pC_m + \ldots\left(k_{fm}C_m + k_{fs}C_s\right)(\xi_0 - \xi_2) - k_t\xi_0\xi_2 \tag{9.19}$$

$$\frac{d\mu_0}{dt} = \left(k_{fm}C_m + k_{fs}C_s\right)\xi_0 + (0.5k_t)\,\xi_0^2 \tag{9.20}$$

$$\frac{d\mu_1}{dt} = \left(k_{fm}C_m + k_{fs}C_s + k_t\xi_0\right)\xi_1 \tag{9.21}$$

$$\frac{d\mu_2}{dt} = \left(k_{fm}C_m + k_{fs}C_s\right)\xi_2 + k_t\xi_0\xi_2 + k_t\xi_1^2 \tag{9.22}$$

$$\frac{dC_i}{dt} = -k_dC_i \tag{9.23}$$

$$\frac{dC_m}{dt} = -\left(k_p + k_{fm}\right)C_m\xi_0 \tag{9.24}$$

$$k_p = \frac{k_{p0}}{1 + \frac{k_{p0}\xi_0}{Dk_{\theta p}}} \qquad k_t = \frac{k_{t0}}{1 + \frac{k_{t0}\xi_0}{Dk_{\theta t}}} \tag{9.25}$$

$$k_{p0} = 7 \times 10^6 \exp\left(\frac{-2.6334 \times 10^4}{R_gT}\right) \tag{9.26}$$

$$k_{t0} = 1.79 \times 10^9 \exp\left(\frac{-1.1704 \times 10^4}{R_g T}\right) \tag{9.27}$$

$$k_{\theta p} = 3.0233 \times 10^{13} \exp\left(\frac{-1.1700 \times 10^5}{R_g T}\right) \tag{9.28}$$

$$k_{\theta t} = 1.454 \times 10^{20} C_{I0} \exp\left(\frac{-1.4584 \times 10^5}{R_g T}\right) \tag{9.29}$$

$$k_d = 1.58 \times 10^{15} \exp\left(\frac{-1.2874 \times 10^5}{R_g T}\right) \tag{9.30}$$

$$k_{fm} = 4.661 \times 10^9 \exp\left(\frac{-7.4479 \times 10^4}{R_g T}\right) \tag{9.31}$$

$$k_{fs} = 1.49 \times 10^9 \exp\left(\frac{-6.6197 \times 10^4}{R_g T}\right) \tag{9.32}$$

$$D = \exp\left(\frac{2.303\left(1 - \phi_p\right)}{D_2}\right) \tag{9.33}$$

$$D_2 = 0.168 - 8.21 \times 10^{-6}(T - 387)^2 + \ldots 0.03(T - \phi_p) \tag{9.34}$$

$$\phi_m = \frac{C_m MW_m}{\rho_m} \qquad \phi_p = \frac{\mu_1 MW_m}{\rho_p} \tag{9.35}$$

$$f = f_0 \exp\left(-C\left(\frac{1}{V_f} - \frac{1}{V_{fcr}}\right)\right) \tag{9.36}$$

$$V_{fcr} = 0.1856 - 2.965 \times 10^{-4}(T - 273) \tag{9.37}$$

$$V_f = 0.025 + 0.001(T - 167)\phi_m + \ldots 0.00048\left((T - 387)\phi_p + (T - 249)\phi_s\right) \tag{9.38}$$

$$R_p = k_p C_m \xi_0 \tag{9.39}$$

$$\rho_{mix} = (\mu_1 + C_m)\, MW_m + C_s MW_s + C_i MW_i \tag{9.40}$$

$$C_p = \frac{\mu_1 C_{p,p} + C_m C_{p,m} + C_s C_{p,s}}{\mu_1 + C_m + C_s} \tag{9.41}$$

$$U = U_0\left(0.2 + 0.8 \exp\left(-5x^5\right)\right) \tag{9.42}$$

$$\eta = \eta_0\left(1 + \eta_{sp}\right) \tag{9.43}$$

$$\eta_{sp} = c\eta_{int}\,(1 + k_h \eta_{int} c) \tag{9.44}$$

$$c = \mu_1 MW_m / 10^3 \tag{9.45}$$

$$\eta_{int} = 6.75 \times 10^{-3}\,(M_w)^{0.72} \tag{9.46}$$

$$M_n = \frac{\mu_1 + \xi_1}{\mu_0 + \xi_0} MW_m \tag{9.47}$$

$$M_w = \frac{\mu_2 + \xi_2}{\mu_1 + \xi_1} MW_m \tag{9.48}$$

$$x = \frac{\mu_1}{C_{m,0}} \tag{9.49}$$

- In Eqs. (9.21) and (9.23), f is used to denote the efficiency of the initiator (which is dependant on viscosity). We calculated this effect in Eqs. (9.36)–(9.38) which were taken from [2].
- As the reaction progresses, the increased viscosity causes a decrease in the heat transfer ability between the jacket and the reactor volume. This is modeled in Eq. (9.42) taken from [3].
- Viscosity values (which can be readily inferred from the torque measurements in the stirrer) were assumed to be measured and related to the process conditions by Eqs. (9.43)–(9.46) taken from [2].
- Equation (9.39) is different from [1] (where a quasi-steady-state assumption is made only for this equation) and uses ξ_0 [8] which is consistent with the use of dynamic equations describing the evolution of the moments.

References

1. Ekpo, E.E., Mujtaba, I.M.: Evaluation of neural networks-based controllers in batch polymerisation of methyl methacrylate. Neurocomputing **71**(7), 1401–1412 (2008)
2. Fan, S., Gretton-Watson, S.P., Steinke, J.H.G., Alpay, E.: Polymerisation of methyl methacrylate in a pilot-scale tubular reactor: modelling and experimental studies. Chem. Eng. Sci. **58**(12), 2479–2490 (2003)
3. Rho, H.J., Huh, Y.J., Rhee, H.K.: Application of adaptive model-predictive control to a batch MMA polymerization reactor. Chem. Eng. Sci. **53**(21), 3729–3739 (1998)
4. Aumi, S., Mhaskar, P.: Integrating data-based modeling and nonlinear control tools for batch process control. AIChE J. (2011). https://doi.org/10.1002/aic.12720
5. Seber, G.A.F.: Multivariate Observations. Wiley, New York (1984)
6. Bezdek, J.: A convergence theorem for the fuzzy ISODATA clustering algorithms. IEEE Trans. Pattern Anal. Mach. Intell. **2**(1), 1–8 (1980)
7. Flores-Cerrillo, J., MacGregor, J.F.: Control of batch product quality by trajectory manipulation using latent variable models. J. Process Control **14**(5), 539–553 (2004)
8. Soroush, M., Kravaris, C.: Nonlinear control of a batch polymerization reactor: an experimental study. AIChE J. **38**(9), 1429–1448 (1992)

Chapter 10
Multi-rate Modeling and Economic Model Predictive Control of the Electric Arc Furnace

10.1 Introduction

The modeling approaches presented in the earlier chapters assumed availability of uniformly sampled measurements. This assumption may not hold for many practical situations, an example being electric arc furnaces (EAFs). EAFs play a prominent role in the steel industry and are widely used for recycling scrap metal. Operated primarily as batch processes (a batch is referred to as a heat), EAFs melt the scrap and adjust the chemical composition of the molten metal to obtain steel of the desired product grade. The required melting of steel scrap results in a highly energy intensive process, necessitating efficient operation. Given that the feed to the EAF, from recycled steel, comes from diverse sources with obscure compositions that vary significantly, efficient operation can only be achieved via an online measurement and feedback control strategy. Closed-loop control of the EAF, however, presents a challenging problem due to the lack of online measurements of key process variables. This complicates both the model development and control implementation steps.

One approach to modeling and control of the EAF process in particular [1–5], and batch processes [6, 7] in general, is to develop first-principles/mechanistic models, and use them for the purpose of optimization. Note that while a first-principles model provides excellent predictive capabilities when sufficient measurements are available to uniquely estimate the associated parameters, the resultant optimization problem is often quite computationally complex and difficult to solve and implement in real-time. To address these challenges, recent results have exploited the structure of the optimization problem to identify the shape of the optimal constraints, which can then be parameterized and readily updated online [8, 9]. More recently, the concept of reachability regions is used to implement model predictive control strategies where the controller, instead of trying to drive the process to the desired end-point at all computation times, guides the process through the reachability regions (computed off-line) [10, 11]. Another effort to shift the computational effort to an off-line step includes design of explicit model predictive control involving multi-parametric programming, where the state of the system is represented as a vector of parameters

P. Mhaskar et al., *Modeling and Control of Batch Processes*,
Advances in Industrial Control, https://doi.org/10.1007/978-3-030-04140-3_10

so that the optimal solution for all possible realizations of the state vector can be precomputed as explicit functions [12, 13]. While these approaches mitigate the online computation aspect of the problem, the problem of developing and implementing a first-principles model-based controller remains a challenging task.

The data-driven model development approaches presented in the previous chapters assume the process measurements to be available at the same sampling rate, motivating the need to generalize these results for the case of the EAF process, where measurements are available at different sampling rates.

Regardless of the nature of the model used, the control problem can benefit from utilizing notions of economic control recently proposed for continuous processes [14–17, 48]. The key idea in these developments is that the controller determines the set-point internally to satisfy the prescribed economic objective and requires a rigorous analysis to ensure stability is preserved. In contrast, the batch control problem requires driving the process to a target that is often not a steady state, but is the desired end-point. Online control of EAF processes therefore stands to gain from incorporation of economic considerations in the control implementation. Motivated by the above considerations, this chapter addresses the problem of economic model predictive control (EMPC) of the electric arc furnace using data-driven models. To this end, we first review the electric arc furnace process and present a first-principles model that we utilize as a test bed to implement and validate the proposed approach. We also review the existing data-driven multi-model approach for batch process control in Sect. 10.2. Subsequently, multi-rate models are proposed in Sect. 10.3.1 that incorporate infrequent and frequent measurements to improve the predictions of the process variables. In Sect. 10.3.2, multi-rate models for the EAF process are computed. In Sect. 10.4.1 a two-tiered economic MPC is developed. In the first tier, the best achievable product (in terms of meeting product specifications) is determined by penalizing the deviation of the end-point variables from the desired target, while accounting for the input constraints in the optimization problem. Then, in the second layer, the optimal inputs are computed where the best achievable end-point (computed using the first layer) is imposed as a constraint, with economic requirements specified in the objective function. The proposed two-tiered economic MPC method and the closed-loop simulation results demonstrating its effectiveness are presented in Sect. 10.4.2. The conclusions of this chapter are summarized in Sect. 10.5.

10.2 Preliminaries

A description of the electric arc furnace process is provided below, followed by the description of the test-bed model, and the aspects of the simulation designed to replicate practical application issues. Then a review of an existing data-driven multi-model approach for batch process control is presented.

10.2.1 Electric Arc Furnace Process

The EAF is a batch process that involves a series of distinct operating phases that include the initial charging of the batch, followed by preheating, melting, and tapping of the furnace. The scrap charge is generally comprised of a variety of sources selected based on a number of factors such as the availability of each scrap source and the desired product grade. Typically, two or three loads of scrap are charged in each batch depending on the bulk density of the scrap and the volume of the furnace. The furnace charge may also be supplemented with some direct reduced iron (DRI) or pig iron for chemical balance and to improve production yields. Once the batch is charged, the EAF is preheated through natural gas combustion to raise the temperature of the steel. Subsequent to preheating, electrodes are lowered into the furnace and the electric power is turned on to an intermediate voltage while the electrodes bore into the scrap. The voltage is increased once a sufficient amount of molten steel is formed at the base of the arc, and the electrodes are submerged into the melt to avoid damage to the furnace walls. During the initial stages of the meltdown, a high voltage is selected that allows for more energy to be transferred to the surrounding scrap. As the batch approaches completion, a lower voltage arc is preferred to avoid damage to the exposed furnace walls. Moreover, slag is foamed during the EAF operation by lancing carbon and oxygen to form carbon monoxide gas that bubbles through the slag layer. The foaming slag cloaks the arc, thereby protecting the furnace walls from arc radiation and improving the energy efficiency. During the batch operation, impurities such as phosphorus, sulfur, aluminum, silicon, manganese, and carbon are removed from the steel as they react with oxygen and float into the slag. After a predefined batch duration, the temperature and carbon content of the steel are measured to determine whether further inputs are needed to reach the desired end-point specifications [4, 5]. Once the desired steel composition and temperature are obtained, the vessel is tapped and the molten steel is poured into a ladle for transport to the downstream units for further processing.

In this chapter, we focus on the melting process (see Fig. 10.1 for a schematic of an EAF during the melting stage). To this end, we utilize a first-principles model as a test bed [2, 18]. The model describes the melting process by using a total of 14 state variables and six manipulated variables. The model parameters were estimated using operating data from an industrial EAF [18]. While the model is focused on the melting process, and does not capture all the details of the EAF process, it is sufficiently detailed and validated through real plant data, making it an excellent candidate to adapt and utilize as a test bed to implement and evaluate the proposed approach.

In the steel industry, accurate, reliable, and low-maintenance sensors for continuous online measurement are typically not available for many of the EAF process variables because of the harsh operating environment, extreme temperatures, and highly corrosive nature of the molten steel. Recognizing the limited availability of process measurements in practice, in this chapter the measurements available to build the data-driven model and to implement the proposed control approach include

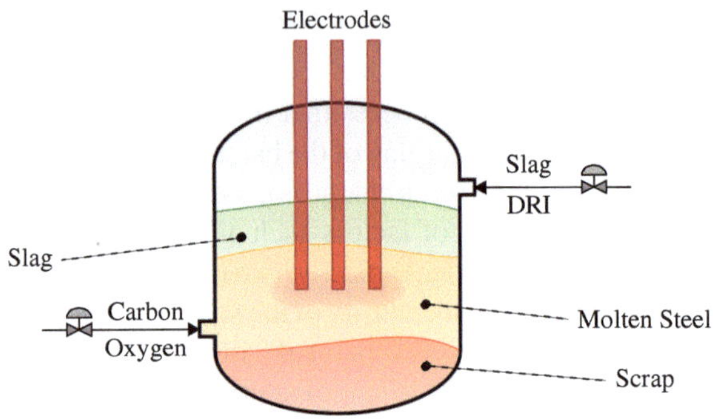

Fig. 10.1 An illustrative diagram of the EAF process

Table 10.1 List of infrequent measurement variables of the EAF process

Variable	Description	Units
T	Temperature of molten steel	K
x_{Fe}	Mass fraction iron in molten steel	kg/kg
x_C	Mass fraction carbon in molten steel	kg/kg
x_{Slag}	Mass fraction lime/dolime in slag	kg/kg
x_{FeO}	Mass fraction iron oxide in slag	kg/kg
x_{SiO_2}	Mass fraction silicon dioxide in slag	kg/kg

Table 10.2 List of frequent measurement variables of the EAF process

Variable	Description	Units
P	Relative pressure	Pa
x_{CO}	Mass fraction carbon monoxide in gas	kg/kg
x_{CO_2}	Mass fraction carbon dioxide in gas	kg/kg
x_{N_2}	Mass fraction nitrogen in gas	kg/kg

infrequent and frequent measurement variables where the infrequent measurements related to the slag and molten steel are available with a sampling time of 11 min while the frequent measurements corresponding to the off-gas composition are available with a sampling time of 1 min. A list of the infrequent and frequent process measurement variables is given in Tables 10.1 and 10.2, respectively, and the manipulated inputs are listed in Table 10.3. In keeping with typical EAF batch runs, each batch of the test-bed simulation is run for a duration of 66 min, which results in 7 infrequent measurements and 67 frequent measurements.

Another important feature necessary to make the test-bed application realistic is to recognize the significant variability of the feed caused by the various scrap sources of diverse steel grades. To replicate this variability across the batches, the initial concentrations of the species within the molten steel are assumed to have a variance of 50% with respect to the nominal values. On the other hand, the EAF processes are run by utilizing lime and dolime that is bought from suppliers, thus the resulting slag formed has lesser variability. Therefore, the initial conditions for the

Table 10.3 List of manipulated variables and corresponding input costs for the EAF process

Variable	Description	Units	Costs	Units of cost
m_{gas}	Off-gas turbine flow	kg/s	4.6296×10^{-5}	$\$ \cdot \text{kg}^{-1}$
m_{O_2}	Oxygen lanced	kg/s	0.01749	$\$ \cdot \text{kg}^{-1}$
m_{DRI}	DRI additions	kg/s	0.0900	$\$ \cdot \text{kg}^{-1}$
m_{Slag}	Slag additions	kg/s	0.0600	$\$ \cdot \text{kg}^{-1}$
E	Electric arc power	kW	0.0625	$\$ \cdot \text{kWh}^{-1}$
m_{C}	Carbon injected	kg/s	0.2500	$\$ \cdot \text{kg}^{-1}$

concentrations of the species within the slag are assumed to have a variance of 10% with respect to the nominal values (note though that as the EAF batch progresses, the slag composition does vary significantly owing to the variance in the molten metal composition). Finally, the initial temperatures of the molten steel and scrap during the nascent period of the EAF are highly dependent on the tightly controlled preprocessing steps taken prior to the commencing of the melting phase. The initial values of the temperatures are thus assumed to have a modest variance of 10% with respect to the nominal initial conditions.

Additionally, to ensure the EAF test bed is characteristic of real industrial processes, the measurements are assumed to be corrupted by uncorrelated noise that is Gaussian distributed with zero-mean and appropriate variances. The noise variances are established such that the average signal-to-noise ratio in the measurements is approximately 30 dB, representative of the typical measurement noise encountered in industrial practice due to random errors in the instrumentation.

From a control standpoint, the objective in EAF processes is to achieve a desired end-point, which like many batch processes typically correspond to a non-equilibrium point, by batch termination. In industrial practice, the EAF process is typically operated until the carbon content is reduced below the desired level and the molten steel temperature is greater than the required temperature for pouring into ingot molds or casting. In addition, it is necessary to maintain the slag at the proper chemical composition to minimize overoxidation of the bath. These primary targets are often used by plant operators to evaluate the EAF batch maturity [19]. In this chapter, we determine the desired end-points based on the attributes of the molten steel product obtained using the EAF simulation test bed and standard operating policies. To this end, we explicitly characterize the end-point targets through the melt temperature ($\text{T}_{t_f} \geq 1830$ K), mass fraction of carbon in the molten steel ($x_{\text{C},t_f} \leq 0.0047$ kg/kg), and the mass fraction of iron oxide in the slag ($x_{\text{FeO},t_f} \leq 0.475$ kg/kg). Moreover, the architecture of industrial EAFs usually incorporates water-cooled furnace walls and roof panels to avoid damage due to overheating. Path constraints that limit the maximum achievable temperature of the molten steel ($\text{T}_t \leq 1950$ K, $t \in [0, t_f]$) ensure that the furnace temperature is within an operating region that takes safety considerations into account. In addition, the relative pressure within EAFs is typically maintained to be negative ($\text{P}_t \leq 0$ Pa, $t \in [0, t_f]$) so as to prevent exposure of harmful chemicals to plant operators and retain the hazardous gaseous emissions within the furnace.

Table 10.4 List of PI control loop-pairings and the corresponding controller parameters for the EAF process

Controlled variable	Manipulated variable	Parameters	
		K_C	τ_I
T	E	50.0	20.0
x_C	m_C	2.0	10.0
x_{FeO}	m_{O_2}	3.0	5.0
P	m_{gas}	0.5	3.0

As with many industrial batch processes, the commonly used operating policy involves limited feedback applied primarily by experienced plant operators and typically does not take into account end-point projections or explicit economic considerations. Another less commonly used operating strategy is through tracking suitable trajectories of the measurable process variables, where the variable trajectories to be tracked are the ones that historically generated on-spec product. As an experienced plant operator is untenable in this study, trajectory tracking control is implemented as a surrogate to the operator in order to mimic typical industrial EAF operation where reference profiles of the measurement variables are closely followed to ensure the constraints are satisfied during normal batch operation. The tracking objective is traditionally accomplished for a set of measurable process variables related to the desired end-point targets using local controllers that invoke classic (linear) control approaches such as local proportional–integral (PI) controllers. In this chapter, to generate the database of historical batches, four PI control loops are used to track the melt temperature, mass fraction of carbon in the molten steel, the mass fraction of iron oxide in the slag, and the relative pressure inside the reactor. In keeping with the frequency of measurement of three of these variables, and for the sake of consistency, PI control calculations are done at the same frequency (i.e., corresponding to the times that the infrequent measurements become available). A detailed breakdown regarding the PI control loops, the manipulated variables that are paired with the controlled variables, and the corresponding controller parameters are given in Table 10.4. The trajectories to be tracked are the conventional operating profiles that if closely followed result in steel product that meets the desired end-point targets. Nevertheless, PI controllers remain inherently based on a decentralized (single-input-single-output) framework that is not well suited to account for the complex multivariable interactions among the various control loops, process constraints, economic objectives, and optimality.

In contrast to trajectory tracking control, an industrially relevant cost functional that explicitly takes the process economics (e.g., operating cost or profit) into consideration can be used to determine the control actions that not only satisfy the desired end-point targets but also obtain economic optimality. As such, the economic objective we formulate in this chapter represents the total operating expense of an EAF run, thus penalizing energy and resource consumption over the entire batch duration to operate the EAF in an economically optimal fashion while ensuring the constraints are satisfied. The nominal costs for the manipulated inputs are given in Table 10.3

[20]. We also take into account input bounds due to the physical limitations of the system, such as the maximum arc voltage or oxygen lanced into the melt.

10.2.2 Data-Based Batch Process Modeling and Control

For the proposed data-driven modeling and control approach, we generalize a recently developed multi-model approach that incorporates multiple local dynamic linear models and fuzzy c-means clustering-based MPC [21, 22], where it is assumed that the measurements are all available with the same frequency. We next review the modeling approach in Sect. 10.2.2.1 and the control design in Sect. 10.2.2.2.

10.2.2.1 Multi-model Approach

The modeling approach in [21, 22] is predicated on the idea that a nonlinear process can be approximated by appropriately combining linear models that are valid around the point of linearization. To this end, local linear models are first developed. As one choice of the local linear models, the following form is used:

$$\hat{y}_k = \sum_{i=1}^{n_y} A_i y_{k-i} + \sum_{j=1}^{n_u} B_j u_{k-j} + \nu \tag{10.1}$$

where A and B are coefficient matrices for the output and input variables denoted by $y \in \mathbb{R}^p$ and $u \in \mathbb{R}^m$ with corresponding lag orders n_y and n_u, respectively, and ν is a bias term. The model can be written in a concatenated form as follows:

$$\hat{y}_k = C \begin{bmatrix} x_k^* \\ 1_{p\times 1} \end{bmatrix} \tag{10.2}$$

where $1_{p\times 1}$ denotes a vector of ones with length p and

$$C = \begin{bmatrix} A_1 & \dots & A_{n_y} & B_1 & \dots & B_{n_u} & \nu \end{bmatrix} \tag{10.3}$$

with

$$x_k^* = \begin{bmatrix} y_{k-1}^T & \cdots & y_{k-n_y}^T & u_{k-1}^T & \cdots & u_{k-n_u}^T \end{bmatrix}^T \tag{10.4}$$

The matrix of model coefficients in Eq. (10.2) can be determined using ordinary least squares (OLS) [21, 22]. However, OLS may not precisely determine robust regression coefficients when the process data is highly correlated or colinear, which leads to ill-conditioning problems and imprecise model coefficients. The drawbacks of OLS regression can be addressed by using latent variable regression techniques such as principal component regression or partial least squares (PLS) regression.

In PLS regression, the regressor matrix (X) of the measurement variables and response matrix (Y) are mean centered and scaled to unit variance and then projected onto orthogonal subspaces comprised of n_A-pairs of latent variables. The preprocessing gives each variable equal importance during model identification while the bias term becomes redundant. Mathematically, PLS regression involves the decomposition of the regressor and response matrices into a summation of the outer products of the n_A score and loading vectors as follows:

$$\begin{aligned} X &= \sum_{i=1}^{n_A} t_i p_i^T + E_X \\ &= T P^T + E_X \end{aligned} \tag{10.5}$$

$$\begin{aligned} Y &= \sum_{i=1}^{n_A} r_i q_i^T + E_Y \\ &= R Q^T + E_Y \end{aligned} \tag{10.6}$$

where t and r are the scores for the projection of the input and output data matrices onto the respective subspaces whose orientation is defined by the vectors p and q. Moreover, the matrices T, P, R, and Q succinctly express the PLS decomposition with E_X and E_Y as the residual matrices, resulting in estimated model coefficients $C = f(P, T, Q)$.

Although PLS-based approaches to building dynamical models enable utilizing sound statistical techniques in the model parameter estimation, as a standalone model, it yields a good, but local linear model. To characterize the process behavior over the entire range of operation, the inclusion of multiple models in an integrated approach is necessary. One such approach involves the partitioning of the historical database of training batches into n_L clusters so that the corresponding linear models can be identified for each cluster [21]. Using fuzzy c-means clustering, the data set can be partitioned into clusters by assigning the i-th data sample the membership values $\mu_{l,i}$, which represents the degree to which the i-th sample belongs to the l-th cluster. The membership values must satisfy the following conditions

$$\mu_{l,i} \in [0, 1] \; \forall \; l \in [1, n_L] \tag{10.7}$$

$$\sum_{l=1}^{n^L} \mu_{l,i} = 1 \; \forall \; i \in [1, n_{\text{obs}}] \tag{10.8}$$

where n_{obs} represents the total number of observations within the training data set. The majority of fuzzy clustering algorithms is based on reducing the total variance in the data from cluster centers, which is expressed by minimizing the following nonlinear objective function

$$\min_{c_l} \mathcal{J} = \sum_{i=1}^{n_{\text{obs}}} \sum_{l=1}^{n_L} \mu_{l,i}^2 \left\| x_i^* - c_l \right\|^2 \tag{10.9}$$

where c_l denote the cluster center vectors to be determined. The membership values $\mu_{l,i}$ are then related to the cluster centers c_l as follows:

$$\mu_{l,i} = \frac{\left\| x_i^* - c_l \right\|^{-2}}{\sum_{l'=1}^{n_L} \left\| x_i^* - c_{l'} \right\|^{-2}} \tag{10.10}$$

which shows that the degree to which the data sample x_i^* belongs to the cluster l is inversely proportional to the squared distance between the respective point and the cluster center. Therefore, a fuzzy c-means clustering-based continuous weighting function can be developed and incorporated with the dynamic modeling approach to integrate multiple local linear models into a global nonlinear model as follows:

$$\hat{y}_k = \sum_{l=1}^{n_L} \omega_{l,k} \left(\sum_{i=1}^{n_y} A_{l,i} y_{k-i} + \sum_{j=1}^{n_u} B_{l,j} u_{k-j} + \nu_l \right) \tag{10.11}$$

$$= \sum_{l=1}^{n_L} \omega_{l,k} C_l \begin{bmatrix} x_k^* \\ 1 \end{bmatrix} \tag{10.12}$$

where the weighting $\omega_{l,k}$ are given by

$$\omega_{l,k} = \frac{\left\| x_k^* - c_l \right\|^{-2}}{\sum_{l'=1}^{n_L} \left\| x_k^* - c_{l'} \right\|^{-2}} \tag{10.13}$$

The final form of the model combines the local linear models with weights to effectively describe the global nonlinear dynamics [21, 22], under the assumption of measurements available at the same frequency. A nonlinear model of this type is well suited to predict the dynamic process behavior and thus can be used for variable trajectory tracking applications in predictive controllers.

10.2.2.2 Trajectory Tracking Predictive Control

A predictive controller for tracking reference trajectories for batch processes is presented in this section. The control action at each sampling instance is computed by solving the following optimization problem

$$\min_{u \in \mathcal{U}} \mathcal{J} = \sum_{k=1}^{n_P} \| \hat{y}_k - y_k^{\text{ref}} \|_{Q_w}^2 + \| \Delta u_k \|_{R_w}^2 \tag{10.14}$$

$$\text{s.t.} \begin{cases} \hat{y}_k = \sum_{l=1}^{n_L} \omega_{l,k} \left(\sum_{i=1}^{n_y} A_{l,i} \hat{y}_{k-i} + \sum_{j=1}^{n_u} B_{l,j} u_{k-j} + \nu_l \right) \\ \hat{y}_0 = y_t \end{cases} \tag{10.15}$$

where $u \in \mathbb{R}^m$ denotes the vector of constrained input variables, taking values in a nonempty convex set $\mathcal{U} \subseteq \mathbb{R}^m$. The first term in the objective function penalizes discrepancies between the predicted output trajectories $\hat{y}$ and the reference trajectories y^{ref} over the prediction horizon n_P, and the second term is a move suppression term that penalizes the magnitude of input changes [21, 22]. Further, Q_w is a positive semi-definite symmetric matrix used to penalize the deviations of the outputs from their nominal values and R_w is a strictly positive-definite symmetric matrix to penalize changes in the manipulated variables. The predictive model in the MPC formulation, specifically the nonlinear weighting function, makes this optimization problem a nonlinear program. While the trajectory tracking control paradigm has been successfully applied in many applications, the primary objective in batch processes is to reach a desired end-point by batch termination. In many instances, the initially optimal state variable trajectories to be tracked may be rendered suboptimal in the presence of measurement noise or process disturbances. On the other hand, directly accounting for process economics to compute the control actions can achieve economic optimality that may not be readily attainable through traditional tracking control techniques.

10.3 Data-Driven Multi-rate Model

In this section, we propose a multi-rate modeling framework and apply it to the EAF process.

10.3.1 Multi-rate Model Formulation

To model multi-rate sampled systems, we propose a novel formulation in which the frequent process measurements are used along with the infrequent measurements for developing effective and reliable multi-rate models. To this end, consider a multi-input multi-output nonlinear system where $y^{\wp_1} \in \mathbb{R}^{p_1}$ and $y^{\wp_2} \in \mathbb{R}^{p_2}$ denote the vector of infrequent and frequent measurement variables, respectively, and $u \in \mathbb{R}^m$ denotes the vector of constrained input variables. Typically, the number of infrequent measurement variables is greater than the frequent measurement variables ($p_1 \geq p_2$). Further, given that in between an infrequent sampling interval, there are γ frequent sampling instances, let k_1 and k_2 be the indexes for the sampling instances of the infrequent and frequent measurements with n_K as the final infrequent sampling instance. Along the lines of Eqs. (10.11) and (10.12), local linear models for the infrequent measurement variables can be developed as follows:

$$\hat{y}^{\wp_1}_{k_1|k_2-1} = \sum_{l=1}^{n_L^{\wp_1}} \omega^{\wp_1}_{l,k_1} \left(\sum_{i=1}^{n_y^{\wp_1}} A^{\wp_1}_{l,i} y^{\wp_1}_{k_1-i} + \sum_{i_s=1}^{\gamma k_1 - k_2} A^{s,\wp_2}_{l,i_s} \hat{y}^{\wp_2}_{\gamma k_1 - i_s} \right.$$
$$\left. + \sum_{i_s'=\gamma k_1 - k_2 - 1}^{\gamma k_1 - n_y^{s,\wp_2}} A^{s,\wp_2}_{l,i_s'} y^{\wp_2}_{\gamma k_1 - i_s'} + \sum_{j=1}^{n_u^{\wp_1}} B^{\wp_1}_{l,j} u_{k_1-j} + \nu^{\wp_1}_l \right) \tag{10.16}$$

where $\hat{y}^{\wp_1}_{k_1|k_2-1}$ denotes the predicted vector of infrequent measurements at the k_1-th sampling instant given the frequent measurement variables until the (k_2-1)-th frequent sampling instance for $k_2 - 1 \in (\gamma k_1, \gamma(k_1-1)]$, u_{k_1} is the vector of inputs, $\omega^{\wp_1}_{l,k_1}$ is the weight given to the l-th model of the $n_L^{\wp_1}$ total models for predicting the infrequent measurements, $A^{\wp_1}_{l,i}$ and $A^{s,\wp_2}_{l,i_s}$ denote the model coefficients for incorporating information from the infrequently and frequently sampled measurements, $B^{\wp_1}_{l,j}$ denotes the model coefficients for the input variables, and $\nu^{\wp_1}_l$ the bias term for the l-th model. Further, $n_y^{\wp_1}$ is the lag order for the infrequent measurement variables, $n_y^{s,\wp_2}$ is the lag order for the supplementary frequent measurement variables used in predicting the infrequent measurement variables, and $n_u^{\wp_1}$ is the lag order for the manipulated variables. Since measurements are only available until the (k_2-1)-th frequent sampling instance, predicting the future infrequently sampled variables requires either estimates or measurements, or both (depending on the respective infrequent and frequent sampling instances in consideration), of the frequently sampled variables, which is incorporated through the respective summations over the estimated and measured frequently sampled variables. The weights $\omega^{\wp_1}_l$ for predicting the infrequent measurements are dependent on the cluster centers $c^{\wp_1}_l$ and given by

$$\omega^{\wp_1}_{l,k_1} = \frac{\left\| x^{\wp_1,*}_{k_1} - c^{\wp_1}_l \right\|^{-2}}{\sum_{l'=1}^{n_L^{\wp_1}} \left\| x^{\wp_1,*}_{k_1} - c^{\wp_1}_{l'} \right\|^{-2}} \tag{10.17}$$

with

$$x^{\wp_1,*}_{k_1} = \left[{y^{\wp_1}_{k_1-1}}^T \ \cdots \ {y^{\wp_1}_{k_1-n_y^{\wp_1}}}^T \ {y^{\wp_2}_{\gamma k_1-1}}^T \ \cdots \ {y^{\wp_2}_{\gamma k_1-n_y^{s,\wp_2}}}^T \ {u_{k_1-1}}^T \ \cdots \ {u_{k-n_u^{\wp_1}}}^T \right]^T \tag{10.18}$$

and the cluster centers can be determined using the $n^{\wp_1}_{\text{obs}}$ samples by solving the nonlinear optimization problem

$$\min_{c_l^{\wp_1}} \mathcal{J} = \sum_{i=1}^{n^{\wp_1}_{\text{obs}}} \sum_{l=1}^{n_L^{\wp_1}} (\mu^{\wp_1}_{l,i})^2 \left\| x^{\wp_1,*}_i - c^{\wp_1}_l \right\|^2 \tag{10.19}$$

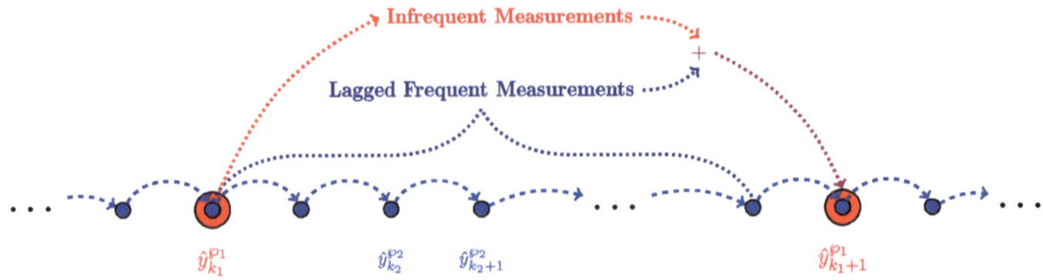

Fig. 10.2 An illustration of the multi-rate modeling approach for infrequent and frequent variables

where the membership of the samples is defined by

$$\mu_{l,i}^{\wp_1} = \frac{\left\| x_i^{\wp_1,*} - c_l^{\wp_1} \right\|^{-2}}{\sum_{l'=1}^{n_L^{\wp_1}} \left\| x_i^{\wp_1,*} - c_{l'}^{\wp_1} \right\|^{-2}} \tag{10.20}$$

Estimates of the infrequent measurements variables, however, require either estimates or measurements of both the infrequent and frequent variables as formalized in Eq. (10.16).

Modeling the infrequent measurements alone without considering the frequent measurements can result in a model with inadequate predictive performance because of the long delays between sampling instances during which the operating dynamics can vary significantly. In this chapter, the proposed data-driven multi-rate models are developed such that all the available process measurements are used to obtain the most accurate predictions. In particular, while past frequently measured variables are used to predict future frequently measured variables, the future infrequently measured variables are predicted using both the past infrequently measured variables and the past frequently measured variables (see Fig. 10.2 for an illustration). Since the key process variables are typically measured infrequently, accurate predictions and effective output-feedback control significantly benefit from the integrated modeling approach.

Therefore, another model that predicts the frequent measurement variables can be developed as follows:

$$\hat{y}_{k_2|k_2-1}^{\wp_2} = \sum_{l=1}^{n_L^{\wp_2}} \omega_{l,k_2}^{\wp_2} \left(\sum_{i=1}^{n_y^{\wp_2}} A_{l,i}^{\wp_2} y_{k_2-i}^{\wp_2} + \sum_{j=1}^{n_u^{\wp_2}} B_{l,j}^{\wp_2} u_{k_2-j} + \nu_l^{\wp_2} \right) \tag{10.21}$$

where $\hat{y}_{k_2|k_2-1}^{\wp_2}$ denotes the predictions of the frequent measurements given past measurements and $\omega_{l,k_2}^{\wp_2}$ is the weight given to the l-th model of the $n_L^{\wp_2}$ total models for predicting the frequent measurements, $A_{l,i}^{\wp_2}$ and $B_{l,j}^{\wp_2}$ are the model coefficients, and $\nu_l^{\wp_2}$ the bias term for the l-th model with $n_y^{\wp_2}$ as the lag order for the frequent measurement variables and $n_u^{\wp_2}$ as the lag order for the manipulated variables. The weights $\omega_l^{\wp_2}$ are dependent on the cluster centers $c_l^{\wp_2}$ and given by

$$\omega_{l,k_2}^{\wp_2} = \frac{\left\| x_{k_2}^{\wp_2,*} - c_l^{\wp_2} \right\|^{-2}}{\sum_{l'=1}^{n_L^{\wp_2}} \left\| x_{k_2}^{\wp_2,*} - c_{l'}^{\wp_2} \right\|^{-2}} \tag{10.22}$$

with

$$x_{k_2}^{\wp_2,*} = \left[y_{k_2-1}^{\wp_2}{}^T \; \dots \; y_{k_2-n_y^{\wp_2}}^{\wp_2}{}^T \; u_{k_2-1}{}^T \; \dots \; u_{k-n_u^{\wp_2}}{}^T \right]^T \tag{10.23}$$

and the cluster centers can be determined using the $n_{\text{obs}}^{\wp_2}$ samples by solving the nonlinear optimization problem

$$\min_{c_l^{\wp_2}} \mathcal{J} = \sum_{i=1}^{n_{\text{obs}}^{\wp_2}} \sum_{l=1}^{n_L^{\wp_2}} \left(\mu_{l,i}^{\wp_2}\right)^2 \left\| x_i^{\wp_2,*} - c_l^{\wp_2} \right\|^2 \tag{10.24}$$

where the membership of the samples is defined by

$$\mu_{l,i}^{\wp_2} = \frac{\left\| x_i^{\wp_2,*} - c_l^{\wp_2} \right\|^{-2}}{\sum_{l'=1}^{n_L^{\wp_2}} \left\| x_i^{\wp_2,*} - c_{l'}^{\wp_2} \right\|^{-2}} \tag{10.25}$$

The model coefficients in Eqs. (10.16) and (10.21) can be determined using PLS regression (see Remark 10.1 on the identification aspects of the model parameters and Remark 10.2 for comparison with principal component analysis-based models). A formulation of a general approach is presented here for handling multi-rate measurements that takes advantage of both infrequent and frequent measurements to improve the prediction accuracy of the models. It should be noted that between two infrequent measurements, several frequent measurements become available and help with the prediction of the infrequent measurement. However, no infrequent measurement becomes available between two frequent measurements; therefore, there is no benefit in incorporating these infrequent measurements in the model for predicting the frequent measurements.

Remark 10.1 The accurate estimation of the model parameters often requires a select number of identification batches where a suitable excitation signal is added onto the nominal inputs to establish a measured response. Due to their prevalence, digitally generated signals that can be readily implemented are attractive for identification purposes. However, the optimal amount of stimulus required to receive an appropriately excited output response and the adequate number of excited response data samples is an important issue for further consideration and outside the scope of the present chapter. In the present manuscript, the predictive ability of models is verified by $\mathcal{k}$-fold cross-validation.

Remark 10.2 In contrast to the principal component analysis-based approaches where a single linear time-varying model is built (around nominal trajectories), the proposed formulation focuses on developing multiple local linear models that are not indexed with time. One benefit of the proposed approach therefore is that batches of the same duration (or identifying a variable that can be used for indexing) are not needed. Furthermore, when identifying linear time varying models (even in the latent variable space), the number of matrices that need to be identified equals the number of sampling instances. In the present approach, on the other hand, the number of models is not necessarily equal to the number of sampling points, but can be picked via the process of cross-validation.

Remark 10.3 The first-principles-based model that is used as a test bed to collect process data only describes the melting phase of the EAF process [2]. Therefore, the model has only one distinct dynamic phase and the state variables evolve throughout the batch operation because of the changing operating dynamics. More detailed models that cover a wider duration of the EAF process operation are currently in development. Future modeling and control work will focus on explicitly handling the discrete nature of the process dynamics and its application to the detailed process model.

10.3.2 Electric Arc Furnace Modeling Results

To develop the multi-rate models for the EAF process measurements, a database of historical batches was initially generated. To this end, the deterministic EAF process model was simulated to generate 23 normal operation batches from different initial conditions. In addition to the normal operation batches, seven identification batches are used to augment the database. For five of these batches, a low-amplitude pseudo-random binary sequence (PRBS) signal is added to the input value computed by the PI controller, except for the last actuation. In the last two, batches, the PRBS signal was added all the way through. In the identification batches, not adding the PRBS to the final actuation allows the last input actuation to be able to compensate for disturbances in the system and achieve product that meets the desired end-point target specifications, which enables many of the identification batches to result in on-spec product. The PRBS signal added to the manipulated inputs, shown in Fig. 10.3, consisted of varying amplitude through the course of the batch. This identification approach was observed to result in good input signal excitation for developing accurate models. It was also verified that all seven identification batches still resulted in on-spec product. Note that unlike conventional batch processes, identification batches of the EAF process do not necessarily result in wasted product because after the identification experiments are completed, further input actions can be made to correct for the end-point targets and recover the product, thus mitigating the drawbacks of running identification batches.

The identification procedure for the local linear models was as follows. For a given lag structure, the number of clusters was varied from $n_L = 1$ to 40. For each choice of n_L, the batch data was partitioned into regressor X and regressor Y matrices, decomposed using PLS, and clustered using fuzzy c-means clustering. Since the model weights were then known, multiple ARX models were estimated simultane-

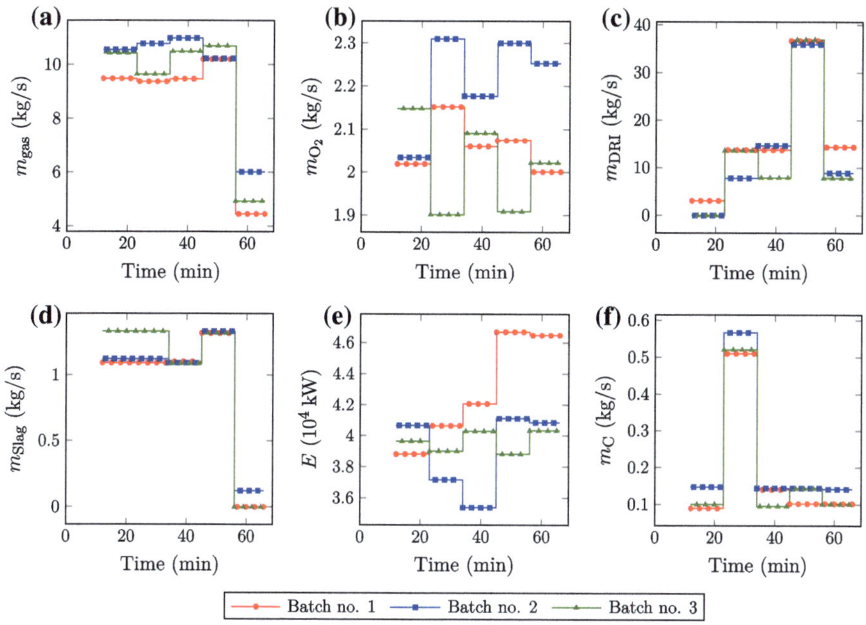

Fig. 10.3 Manipulated variables of the EAF process for selected identification batches with a low-amplitude pseudo-random binary sequence signal

ously using PLS. The models were built using k-fold cross-validation with $k = 7$. The goodness of each fit was judged using the weighted root-mean-square error (RMSE) in predicting back the validation samples, where the weights for the controlled variables (variables with end-point constraints) are twice the weights for the other variables.

The proposed multi-rate models are compared against a single model, identified using the multi-model approach reviewed in Sect. 10.2.2.1, that predicts all available process variables at only the infrequent sampling instances. The minimum weighted RMSE values obtained using the cross-validation are provided in Table 10.5. The lag structure, the number of clusters, and the number of latent variables (n_A) for the multi-rate models and the single model that yielded the lowest weighted RMSE values are also presented. It should be noted that all the frequent measurements available between infrequent sampling instances are used to augment the regressor matrix for predicting the infrequent variables. In contrast, the single model only predicts future variables using the past variables at only the infrequent sampling instances without incorporating any of the frequent measurements available between infrequent sampling instances. Moreover, 10 new batches (not used during the model development) are simulated for validation purposes. The model validation results for the infrequent and frequent variables are shown in Tables 10.6 and 10.7 (respectively), and the corresponding output of the multi-rate models, single model and actual profile

Table 10.5 Data-driven modeling results for both single and multi-rate model

Model	Type	Weighted RMSE	Variable lags			n_L	n_A
			Infrequent	Frequent	Manipulated		
Single	–	0.3721	1	1	1	22	7
Multi-rate	Infrequent	0.2882	1	1	1	29	5
	Frequent	0.0150	–	1	1	3	13

Table 10.6 Model validation results for the infrequent measurement variables of the EAF process

Variable	Weighted RMSE		Units
	Single model	Multi-rate model	
T	19.0585	12.6981	K
x_{Fe}	0.0016	0.0012	kg/kg
x_{C}	0.0027	0.0024	kg/kg
x_{Slag}	0.0124	0.0087	kg/kg
x_{FeO}	0.0296	0.0133	kg/kg
x_{SiO_2}	0.0033	0.0024	kg/kg

trajectories from the mechanistic model for a set of new initial conditions are shown in Figs. 10.4 and 10.5.

The prediction by the data-based model is an open-loop prediction where the predictions throughout the batch are not corrected and the model errors are allowed to accumulate over the entire batch duration. It is readily observed that the single model that predicts all the process variables at only the infrequent sampling instances does not predict the variables well, resulting in higher weighted RMSE values. Additionally, since the model validation results involved open-loop predictions, the multi-rate models are expected to further improve accuracy over the single model when predictions of the frequent variables are substituted for the preferred available measurements. Accordingly, the single model will not be considered further as its capability, and that of a predictive controller derived from it, will falter in relation to the multi-rate models. Overall, the proposed multi-model approach captured the nonlinear nature of the EAF and provided relatively reliable predictions.

10.4 Economic Model Predictive Control

In this section, we propose a two-tiered economic model predictive control algorithm and then implement it on the EAF process test bed.

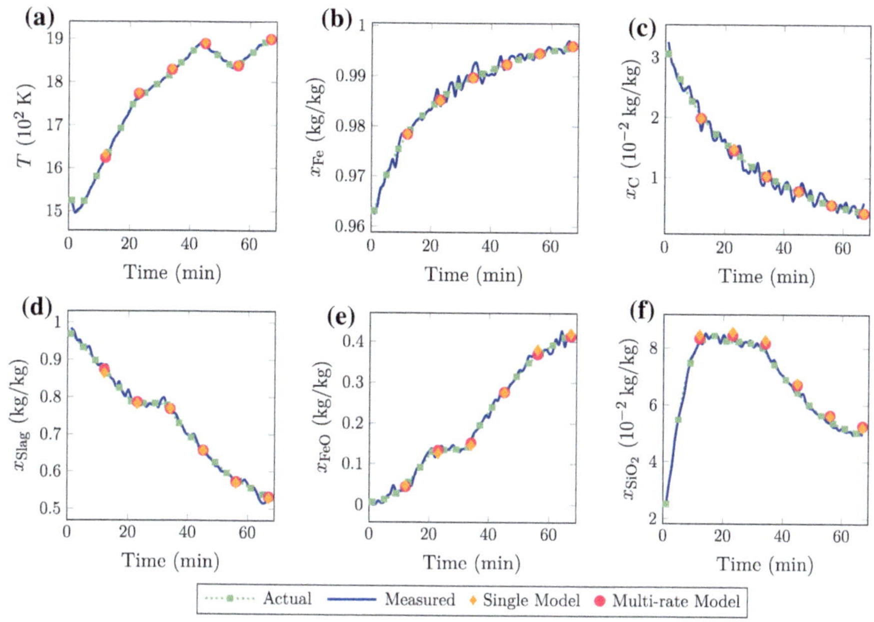

Fig. 10.4 Model validation results for the infrequent measurement variables of the EAF process

Fig. 10.5 Model validation results for the frequent measurement variables of the EAF process

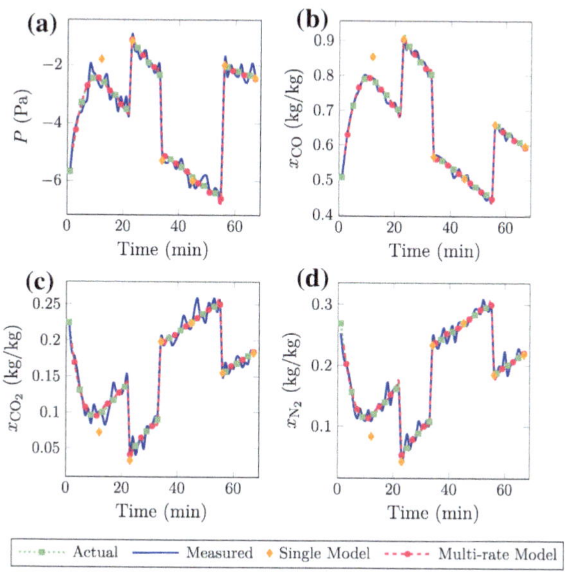

Table 10.7 Model validation results for the frequent measurement variables of the EAF process

Variable	Weighted RMSE		Units
	Single model	Multi-rate model	
P	3.6173	0.2269	Pa
x_{CO}	0.0404	0.0092	kg/kg
x_{CO_2}	0.0219	0.0103	kg/kg
x_{N_2}	0.0232	0.0103	kg/kg

10.4.1 Economic Model Predictive Control Formulation

The key idea of the proposed economic MPC is to use a tiered economic model predictive control framework to achieve an acceptable (on-spec) end-point while optimizing an economics-based cost function. To this end, let y_{des} be the desired product end-point. In the first tier, the EMPC computes the best (in terms of satisfaction of product specifications) achievable end-point at each infrequent sampling instance by solving the following optimization problem

$$\min_{u \in \mathcal{U}} \ \mathcal{J}_1 = \left\| \hat{y}_e - y_{\text{des}} \right\|_{\Psi}^2 \tag{10.26}$$

$$\text{s.t.} \begin{cases} \hat{y}^{\wp_1}_{k_1|k_2-1} = \sum_{l=1}^{n_L^{\wp_1}} \omega^{\wp_1}_{l,k_1} \left(\sum_{i=1}^{n_y^{\wp_1}} A^{\wp_1}_{l,i} y^{\wp_1}_{k_1-i} + \sum_{i_s=1}^{\gamma k_1 - k_2} A^{s,\wp_2}_{l,i_s} \hat{y}^{\wp_2}_{\gamma k_1 - i_s} \right. \\ \qquad \left. + \sum_{i'_s=\gamma k_1-k_2-1}^{\gamma k_1 - n_y^{s,\wp_2}} A^{s,\wp_2}_{l,i'_s} y^{\wp_2}_{\gamma k_1 - i'_s} + \sum_{j=1}^{n_u^{\wp_1}} B^{\wp_1}_{l,j} u_{k_1-j} + \nu^{\wp_1}_l \right) \\ \hat{y}^{\wp_2}_{k_2|k_2-1} = \sum_{l=1}^{n_L^{\wp_2}} \omega^{\wp_2}_{l,k_2} \left(\sum_{i=1}^{n_y^{\wp_2}} A^{\wp_2}_{l,i} y^{\wp_2}_{k_2-i} + \sum_{j=1}^{n_u^{\wp_2}} B^{\wp_2}_{l,j} u_{k_2-j} + \nu^{\wp_2}_l \right) \\ \hat{y}^{\wp_1} \in \mathcal{Y}_{\wp_1}, \quad \hat{y}^{\wp_2} \in \mathcal{Y}_{\wp_2} \\ \hat{y}_e = g\left(\hat{y}^{\wp_1}_{n_K}\right) \\ y^{\wp_1}_{k_1-1} = y^{\wp_1}_t, \quad y^{\wp_2}_{k_2-1} = y^{\wp_2}_t \end{cases} \tag{10.27}$$

where $\hat{y}_e = g\left(\hat{y}^{\wp_1}_{n_K}\right)$ defines the batch end-point characteristics through the estimates of the measurement variables at the last sampling instant, $u \in \mathbb{R}^m$ denotes the vector of constrained input variables, taking values in a nonempty convex set $\mathcal{U} \subseteq \mathbb{R}^m$ with $\mathcal{U} := \{u \in \mathbb{R}^m : u_{\min} \leq u \leq u_{\max}\}$, $u_{\min} \in \mathbb{R}^m$ and $u_{\max} \in \mathbb{R}^m$ denote the lower and upper bounds on the manipulated input, respectively, and Ψ is a positive-definite matrix used to penalize the squared deviation of the end-point variables with respect to the desired end-point values. Further, $\mathcal{Y}_{\wp_1} \in \mathbb{R}^{\wp_1}$ and $\mathcal{Y}_{\wp_2} \in \mathbb{R}^{\wp_2}$ denote the constraints on the infrequent and frequent measurement variables, respectively, and $y^{\wp_1}_t$ and $y^{\wp_2}_t$ provide the initialization of the optimization problem at the current process conditions. The multi-rate models utilize the available process measurements to predict the variables at the next sampling instance, while latter sampling instances incorporate estimates of the process measurements to predict the entire process variable trajectories over the prediction horizon.

Subsequent to obtaining the end-point that best satisfies the product specifications, $\hat{y}_e^*$, the next tier computes the optimal (from an economic perspective) input moves, albeit with the constraint that the best product specifications should still be achieved. The second tier of the EMPC solves an optimization problem to determine the optimal input trajectory by minimizing an economic objective function at the infrequent sampling instances as follows:

$$\min_{u \in \mathcal{U}} \mathcal{J}_2 = \delta \cdot \sum_{i=k_1-1}^{n_K-1} \lambda^T u_i \tag{10.28}$$

$$\text{s.t.} \begin{cases} \hat{y}^{\wp_1}_{k_1|k_2-1} = \sum\limits_{l=1}^{n_L^{\wp_1}} \omega^{\wp_1}_{l,k_1} \left(\sum\limits_{i=1}^{n_y^{\wp_1}} A^{\wp_1}_{l,i} y^{\wp_1}_{k_1-i} + \sum\limits_{i_s=1}^{\gamma k_1-k_2} A^{s,\wp_2}_{l,i_s} \hat{y}^{\wp_2}_{\gamma k_1-i_s} \right. \\ \qquad \left. + \sum\limits_{i'_s=\gamma k_1-k_2-1}^{\gamma k_1-n_y^{s,\wp_2}} A^{s,\wp_2}_{l,i'_s} y^{\wp_2}_{\gamma k_1-i'_s} + \sum\limits_{j=1}^{n_u^{\wp_1}} B^{\wp_1}_{l,j} u_{k_1-j} + \nu^{\wp_1}_l \right) \\ \hat{y}^{\wp_2}_{k_2|k_2-1} = \sum\limits_{l=1}^{n_L^{\wp_2}} \omega^{\wp_2}_{l,k_2} \left(\sum\limits_{i=1}^{n_y^{\wp_2}} A^{\wp_2}_{l,i} y^{\wp_2}_{k_2-i} + \sum\limits_{j=1}^{n_u^{\wp_2}} B^{\wp_2}_{l,j} u_{k_2-j} + \nu^{\wp_2}_l \right) \\ \hat{y}^{\wp_1} \in \mathcal{Y}_{\wp_1}, \quad \hat{y}^{\wp_2} \in \mathcal{Y}_{\wp_2} \\ h\left(\hat{y}^{\wp_1}_{n_K}, \hat{y}^*_e\right) \le 0 \\ y^{\wp_1}_{k_1-1} = y^{\wp_1}_t, \quad y^{\wp_2}_{k_2-1} = y^{\wp_2}_t \end{cases} \tag{10.29}$$

where $h\left(\hat{y}^{\wp_1}_{n_K}, \hat{y}^*_e\right) \le 0$ denotes end-point constraints that are appropriately relaxed as an interval with $\hat{y}_e^*$ as a bound (see Remark 10.4 for more details on the end-point constraint relaxation), δ is the hold time for the control action, and $\lambda \in \mathbb{R}^m$ are the operating costs associated with the manipulated variables. The two-tiered economic MPC approach allows for a framework that can handle distinct and competing objectives and deal with the trade-off between them without resorting to using arbitrary penalty weights (see Remark 10.5). In both tiers of the proposed approach, the models used and the process constraints are identical, while the objective function for the first tier is end-point optimization and for the second tier is economic MPC with the optimal end-point as an additional constraint. The feasibility of this optimization problem is in general not guaranteed, but hard constraints are incorporated so even if the target end-point may not be reached, no safety violations will occur. Additionally, if the optimization problem is infeasible, it suggests that for the current batch, safety constraints cannot be satisfied and thus may require an earlier termination of the batch to avoid safety issues. On the other hand, if it is possible to meet the safety requirements, the first tier computes the solution that best achieves the desired target. The second tier is then guaranteed to be feasible, and in the worst-case scenario be as economic as the first-tier solution.

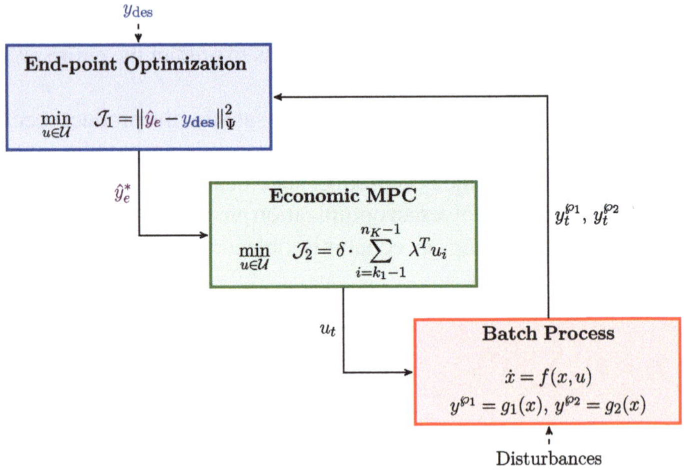

Fig. 10.6 A flow diagram of the two-tiered economic model predictive control approach

In summary, the two-tiered EMPC strategy uses the data-driven multi-rate models to predict the behavior of the batch dynamics over a receding horizon. The control problem is solved at each infrequent sampling instance to determine the best achievable end-point target (the nearest to the desired end-point) and the economically optimal path to reach the optimal end-point using the identified multi-rate models and the process constraints, and only the control action of the current infrequent sampling interval is implemented with the calculation repeated at subsequent sampling instances using new measurements and updated plant information. A block diagram of the two-tiered economic MPC approach is shown in Fig. 10.6.

Remark 10.4 Recognizing that the optimal end-point from the first layer is the best end-point that we can reach, we implement the constraint as an interval that is bounded on one end by the best achievable end-point (as given by the first tier) and bound on the other end by the extreme limit the variable can obtain. Thus, the constraints always include a target known to be feasible. The added flexibility of the relaxed end-point constraints, however, allows achieving a better economic optimum if at all possible while ensuring the constraints are satisfied.

Remark 10.5 The typical control objective in batch processes is to minimize the deviation between the product end-point and the desired target within the finite batch duration. Another control objective routinely involved in batch process control is to optimize the process economics while constraining the end-point to satisfy the desired specifications. These two competing objectives (minimize end-point deviation and optimize economics) when considered simultaneously result in a multi-objective optimization problem with inherent complexities in negotiating the possible trade-offs between the competing objectives. To address this issue without resorting to the compromise of using arbitrary penalty weights, the two-tiered EMPC approach is proposed that is capable of first determining the best attainable end-point and then subsequently optimizing the operating economics of reaching that optimal achievable end-point. Similar tiered optimization approaches were used in [23] in a hierarchical supervisory control scheme and in [24] in optimal operation under partial plant shutdown.

Remark 10.6 The two-tiered EMPC framework using the proposed data-driven multi-rate models is relatively tractable in contrast to the first-principles model-based control designs. Nevertheless, for embedded systems with fast dynamics where the sampling times are in milli- or nanoseconds and thus the computational time can be prohibitive, the two-tiered EMPC can be implemented such that the first tier is only solved occasionally in order to update the best achievable end-point, as appropriate, to assist with the computation time.

Remark 10.7 Existing EMPC formulations [14–17, 48] have shown promising results in the context of continuous systems. However, the fact that the desired end-point in batch processes is typically not an equilibrium point precludes the direct implementation of these EMPC techniques. Note also that any economic benefits achieved through the proposed two-tier formulation rely on the ability to reach (or get close to) the process specification by following different paths (and thus following different manipulated input trajectories). If the process dynamics are such that there exists only one unique path to the desired end-point, then the first- and second-tier optimization problems will both yield identical results, but no worse than the solution obtained at the first tier.

10.4.2 Economic Model Predictive Control Results

Closed-loop simulations for ten new initial conditions were performed using the proposed two-tired economic MPC design, and the performance was compared against the standard EAF operating approach involving PI controllers described in Sect. 10.2. The two layers of the EMPC were executed consecutively at every infrequent sampling instance. All new initial conditions for the closed-loop results were obtained from the same distribution as the initial conditions in the training data. A representative set of closed-loop simulation results (batch no. 3) is presented in Figs. 10.7 and 10.8 for the infrequent and frequent process measurements, respectively. The closed-loop trajectories of the end-point constrained variables and the corresponding intervals of the end-point constraints are shown in Fig. 10.7 as well. For instance, the end-point constraint for the temperature in Fig. 10.7a is represented as an interval with the lower limit as the optimal end-point obtained from the first tier of the EMPC and the upper limit as the path constraint for the temperature imposed for safety considerations. Similarly, the end-point constraints for the mass fractions of carbon and iron oxide in Fig. 10.7c and e, respectively, are represented as intervals with the upper limits obtained from the first tier of the EMPC and the lower limits as zero,

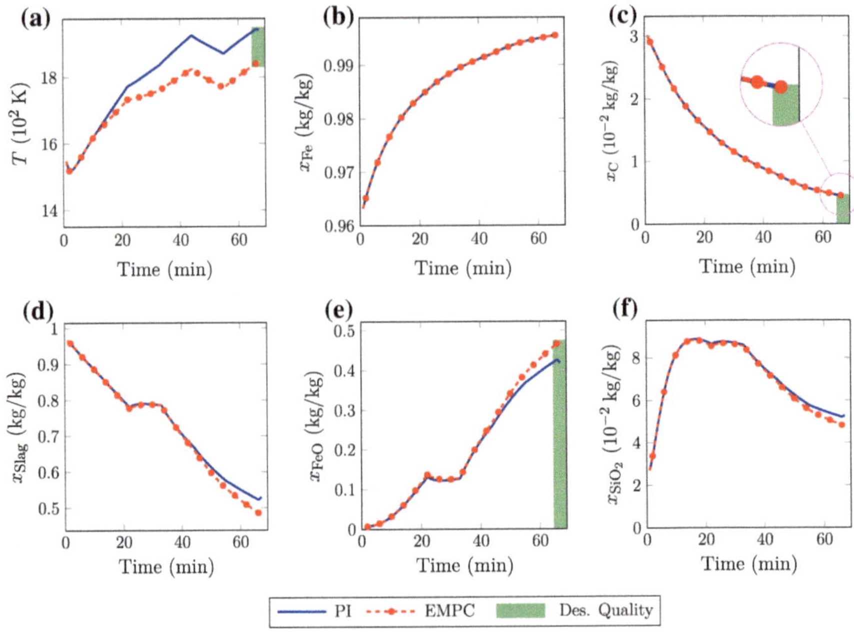

Fig. 10.7 Comparison of the trajectories for the infrequent measurement variables obtained from the proposed economic MPC and conventional method (The inset in subplot (c) zooms in to show the end-point constraint is satisfied)

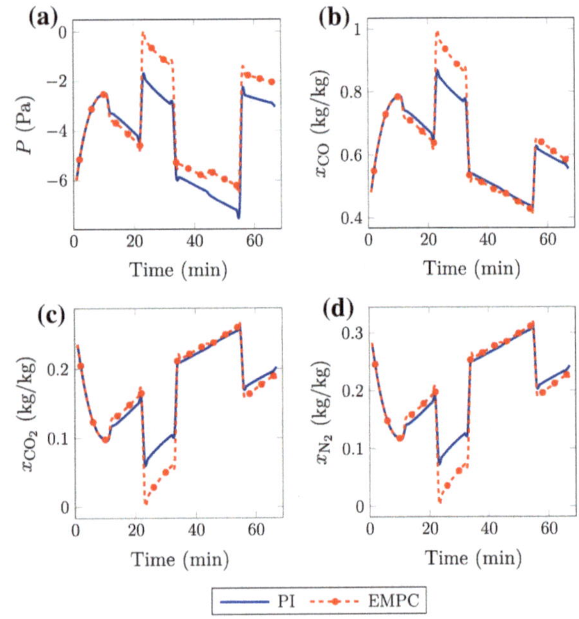

Fig. 10.8 Comparison of the trajectories for the frequent measurement variables obtained from the proposed economic MPC and conventional method

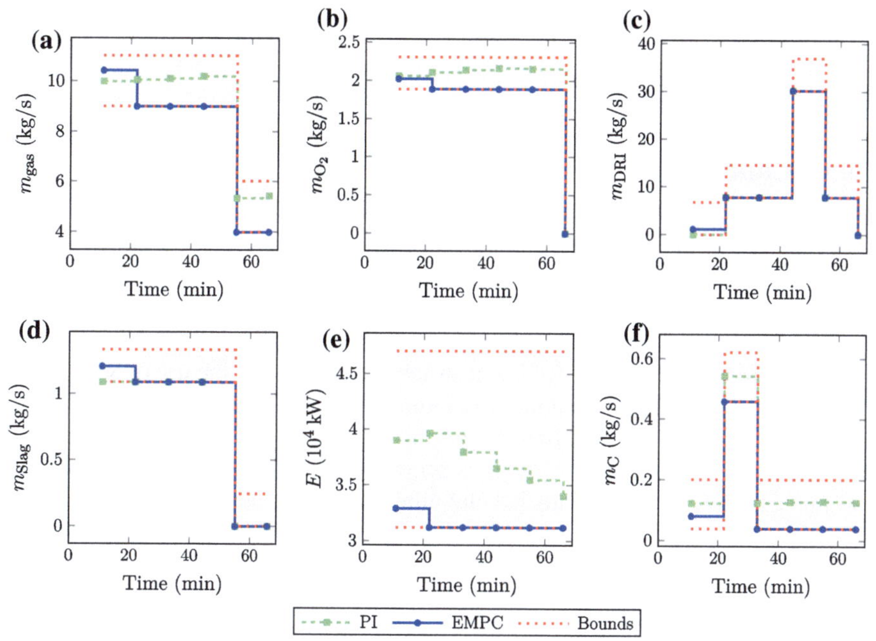

Fig. 10.9 Closed-loop profiles of the manipulated variables obtained from the proposed economic MPC and conventional method for a selected batch of the EAF process

which are the minimum physically realizable values of the concentration measurements. The desired end-points were selected based on desired attributes of the steel product as discussed in Sect. 10.2.1. Although both the EMPC and the standard control approaches are able to meet the desired product quality, the proposed economic MPC controller offered a significant cost advantage of approximately \$5.7372 $\times$ 10^3 compared to \$6.2020 $\times$ 10^3 for the standard approach, an average savings of \$464.75 per batch or approximately 7.49% improvement per batch. The poor performance of the PI controllers can be attributed to the fact that they are inherently based on a decentralized (single-input, single-output) framework that does not account for interactions among the various control loops. Furthermore, the conventional PI control also does not explicitly take into consideration process constraints and optimality.

The proposed economic MPC design was efficiently solvable with the average CPU time required to solve the two-tired MPC optimization problem with the longest prediction horizon as 3.9983 s (maximum 9.2665 s) using `GAMS` with `IPOPT` as the solver on an Intel Dual Core machine with 4 GB RAM. The closed-loop input profiles are shown in Fig. 10.9, and it is readily observed that the EMPC approach recognizes that in certain instances, the end-point constraints can still be achieved while maintaining the manipulated variables close to their lower bounds, due to which the operating costs of the EMPC approach are significantly lower. Overall, the simulation results demonstrated the advantages of implementing the proposed two-tiered economic predictive controller over standard operating policies.

10.5 Conclusion

This chapter considers the problem of data-based model development and economic model predictive control of the electric arc furnace. To this end, a multi-rate modeling approach is developed for the infrequent and frequent process measurements that efficiently incorporate the available measurements, the simplicity of local linear models, the data extraction capabilities of latent variable methods, and the use of appropriate clustering and weighting techniques to capture the nonlinear time-varying dynamics of the EAF process. By exploiting the enhanced capability of multi-rate data-driven models, a two-tiered economic MPC is proposed that ensures the desired product end-point specifications are satisfied and minimizes an economics-based cost function. The implementation of the multi-rate data-driven modeling approach and the economic MPC framework subject to measurement noise, constraints, and desired end-point targets is illustrated on a electric arc furnace test-bed simulation process.

References

1. Gosiewski, A., Wierzbicki, A.: Dynamic optimization of a steel-making process in electric arc furnace. Automatica **6**, 767–778 (1970)
2. Bekker, J.G., Craig, I.K., Pistorius, P.C.: Modelling and simulation of an electric arc furnace process. ISIJ Int. **39**, 23–32 (1999)
3. Oosthuizen, D.J., Craig, I.K., Pistorius, P.C.: Economic evaluation and design of an electric arc furnace controller based on economic objectives. Control Eng. Pract. **12**, 253–265 (2004)
4. MacRosty, R.D.M., Swartz, C.L.E.: Dynamic modeling of an industrial electric arc furnace. Ind. Eng. Chem. Res. **44**, 8067–8083 (2005)
5. MacRosty, R.D.M., Swartz, C.L.E.: Dynamic optimization of electric arc furnace operation. AIChE J. **53**, 640–653 (2007)
6. Shi, D., El-Farra, N.H., Li, M., Mhaskar, P., Christofides, P.D.: Predictive control of particle size distribution in particulate processes **61**, 268–281 (2006)
7. Christofides, P.D., El-Farra, N.H., Li, M., Mhaskar, P.: Model-based control of particulate processes. Chem. Eng. Sci. **63**, 1156–1172 (2008)
8. Chachuat, B., Marchetti, A., Bonvin, D.: Process optimization via constraints adaptation. J. Proc. Cont. **18**, 244–257 (2008)
9. Chachuat, B., Srinivasan, B., Bonvin, D.: Adaptation strategies for real-time optimization. Comput. Chem. Eng. **33**, 1557–1567 (2009)
10. Aumi, S., Mhaskar, P.: Safe-steering of batch process systems. AIChE J. **55**, 2861–2872 (2009)
11. Aumi, S., Mhaskar, P.: Robust model predictive control and fault handling of batch processes. AIChE J. **57**, 1796–1808 (2011)
12. Rivotti, P., Pistikopoulos, E.N.: Constrained dynamic programming of mixed-integer linear problems by multi-parametric programming. Comput. Chem. Eng. **70**, 172–179 (2014)
13. Rivotti, P., Pistikopoulos, E.N.: A dynamic programming based approach for explicit model predictive control of hybrid systems. Comput. Chem. Eng. **72**, 126–144 (2015)

14. Heidarinejad, M., Liu, J., Christofides, P.D.: Economic model predictive control of nonlinear process systems using Lyapunov techniques. AIChE J. **58**, 855–870 (2012)
15. Amrit, R., Rawlings, J.B., Biegler, L.T.: Optimizing process economics online using model predictive control. Comput. Chem. Eng. **58**, 334–343 (2013)
16. Ellis, M., Durand, H., Christofides, P.D.: A tutorial review of economic model predictive control methods. J. Proc. Cont. **24**, 1156–1178 (2014)
17. Ellis, M., Christofides, P.D.: On finite-time and infinite-time cost improvement of economic model predictive control for nonlinear systems. Automatica **50**, 2561–2569 (2014)
18. J.G. Bekker, Modelling and control of an electric arc furnace off-gas process. Master's thesis, University of Pretoria, South Africa (1999)
19. Clerici, P., Dell'Acqua, F., Maiolo, J., Vittorio, S.: Tenova's intelligent arc furnace 'iEAF' – Concept and technical overview. Steel Times Int. (2008)
20. Wang, C., Larsson, M., Ryman, C., Grip, C.-E., Wikström, J.-O., Johnsson, A., Engdahl, J.: A model on CO_2 emission reduction in integrated steelmaking by optimization methods. Int. J. Energy Res. **32**, 1092–1106 (2008)
21. Aumi, S., Mhaskar, P.: Integrating data-based modeling and nonlinear control tools for batch process control. AIChE J. **58**, 2105–2119 (2012)
22. Aumi, S., Corbett, B., Clarke-Pringle, T., Mhaskar, P.: Data-driven model predictive quality control of batch processes. AIChE J. **59**, 2852–2861 (2013)
23. Swartz, C.L.E.: An algorithm for hierarchical supervisory control. Comput. Chem. Eng. **19**, 1173–1180 (1995)
24. Chong, Z., Swartz, C.L.E.: Optimal operation of process plants under partial shutdown conditions. AIChE J. **59**, 4151–4168 (2013)
25. Flores-Cerrillo, J., MacGregor, J.F.: Latent variable MPC for trajectory tracking in batch processes. J. Proc. Cont. **15**, 651–663 (2005)
26. Golshan, M., MacGregor, J.F., Bruwer, M.J., Mhaskar, P.: Latent variable model predictive control (LV-MPC) for trajectory tracking in batch processes. J. Proc. Cont. **20**, 538–550 (2010)
27. Yu, H., Flores-Cerrillo, J.: Latent variable model predictive control for trajectory tracking in batch processes: internal model control interpretation and design methodology. Ind. Eng. Chem. Res. **52**, 12437–12450 (2013)
28. Rashid, M.M., Mhaskar, P., Swartz, C.L.E.: Multi-rate modeling and economic model predictive control of the electric arc furnace. J. Proc. Cont. **40**, 50–61 (2016)
29. Moonen, M., De Moor, B., Vandenberghe, L., Vandewalle, J.: On- and off-line identification of linear state-space models. Int. J. Control **49**, 219–232 (1989)
30. Jansson, M., Wahlberg, B.: On consistency of subspace methods for system identification. Automatica **34**, 1507–1519 (1998)
31. Shi, R., MacGregor, J.F.: Modeling of dynamic systems using latent variable and subspace methods. J. Chemom. **14**, 423–439 (2000)
32. Qin, S.J.: An overview of subspace identification. Comput. Chem. Eng. **30**, 1502–1513 (2006)
33. Corbett, B., Mhaskar, P.: Subspace identification for data-driven modeling and quality control of batch processes. AIChE J. **62**, 1581–1601 (2016)
34. Van Overschee, P., De Moor, B.: Subspace Identification for Linear Systems: Theory, Implementation, Applications. Kluwer Academic Publishers, Norwell (1996)
35. Ljung, L.: System Identification: Theory for the User. Prentice-Hall, Inc., Englewood Cliffs (1999)
36. Liu, Z., Hansson, A., Vandenberghe, L.: Nuclear norm system identification with missing inputs and outputs. Syst. Control Lett. **62**, 605–612 (2013)
37. Liu, Z., Vandenberghe, L.: Interior-point method for nuclear norm approximation with application to system identification. SIAM J. Mat. Anal. Appl. **31**, 1235–1256 (2009)
38. Gibson, S., Ninness, B.: Robust maximum-likelihood estimation of multivariable dynamic systems. Automatica **41**, 1667–1682 (2005)
39. Srinivasan, B., Bonvin, D.: Real-time optimization of batch processes by tracking the necessary conditions of optimality. Ind. Eng. Chem. Res. **46**, 492–504 (2007)

40. Bonvin, D., Srinivasan, B.: On the role of the necessary conditions of optimality in structuring dynamic real-time optimization schemes. Comput. Chem. Eng. **51**, 172–180 (2013)
41. Srinivasan, B., Palanki, S., Bonvin, D.: Dynamic optimization of batch processes: I. Characterization of the nominal solution. Comput. Chem. Eng. **27**, 1–26 (2003)
42. Srinivasan, B., Bonvin, D., Visser, E., Palanki, S.: Dynamic optimization of batch processes: II. Role of measurements in handling uncertainty. Comput. Chem. Eng. **27**, 27–44 (2003)
43. Van Overschee, P., De Moor, B.: A unifying theorem for three subspace system identification algorithms. Automatica **31**, 1853–1864 (1995)
44. Raghavan, H., Tangirala, A.K., Gopaluni, R.B., Shah, S.L.: Identification of chemical processes with irregular output sampling. Control Eng. Pract. **14**, 467–480 (2006)
45. Gopaluni, R.B.: Nonlinear system identification under missing observations: the case of unknown model structure. J. Proc. Control **20**, 314–324 (2010)
46. Balzano, L., On, Wright S.J.: GROUSE and incremental SVD. In IEEE 5th International Workshop on Computational Advances in Multi-Sensor Adaptive Processing (CAMSAP), pp. 1–4. St. Martin. (2013)
47. Balzano, L., Wright, S.J.: Local convergence of an algorithm for subspace identification from partial data. Found. Comput. Math. **15**, 1279–1314 (2015)
48. Diehl, M., Amrit, R., Rawlings, J.B.: A Lyapunov function for economic optimizing model predictive control. IEEE Trans. Auto Cont. **56**, 703–707 (2011)
49. Bemporad, A., Morari, M., Dua, V., Pistikopoulos, E.N.: The explicit linear quadratic regulator for constrained systems. Automatica **38**, 3–20 (2002)
50. Yang, X., Biegler, L.T.: Advanced-multi-step nonlinear model predictive control. J. Proc. Cont. **23**, 1116–1128 (2013)
51. Jäschke, J., Yang, X., Biegler, L.T.: Fast economic model predictive control based on NLP-sensitivities. J. Proc. Cont. **24**, 1260–1272 (2014)
52. Diehl, M., Bock, H.G., Schlöder, J.P.: A real-time iteration scheme for nonlinear optimization in optimal feedback control. SIAM J. Cont. Opt. **43**, 1714–1736 (2005)
53. Kozub, D.J., MacGregor, J.F.: State estimation for semi-batch polymerization reactors. Chem. Eng. Sci. **47**, 1047–1062 (1992)
54. Washington, I.D., Swartz, C.L.E.: Design under uncertainty using parallel multiperiod dynamic optimization. AIChE J. **60**, 3151–3168 (2014)

Part IV
Subspace Identification Based Modeling Approach for Batch Processes

Chapter 11
Subspace Identification for Data-Driven Modeling and Quality Control of Batch Processes

11.1 Introduction

In the previous part of the book, we presented a novel dynamic model building approach and integrated it with PLS. Specifically, partway through the batch, only process measurements from the past and current sampling instant are available. To solve this problem, we built dynamic models [1] to estimate the necessary values. Note that the problem of estimating future process measurements is essentially a dynamic modeling problem.

The primary contribution of the current chapter is the use of subspace identification to model the dynamic evolution of the process and relate available process measurements to quality. Subspace identification is a class of linear, time-invariant, dynamic model identification techniques in which the first step is to identify a valid state trajectory from training data before fitting model parameters. Early contributions in this field established the broad success of these methods for identifying LTI models of continuous processes [2–5]. However, very little work has been carried out on applying subspace identification to batch processes. In [6], CVA (a class of subspace identification algorithms) is used to build a model for monitoring of batch processes, but the issue of control is not addressed. In [7], subspace identification is used for batch-to-batch control. Neither of these contributions treat the use of subspace identification for within-batch control.

This chapter presents a novel approach to within-batch quality modeling and control with the ability to handle variable batch durations. To this end, first a dynamic model of the process is built by adapting existing subspace identification methods for batch data. Next, a quality model is built to relate subspace states to product quality. These two models are used in conjunction to make predictions for new batches. To make control decisions, a model predictive control (MPC) scheme is designed which uses the two models to proscribe appropriate input moves. The efficacy of this approach is demonstrated by comparison with a state-of-the-art latent variable approach and traditional PI trajectory tracking control. The chapter is organized as follows: In the next section, we define the class of batch processes studied and

P. Mhaskar et al., *Modeling and Control of Batch Processes*,
Advances in Industrial Control, https://doi.org/10.1007/978-3-030-04140-3_11

review the literature results on which the proposed method is based. We also review a state-of-the-art latent variable model predictive control scheme which we use as a benchmark in this chapter. Following the preliminaries, we present the batch reaction of polymethyl methacrylate (PMMA) as a motivating example process. Next, we present an analysis of latent variable modeling approaches from a state-space perspective. In the following section, we present the novel subspace quality model predictive control (SQMPC) design. The subsequent section provides application results in which the proposed SQMPC, LVMPC, and traditional PI trajectory tracking are compared. Finally, we draw conclusions and discuss areas for future work.

11.2 Preliminaries

In this section, we review the fundamental principles and methodologies that lay the groundwork for the novel elements of this chapter. To begin, in the first subsection, we define the class of system studied in this chapter. Next, we briefly review latent variable methods which have traditionally played a large role in batch modeling and control. In the final subsection, we briefly review subspace identification, a dynamic modeling approach which we heavily utilize in the proposed approach.

11.2.1 Class of Processes

In this chapter, we consider batch processes, described mathematically:

$$\dot{\boldsymbol{x}} = \boldsymbol{f}(\boldsymbol{x}, \boldsymbol{u}) \tag{11.1}$$

$$\boldsymbol{y} = \boldsymbol{g}(\boldsymbol{x}, \boldsymbol{u}) \tag{11.2}$$

$$\boldsymbol{q} = \boldsymbol{h}(\boldsymbol{x}, \boldsymbol{u}) \tag{11.3}$$

where $\boldsymbol{x}$ is the state variable, and $\boldsymbol{f}$, $\boldsymbol{g}$, and $\boldsymbol{h}$ are nonlinear functions that relate the evolution of the process, output, and quality, respectively, to the state and input. The outputs are denoted by $\boldsymbol{y}[k]$ where k indexes the sample. In general, output variables are assumed to have minimal to zero delay in availability. Examples of typical output variables include temperature and volume. We denote the input by $\boldsymbol{u}(t)$. In this chapter, we consider inputs with a zero-order hold with the same frequency as the sample frequency of the output. We represent this as $\boldsymbol{u}[k]$ corresponding to the input implemented when measurement $\boldsymbol{y}[k]$ becomes available. Quality variables are aspects of the final product that are important to the consumer of the product (either for downstream processing or commercially). Quality variables may be difficult to measure, involving long measurement delays, and typically only become available after the completion of the batch. In this chapter, we denote quality measurements with the vector $\boldsymbol{q}[T]$ where T is the duration of the batch.

11.2.2 Latent Variable Methods

One popular approach to modeling and control of batch processes involves statistical techniques, or latent variable methods, for model building. Before we analyze an existing approach and integrate some of these concepts into a state-space-based modeling and control approach, in this section we review some of the key concepts behind these methods. The underpinning concept of latent variable methods is the projection of data from a high dimension (measurement space) in which the independence of variables is uncertain into a lower dimensional (latent space) with guaranteed variable independence. In this subsection, we present the fundamentals of the latent variable methods: principal component analysis (PCA), principal component regression (PCR), and projection to latent spaces (PLSs).

PCA is, in some sense, the most basic latent variable method. PCA is concerned only with the representation of a single measurement space (the $\mathbf{X}$ block) in a reduced dimensional space. Mathematically, this is represented in the following form:

$$\mathbf{X} = \mathbf{T}\mathbf{P}' \tag{11.4}$$

where $\mathbf{X}$ is the measurement space with columns containing variables and rows' observations. $\mathbf{T}$ is the latent space, referred to as the score matrix, with the same interpretation as $\mathbf{X}$. $\mathbf{P}$ defines the orientation of the latent variable space and is referred to as the loading matrix.

PCR is an extension of PCA and ordinary least squares regression. The objective of PCR is to regress dependent variables referred to as the y block, $\mathbf{Y}$, onto a set of possibly correlated independent variables $\mathbf{X}$. Ordinary least squares regression fails in this case because $\mathbf{X}$ is column-rank deficient. PCR is accomplished by first using PCA to project the $\mathbf{X}$ space variables into an uncorrelated latent (score) space. Ordinary least squares (OLS) regression is then carried out; the columns of $\mathbf{Y}$ are regressed onto the scores. It is worth noting that the resulting model is nevertheless a linear model relating $\mathbf{X}$ and $\mathbf{Y}$.

In contrast to PCR, when using PLS both the $\mathbf{x}$ and $\mathbf{Y}$ blocks are projected into corresponding latent variable spaces. This is accomplished in practice by evaluating a three-component objective: maximizing the explained variance in the $\mathbf{X}$ space, maximizing the explained variance in the $\mathbf{Y}$ space, and maximizing the correlation between the two. Mathematically:

$$\mathbf{T} = \mathbf{X}\mathbf{W}^* \tag{11.5}$$

$$\mathbf{Y} = \mathbf{U}\mathbf{C}' \tag{11.6}$$

where $\mathbf{T}$ and $\mathbf{U}$ are the score spaces of the $\mathbf{X}$ and $\mathbf{Y}$ blocks, respectively. In application, it is assumed that $\mathbf{T} = \mathbf{U}$ which is valid because of the maximized correlation between the two score spaces. The resulting PLS model takes the same form as an OLS or PCR model; namely, it is a linear model relating the $\mathbf{X}$ and $\mathbf{Y}$ blocks. We next describe an existing approach that applies the idea behind PLS for the purpose of batch process modeling and control.

11.2.3 LVMPC

In this subsection, we will review the latent variable model predictive control [8]. This work addresses the problem of direct quality control by inverting a PLS model to determine the input move and has found widespread application and success. We review the approach here to enable its subsequent analysis from a state-space-based perspective and also to serve as an excellent benchmark for our proposed approach that utilizes an altogether different perspectives to the quality control problem.

All modeling approaches inherently assume a certain model structure and then determine the best fit values of the parameters. To do this, the data must be arranged appropriately for parameter estimation. The LVMPC proposed utilizes batch-wise unfolding to build and implement the model (see, e.g., [8] and many subsequent contributions). Batch-wise unfolding is accomplished by re-arranging the training data such that each row of the $\mathbf{X}$ matrix contains the input and output trajectories from a complete batch. This is represented as

$$\mathbf{X} = \begin{bmatrix} \boldsymbol{y}^{(1)}[1] & \boldsymbol{u}^{(1)}[1] & \cdots & \boldsymbol{y}^{(1)}[k_f] & \boldsymbol{u}^{(1)}[k_f] \\ \vdots & \vdots & & \vdots & \vdots \\ \boldsymbol{y}^{(B)}[1] & \boldsymbol{u}^{(B)}[1] & \cdots & \boldsymbol{y}^{(B)}[k_f] & \boldsymbol{u}^{(B)}[k_f] \end{bmatrix} \tag{11.7}$$

where $\boldsymbol{y}^{(b)}[k]$ and $\boldsymbol{u}^{(b)}[k]$ are vector output measurements and inputs, respectively, for batch b at sampling instant k. In the $\mathbf{Y}$ matrix, each row is comprised of the corresponding quality measurements.

$$\mathbf{Y} = \begin{bmatrix} \boldsymbol{q}^{(1)}[k_f] \\ \vdots \\ \boldsymbol{q}^{(B)}[k_f] \end{bmatrix} \tag{11.8}$$

Model fitting is accomplished by applying PLS to the $\mathbf{X}$ and $\mathbf{Y}$ blocks (as described in the previous subsection).

Remark 11.1 Batch-wise unfolding necessitates that each batch in the training data has the same length so that complete columns of the $\mathbf{X}$ block can be populated. In practice, it is common for batches to have varying duration. In many cases, this is due to operating policies, whereby the reaction is stopped when a condition is met (i.e., when a sample with high enough conversion is taken). One commonly adopted solution in the literature [9–13] is to 'align' batches based on a measured monotonically increasing variable. Under batch alignment, columns of the $\mathbf{X}$ block correspond to equal values of the alignment variable rather than equal elapsed time.

Once a PLS model is identified, LVMPC can be used at decision points throughout the batch to adjust the input trajectory and drive the process to the desired quality. In the interest of brevity, only an outline of and the most simple form of the LVMPC calculation will be presented here. The method can be thought of in three steps.

First, the current score is calculated from the existing measurements and nominal input trajectory. At each decision point, only a partial row of the **X** space can be constructed because output measurements following the decision point are not yet available. Therefore, the score is calculated using a missing data algorithm. These algorithms utilize the observed correlation in the **X** space to calculate the score for the unavailable values. In the next step, a quadratic program is evaluated to determine a move in the score space. The move is calculated such that the resulting score corresponds to a predicted quality close to the quality set-point. Finally, the move in the score space is translated into an input move trajectory to be implemented for the remainder of the batch. The inquiring reader is referred to [8] for details. We next review an alternate modeling approach.

11.2.4 Subspace Identification

In this subsection, we will give a brief overview of subspace identification. While we refer the reader to the literature for details (see [2, 3, 5, 14, 15]), we provide enough nomenclature to describe the application of subspace identification in this chapter.

In these approaches, the model parameters are identified assuming a discrete linear model of the form:

$$\mathbf{x}[k+1] = \mathbf{A}\mathbf{x}[k] + \mathbf{B}\boldsymbol{u}[k] \tag{11.9}$$

$$y[k] = \mathbf{C}\mathbf{x}[k] + \mathbf{D}\boldsymbol{u}[k] \tag{11.10}$$

where $\mathbf{x}$ denotes the subspace states (note that the distinction between subspace states denoted $\mathbf{x}$ and first-principles states $\boldsymbol{x}$ will be discussed in detail later in the chapter). The objective of subspace identification is to identify the system matrices **A**, **B**, **C**, and **D** utilizing the input and output data $\boldsymbol{u}$ and $\boldsymbol{y}$. To do this, first a valid state trajectory is identified by finding a common basis between lagged input and output data. Once a state trajectory is obtained, the system matrices can be easily determined (e.g., by OLS).

One important tool that is utilized in the determination of the state trajectories is the Hankel matrices of input and output data. In this chapter, we will adopt the standard notation of these matrices where $\mathbf{Y}_{1|i}$ and $\mathbf{U}_{1|i}$ represent Hankel matrices of output and input data, respectively. Under this notation, the subscripts denote the range of sampling instances in the first column. For instance:

$$\mathbf{Y}_{1|i} = \begin{bmatrix} y[1] & y[2] & \cdots & y[j] \\ \vdots & \vdots & & \vdots \\ y[i] & y[i+1] & \cdots & y[i+j-1] \end{bmatrix} \tag{11.11}$$

Once the data has been arranged into Hankel matrices, a wide variety of subspace identification methods can be used. Subspace identification methods can be broadly categorized into deterministic, stochastic, and combined deterministic–stochastic methods. The key difference between these methods is their handling of process noise (i.e., noise in the states), resulting from the model structure that is assumed. In essence, utilizing of combined deterministic–stochastic methods (under assumptions of a specific structure of the data) also enables estimating the process and measurement noise covariance matrices that can then be utilized within say a Kalman filter for improved state estimation. Three of the most prominent combined deterministic–stochastic methods are numerical subspace state-space system identification (N4SID), multivariable output error state space (MOESP), and canonical variate analysis (CVA) [4]. In [16], a unifying framework is presented for these three methods where the principle difference is the selection of two weighting matrices which determine the orientation of the resulting state space. For simplicity of use, and for the purpose of illustration of our proposed approach, in this chapter we will use the simple deterministic algorithm presented by Moonen et al. [3]. Note that the proposed approach can be readily generalized to utilize the combined deterministic–stochastic methods.

11.3 Motivating Example: Polymethyl Methacrylate Process

In this section, we introduce the polymethyl methacrylate (PMMA) process, an industrially relevant batch process, under traditional trajectory tracking control, to motivate the proposed approach. PMMA is a staple of the polymer industry. Originally marketed as plexiglass, PMMA now has applications in many products ranging from its traditional application as a glass substitute for medical implants. For high-end applications, quality is a key factor. As with many products, when quality is of paramount importance, it is preferable to carry out the polymerization as a batch process.

The quality of a batch of PMMA can be characterized by the molecular weight distribution of the polymer. Since measurement of the complete molecular weight distribution is often not practicable, it is common to summarize the key characteristics of the distribution. The number and weight average molecular weights provide sufficient information to classify the properties of the product. Therefore, the objective for the batch production of PMMA is to reach the desired number and weight average molecular weight at batch termination.

Batch, free-radical polymerization of PMMA is carried out in a jacketed reactor with a continuous stirrer. The reactor is charged with methyl methacrylate (monomer), AIBN (initiator), and toluene (solvent). The heat addition to the reactor is manipulated via the coolant/heating fluid flow rate. The nominal recipe for a batch (in terms of concentration) is presented in Table 11.1. In this chapter, we use a first-principles dynamic model of the PMMA process to generate historical

Table 11.1 Initial batch conditions for PMMA reaction

Variable	Value	σ^2	Units
Initiator concentration	2.06×10^{-2}	1.0×10^{-3}	kg/m^3
Monomer concentration	4.57	0.5	kg/m^3
Temperature	61	2.5	°C

and identification data, and as a test bed to evaluate control designs. This model includes nine states dictated by nonlinear algebraic and differential equations. These states include six molecular weight distribution moments, concentration of monomer and initiator, and reactor temperature. Throughout this study, we regarded the first-principles model as a black box model. This first-principles model is adapted from [17] while making suitable adaptations [8, 18]. More discussion of these adaptations is provided in [1].

As one example of a traditional control implementation, PI trajectory tracking of a nominal temperature profile is used, where a tightly tuned controller rejects deviations in the measured temperature from the nominal trajectory by manipulating the jacket temperature. The key idea is that the nominal temperature trajectory, shown to provide good quality in historical batches, should continue to provide good quality in subsequent batches. Once the batch has begun, reaction is carried until the density reaches a predefined level.

To reflect common variation in the feed, each of these initial conditions is independently drawn from a normal distribution centered around the nominal initial condition. The standard deviations for these initial conditions are given in Table 11.1. In this chapter, we take the measured process variables to be temperature within the reactor, volume of the reactants, and stirrer torque. From the volume measurement, we assume that the density can be determined. For the PMMA process, density is a monotonically increasing variable which reflects, in some sense, the extent of the reaction. Therefore, it provides a crude indication of when the target product has been reached. By terminating the batch based on density rather than at a specific batch duration, the resulting quality variation is reduced. An example of the resulting PI jacket and reactor temperature trajectories is shown in Figs. 11.1 and 11.2, respectively. The bottom row of plots in Fig. 11.3 shows histograms of the resulting number and weight average molecular weight as well as conversion in terms of percent error from the desired values. While the temperature trajectory plot demonstrates that the PI control is able to effectively track the desired temperature trajectory, clearly the desired quality is not always achieved. This is due to the fact that the expected (or past) relationship between a temperature profile and the final quality does not continue to hold. Note that in the present process, viscosity can also be calculated from the stirrer torque. Therefore, there exists a possibility to implement advanced control approaches that utilize all the available measurements for improved quality control.

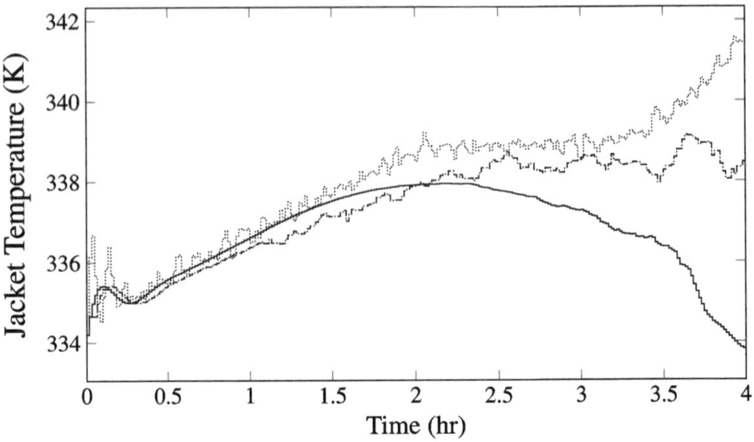

Fig. 11.1 Jacket temperature trajectory for a characteristic closed-loop batch under PI (............), LVMPC (_._._..), and the proposed SQMPC (_____)

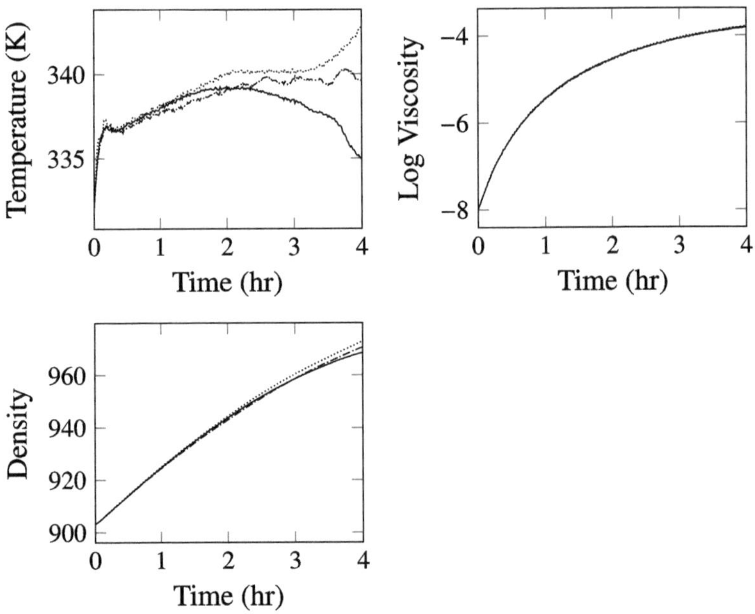

Fig. 11.2 Measured outputs from a characteristic closed-loop batch under PI (............), LVMPC (_._._..), and the proposed SQMPC (_____)

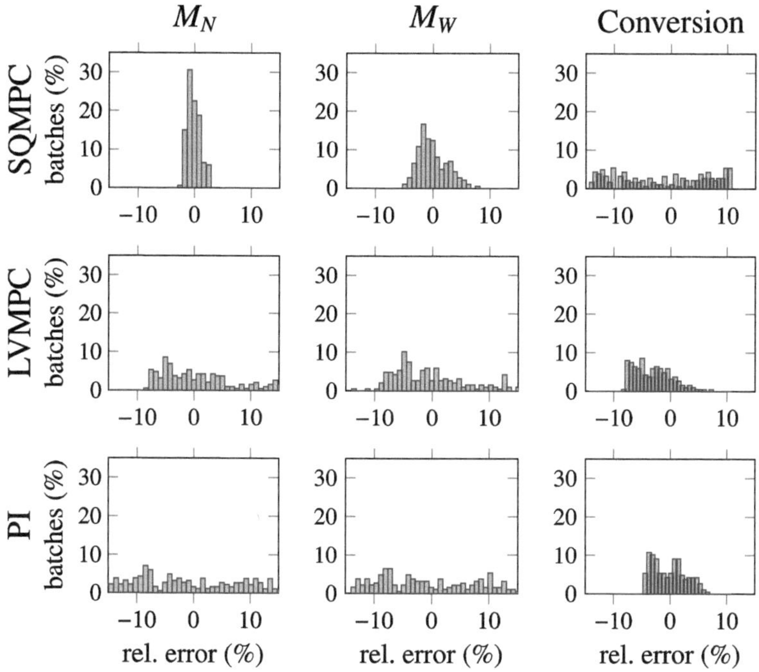

Fig. 11.3 Histograms of case 1 qualities resulting from 186 batches operated under SQMPC, LVMPC, and PI control

11.4 Latent Variable Control Analysis

In this section, we will conduct a brief analysis of latent variable methods. In particular, we focus our discussion on the model form of latent variable models when built on batch-wise unfolded data to reconcile the inherent requirements of equal batch duration in these approaches.

Latent variable methods have proven successful for data-based MPC of batch processes [8, 20–24]. A key issue in these contributions is the arrangement of batch data into matrices for latent variable analysis. In many of these contributions, a latent variable model is built by unfolding batch data in the batch-wise manner described in Eqs. (11.7) and (11.8). Once identified, the models are applied in a controller to control a new batch. Note that at the decision points, partway through the batch, only the past and current measurements are available. Therefore, one important aspect of these methods is to use the identified model to predict the future batch measurement trajectory. This prediction is accomplished using missing data algorithms. In some sense, the missing data algorithms turn the latent variable models into dynamic models of the process. A key to understanding the success of LVMPC is to understand the structure and functioning of latent variable models used in this fashion.

To begin, we will examine the structure of latent variable models. Particularly, consider applying the previously identified model to data from a new batch. To do this, the data is arranged into a row, $\boldsymbol{x}_{new}$ of the same format as the rows of the $\mathbf{X}$ training matrix. The first step in applying the model is to center and scale the row in the same way as was done in the training data. This can be represented

$$\bar{\boldsymbol{x}}_{new} = (\boldsymbol{x}_{new} - \mu_x) \odot \left[\frac{1}{\sigma_{x_1}} \cdots \frac{1}{\sigma_{x_{k_f}}}\right] \tag{11.12}$$

where μ_x represents a vector of the column means of the training data and σ_{x_i} represents the standard deviation of the ith column. Because of the unfolding policy described by Eq. (11.7), this implies that μ_x contains the time-indexed average trajectory of the inputs and outputs in the training data. Likewise, σ_{x_i} represents the time-indexed standard deviation of the inputs and outputs in the training data. After centering and scaling, the score is calculated as

$$\boldsymbol{t} = \bar{\boldsymbol{x}}_{new}\mathbf{w}^* \tag{11.13}$$

where $\mathbf{w}^*$ is the coefficient matrix identified in the modeling step. To clarify structure of this model, we can substitute the form from Eq. (11.7). Doing this, we get

$$\boldsymbol{t} = \left[\boldsymbol{y}[1]\; \boldsymbol{u}[1] \cdots \boldsymbol{y}\left[k_f\right]\; \boldsymbol{u}\left[k_f\right]\right] \tilde{\mathbf{w}}^* - \mu_x \tilde{\mathbf{w}}^* \tag{11.14}$$

where $\tilde{\mathbf{w}}^*$ is the result of taking the appropriate Hadamard product of the coefficients and the standard deviation vector. Next, we can partition the matrix $\tilde{\mathbf{w}}^*$ into submatrices based on the sampling instant of the output measurement or input they multiply. Then, Eq. (11.14) can be written as:

$$\boldsymbol{t} = \sum_{k=1}^{k_f} \left(\left[\boldsymbol{y}[k]\; \boldsymbol{u}[k]\right] - \mu_x[k]\right) \tilde{\mathbf{w}}^*[k] \tag{11.15}$$

where $\mu_x[k]$ is the subvector of the mean input and output trajectories corresponding to the kth sampling instant. This reformulation clearly demonstrates that latent variable models built on batch-wise unfolded data yield time-dependent linear models of the process.

Next, we consider how this structure is utilized by missing data algorithms. There are a large variety of missing data algorithms that have been discussed in the literature each with its own advantages [25, 26]. While each missing data algorithm is unique in its methodology, the basic steps are similar between algorithms. Specifically, first a score is calculated from the available data, and then the missing trajectory is filled in with values consistent with the calculated score. Regardless of the algorithm, both of these steps involve using the relevant parts of the weighting matrix $\tilde{\mathbf{w}}^*$. Therefore, these missing data algorithms inherit the time dependency of the latent variable model structure. In this way, using a latent variable model and missing data algorithms to make trajectory predictions is equivalent to using a linear time-dependent dynamic model in which the dynamic model coefficients depend on sampling instant.

To illustrate the time-dependent indexed nature of latent variable methods when used as dynamic models, we will demonstrate the use of the projection to model plane (PMP) missing data algorithm for estimating the score using measurements available midway through a batch. In many latent variable batch approaches, this score value forms the basis for subsequent control decisions and output prediction. Consider that output data is available up until the current sampling instant, k_c, and the complete input trajectory (past inputs and future candidate trajectory) is also available. Using the PMP method, the estimated score, $\hat{t}$, can be obtained as:

$$\hat{\boldsymbol{t}} = \left(\mathbf{w}^{*'}_{\text{known}}\mathbf{w}^{*}_{\text{known}}\right)^{-1}\mathbf{w}^{*'}_{\text{known}}$$
$$\left(\left[\boldsymbol{y}[1]\ \boldsymbol{u}[1]\ \cdots\ \boldsymbol{y}[k_c]\ \boldsymbol{u}[k_c]\ \boldsymbol{u}[k_c+1]\ \cdots\ \boldsymbol{u}\left[k_f\right]\right] - \mu_{x,known}\right) \odot \left[\frac{1}{\sigma_{x,known}}\right] \tag{11.16}$$

where $\mathbf{w}^{*'}_{\text{known}}$ is a submatrix of the projection matrix, $w^{*'}$, formed by selecting only columns corresponding to known information at instant k_c. Similarly, $\mu_{x,known}$ and $\sigma_{x,known}$ are, respectively, the mean and standard deviation values corresponding to the known information. Note that for all three of these matrices, their contents depend on k_c and therefore are time dependent. Once $\hat{\boldsymbol{t}}$ has been calculated, the predicted output trajectory is given by:

$$\left[\boldsymbol{y}[k_c+1]\ \cdots\ \boldsymbol{y}\left[k_f\right]\right] = \hat{\boldsymbol{t}}\mathbf{P}'_{unknown} \tag{11.17}$$

where $\mathbf{P}_{unknown}$ is a submatrix of the loading matrix, $\mathbf{P}$. Once again, the contents of the submatrix $\mathbf{P}$ are time dependent. By substituting Eq. (11.16) into (11.17), the overall model is a time-varying linear combination of time-indexed past outputs, inputs, and the candidate input trajectory.

The advantage of using time-dependent models for batch processes is that they provide a way to account for nonlinear batch behavior. Batches that are initialized consistently from the same initial conditions will follow the same state trajectories. In this case, a time-dependent model is capable of capturing the dynamic behavior of the process because at corresponding batch times, the process is at corresponding states and therefore has corresponding dynamic behavior. However, for batches that are initialized from a range of initial conditions, the state trajectories do not correspond to corresponding times. More importantly, the time indexing then necessitates the training of the model from batches of equal length (or necessitates the search for an alignment variable).

11.5 Subspace Quality Model Predictive Control (SQMPC)

In this section, we present a novel data-driven model predictive control scheme for batch quality control. The model building approach combines subspace identification to obtain a dynamic model and principal component regression (PCR) to relate process states and product quality. Together, these two models provide a causal model relating a candidate input trajectory to the resulting quality. The section is organized as follows: First, we will present an adaptation of subspace identification for batch processes that allows its application for batch data. Next, we will discuss the use of PCR to relate state and quality. Finally, we will demonstrate how the resulting models can be incorporated in a model predictive control scheme to directly control product quality.

11.5.1 Subspace Identification of Batch Dynamics

In this subsection, we present an adaptation of subspace identification for batch processes. Out of the wide variety of subspace algorithms, one commonality is the re-arrangement of training data into Hankel matrices. In order to use subspace identification for batch processes, the formation of the Hankel matrices from training data has to be carefully considered. In order to understand the necessary modifications, we first consider the form of training data for subspace identification of continuous processes. To carry out subspace identification for a continuous process, the process is first brought to the desired steady-state condition. Next, perturbations (e.g., by PRBS) are made to the inputs over some duration of time. These inputs and the corresponding outputs form a contiguous block of data. The appropriate Hankel matrices for a given subspace identification algorithm can be readily formed from this contiguous block of data in the manner presented in the preliminary section.

In contrast to continuous processes, batch processes, by their very definition, do not possess meaningful steady-state conditions around which to identify a model. Instead, the objective is to develop a model for the transient dynamic behavior of the process. To achieve this, the entire range of dynamic behavior for which a model is desired must be present in the training data. This can be partially accomplished by using data from the complete duration of a batch. However, owing to the variation in initial conditions, it is unlikely that any single batch (even with experimental augmentation) would cover the range of operating conditions necessary to build a reliable, general model. This motivates the need to include multiple batches in the training data. However, since each batch is independent of the batches before or after it, this results in a database with multiple contiguous blocks of data rather than one single contiguous block. One approach would be to build separate Hankel matrices for each of these blocks and then modify the chosen subspace identification algorithm to incorporate the multiple batches. In this chapter, however, we took the simpler approach of modifying the form of the Hankel matrices themselves. This approach has the benefit of maintaining generality so that any subspace identification algorithm can be applied without further modification.

To incorporate multiple independent blocks of training data, corresponding to multiple batches, we horizontally concatenate Hankel submatrices formed from each block. Consider the notation $\boldsymbol{y}^{(b)}[k]$ for the vector of output measurements where the superscript, (b), denotes the database batch index and k denotes the sampling instant (indexed from batch initiation). Then, the Hankel submatrix for any batch b is formed as:

$$\mathbf{Y}_{1|i}^{(b)} = \begin{bmatrix} \boldsymbol{y}^{(b)}[1] & \boldsymbol{y}^{(b)}[2] & \cdots & \boldsymbol{y}^{(b)}\left[j^{(b)}\right] \\ \vdots & \vdots & & \vdots \\ \boldsymbol{y}^{(b)}[i] & \boldsymbol{y}^{(b)}[i+1] & \cdots & \boldsymbol{y}^{(b)}\left[i+j^{(b)}-1\right] \end{bmatrix} \tag{11.18}$$

Once the submatrices have been formed for each batch, they can be concatenated to form the overall Hankel matrix as follows:

$$\mathbf{Y}_{1|i} = \left[\mathbf{Y}_{1|i}^{(1)}\ \mathbf{Y}_{1|i}^{(2)} \cdots \mathbf{Y}_{1|i}^{(B)}\right] \tag{11.19}$$

where B is the number of batches in the training data. Hankel matrices of input data can be formed in exactly the same fashion. The resulting matrices elegantly handle the blocks of data corresponding to independent batches while maintaining the properties required for subspace identification.

Remark 11.2 Notice that in Eq. (11.18) the final time index j is indexed itself by b. This is an important and valuable distinction of the proposed method. The index j corresponds to the final sampling instant at which there is sufficient past and future data to fulfill the required lag structure. In this way, the value of j reflects the length of the batch. Because the Hankel matrices for each batch are concatenated horizontally, only the number of rows in each submatrix needs to match. In other words, the number of columns can vary for each submatrix corresponding to each batch. Therefore, the length of each batch in the training data can be different. This is markedly different from time-dependent models that inherently need training batches of equal length (or aligned to have equal length).

The proposed form of the Hankel matrices is permissible because it maintains the row-space structure of the traditional form. Specifically, the row-spaces of the Hankel matrices $\mathbf{Y}_{1|i}$ and $\mathbf{Y}_{i+1|2i}$ still share a common basis related to the state of the system at time i. Subspace algorithms determine viable states by exploiting this property of the row-space. If instead the Hankel matrix was constructed by first concatenating the outputs from each batch, the discontinuity between batches would violate the row-space relationship in the corresponding columns in the Hankel matrix. Because the proposed form of the Hankel matrices preserves the row-space relationship, any desired subspace identification algorithm can be applied without modification.

Remark 11.3 It is well recognized that a common characteristic of industrial batch processes is sparsity of available measurements. This sparsity does not hinder the application of the subspace identification methods due to unobservability limitations since by construction the LTI system that is identified is naturally observable. The key implication of limited measurements is if they prohibit the identification of a system structure with (the entire set of) states that impact the final product quality. That, however, would be a fundamental limitation of the process instrumentation, and no other approach would be able to capture the process quality from the available measurements either. The construction of the quality model in the proposed approach readily allows diagnosis of a problem with measurement sparsity and can be used to motivate improved instrumentation. The response of the proposed method to sparse measurements is demonstrated in the subsection on quality model fitting in the simulation study later in this chapter.

Remark 11.4 Many subspace identification methods make an assumption about the stationarity (or quasi-stationarity) of the processes. These assumptions are important when the identified model structure is a combined deterministic–stochastic model. In these cases, stationarity has an important impact on the stochastic part of the identified model. While the current work could readily be extended to utilize a combined deterministic–stochastic model, the current focus is on deterministic subspace identification methods for which stationarity is not a concern. Therefore, discussion of the stationarity of the unfolded training data will be postponed to future work, and instead, the applicability of the proposed approach is demonstrated through application.

The result of subspace identification is a single, linear, time-invariant model of the form presented in Eqs. (11.9) and (11.10). For clarity, we will refer to the identified dynamic model using the following nomenclature:

$$\mathbf{x}[k+1] = \mathbf{A}\mathbf{x}[k] + \mathbf{B}\boldsymbol{u}[k] \tag{11.20}$$

$$\boldsymbol{y}[k] = \mathbf{C}\mathbf{x}[k] + \mathbf{D}\boldsymbol{u}[k] \tag{11.21}$$

where the subspace state is denoted as $\mathbf{x}$ to avoid confusion with $\boldsymbol{x}$ which represents the first-principles states from the process Eqs. (11.1)–(11.3).

To some readers, the linear form of Eqs. (11.20) and (11.21) will seem incongruent with the emphasis placed on the unavoidable, inherent nonlinearity of batch systems. Indeed, the nonlinear dynamics of batch processes when expressed in terms of meaningful process variables (such as concentrations and temperature) is an important consideration in batch processes. However, there is no meaningful way to linearize a nonlinear representation of the process. Consider that a first principles model of the form of Eqs. (11.1)–(11.3) was available (note that the present formulation does not require knowledge of this model), in terms of the original state variables, $\boldsymbol{x}$. Since the process is inherently transient, any selected center (state) of linearization is departed quickly rendering the linearization around that point useless.

The key distinction of the proposed approach is that we are not linearizing an existing nonlinear state-space model. Instead, we are identifying model parameters for a LTI model that describes the evolution of the process outputs. This is fundamentally different from linearizing a nonlinear representation because the system order is determined during the identification. One of the reasons that subspace methods have been met with such success is their ability to determine system order as a first step of the identification. Specifically, the system order is determined based on the order needed to adequately describe the dynamics observed in the data. It is a well-known result that any discrete series can be exactly represented in the frequency domain in a finite number of frequency bands using a discrete Fourier transform. It follows that a finite number of observations can be represented exactly by LTI with a finite system order. Obviously, an exact fitting of the data would result in dramatic overfitting. Fortunately, LTI models identified by subspace methods fit states in an order that maximizes the amount of variance explained by each new state. Therefore, it is practicable to fit LTI models of a reasonable order that adequately predict the evolution of the process over an appropriate region in state space.

Remark 11.5 Clearly, the system order is of particular importance for fitting an LTI model to batch data. In subspace identification, the system order is often selected such that the explained variance falls above a specified threshold. This threshold is a key modeling parameter and is especially important given the potentially high-order models needed to explain transient batch behavior. One robust threshold selection method is to test the predictive ability of the resulting dynamic model on validation batches. Doing this, the threshold should be set at the value which minimizes the prediction error for the validation batches. The primary downside to this approach is the need to set aside valuable training data to use instead for validation. One solution to address this problem would be the use of cross-validation; however, this approach was not implemented in this chapter. Better methods for determining the explained variance threshold will be the topic of future work.

The identified model provides a theoretically sound way to predict the evolution of the process outputs (and the associated states) for a candidate input trajectory. In order to be useful for quality control, a model between the subspace states and the final quality must be determined. In the next subsection, we will present a method for modeling this relationship.

11.5.2 PCR Quality Model

In the last subsection, we demonstrated that subspace identification can be used to obtain a LTI dynamic model from batch data. In this subsection, we motivate and present a modeling approach for relating subspace states, $\mathbf{x}$, to batch quality. The key idea in building this model can be described in two logical steps. First, since both the subspace states, $\mathbf{x}$, and the first-principles states, $\boldsymbol{x}$, capture the instantaneous condition of the process, there must be a mapping from $\mathbf{x}$ to $\boldsymbol{x}$. Next, since quality is known to depend on process state, as described by Eq. (11.3), there must be a causal relationship between $\mathbf{x}$ and quality. In this subsection, we will present a method for first obtaining state estimates (we utilize a Luenberger observer for the purpose of illustration), then relating the terminal state of the batch to quality.

The first step in the proposed quality model methodology is to estimate the subspace states $\mathbf{x}$ at the end of each training batch. One of the intermediate results of subspace identification is a vector of state estimates for the training data. However, this vector does not contain states corresponding to the last sampling instant in each batch because lagged future outputs are needed to obtain each state. Furthermore, the state estimates obtained during subspace identification are based on a limited number of lagged past and future outputs. To build an accurate quality model, we require the best possible terminal state estimate, $\hat{\mathbf{x}}[k_f]$, obtainable from the complete trajectories of measured inputs and outputs.

To obtain an estimate of the terminal subspace state for model building, $\hat{\mathbf{x}}[k_f]$, a Luenberger observer is used. The observer can easily be designed based on the identified model such that the state estimates converge after being updated with measurements from the beginning of the batch. By updating the observer with outputs from the duration of the batch, the best possible estimate of $\mathbf{x}[k_f]$ is obtained. By repeating this procedure, terminal subspace state estimates for each training batch are obtained. These estimates are then used in the next stage of building the quality model.

Once estimates of the terminal subspace states are obtained, what remains is to identify the correlation between these states and quality. In this chapter, we use principal component regression (PCR) for this task. As discussed in the preliminaries, PCR is a method for regressing observations $\mathbf{X}$ onto dependent variables $\mathbf{Y}$. In this application, the $\mathbf{X}$ and $\mathbf{Y}$ blocks take the following form:

$$\mathbf{X} = \left[\hat{\mathbf{x}}^{(1)}\left[k_f^{(1)}\right] \cdots \hat{\mathbf{x}}^{(B)}\left[k_f^{(B)}\right]\right]^T \tag{11.22}$$

$$\mathbf{Y} = \left[\boldsymbol{q}^{(1)} \cdots \boldsymbol{q}^{(B)}\right]^T \tag{11.23}$$

where k_f is the final sampling instant. After completing PCR, the quality model takes the linear form:

$$\boldsymbol{q} = \mathbf{Q}\hat{\mathbf{x}}[k_f] + \boldsymbol{c}_q \tag{11.24}$$

where $\mathbf{Q}$ is a matrix of model coefficients and $\boldsymbol{c}_q$ is a vector of constants. Using this linear relationship, any terminal subspace state estimate can be used to predict quality.

Remark 11.6 Note that in Eq. (11.22), k_f is indexed by batch. This is to denote the fact that there is no necessity for batches to share the same length (be aligned) using this method. Any mismatch in batch length is elegantly and appropriately dealt with by the Luenberger observer. This reflects our understanding that quality is determined by state and therefore should not be dependent on time. Taken together, both the dynamic and quality modeling are independent of batch duration meaning that the overall modeling approach is equally valid for training data with mismatched batch length. There is no need to address the alignment problem. This is particularly beneficial because available historical operation data often contains batches of varying lengths, and a monotonically behaving 'alignment' variable might not be readily available.

Remark 11.7 PCR is used to build the quality model to address correlation in the **X** space or equivalent states that appear correlated. In principle, all states identified by subspace identification are independent. Therefore, one might think that there is no correlation in the **X** space of this regression. However, while the states are all uncorrelated, it might be possible that the existing batches evolve in a manner where by the end of the batch some states may be correlated. In this case, using PCR improves the performance of the identified model as opposed to models built using ordinary least squares (OLS) regression. In the case where there is no correlation in the terminal states, PCR will identify the same model as OLS so there is no loss of generality.

11.5.3 SQMPC—Formulation

In the previous two subsections, we described procedures for identifying a dynamic model and quality model from batch data. When combined, these two models provide an overall, causal model of batch quality. In this subsection, we first demonstrate the manner in which these models can be used to make quality predictions based on available process measurements and a candidate input trajectory. Next, we present a model predictive control (MPC) formulation that uses these models to directly control batch quality.

For any point in time sufficiently after the initiation of a new batch, the problem of predicting final quality for a candidate input trajectory is addressed as follows: At batch initiation, the Luenberger observer from the quality model development is initialized. As the batch progresses, the observer is updated with each new output measurement. Convergence of the observer is indicated by the innovation term. Once the innovation has fallen below a selected threshold, the state estimate is considered converged and quality predictions can commence.

To make a quality prediction from the available models, first the terminal subspace state for the batch must be determined. Naturally, the terminal state depends on both the latest estimated state of the process and the future inputs, and can be computed by using the identified dynamic model, written succinctly as:

$$\hat{\mathbf{x}}\left[k_f\right] = \mathbf{A}^{k_f-k}\hat{\mathbf{x}}[k] + \left[\mathbf{A}^{k_f-k-1}\mathbf{B}\ \mathbf{A}^{k_f-k-2}\mathbf{B}\ \cdots\ \mathbf{B}\right]\mathbf{v}[k] \tag{11.25}$$

where k is the current sampling instant, k_f is the final sampling instant in the batch, and the vector $\mathbf{v}[k]$ contains the candidate input trajectory for the remaining batch duration:

$$\mathbf{v}[k] = \begin{bmatrix} \boldsymbol{u}[k] \\ \boldsymbol{u}[k+1] \\ \vdots \\ \boldsymbol{u}\left[k_f - 1\right] \end{bmatrix} \tag{11.26}$$

From this terminal state estimate, we next use the quality model to predict the resulting quality. Specifically:

$$\hat{q} = \mathbf{Q}\hat{\mathbf{x}}[k_f] + \boldsymbol{c_q} \tag{11.27}$$

where $\mathbf{Q}$ and $\boldsymbol{c_q}$ are the model coefficients from the quality model identification.

Direct quality control can be achieved by (in some sense) inverting the relationship between candidate input trajectory and quality described by Eqs. (11.25) and (11.27). One way to carry out this inversion, which provides a number of practical benefits, is to formulate the problem in a shrinking horizon, model predictive control framework. Mathematically, this approach takes the form of a quadratic programming problem defined by:

$$\min_{\mathbf{v}[k]} \quad (\hat{q} - \boldsymbol{q}_d)^T \mathbf{M} (\hat{q} - \boldsymbol{q}_d) + (\Delta\mathbf{v}[k])^T \mathbf{P} (\Delta\mathbf{v}[k]) \tag{11.28}$$

$$s.t. \quad \hat{q} = \mathbf{Q}\left(\mathbf{A}^{k_f-k}\hat{\mathbf{x}}[k] + \left[\mathbf{A}^{k_f-k-1}\mathbf{B}\ \mathbf{A}^{k_f-k-2}\mathbf{B} \cdots \mathbf{B}\right]\mathbf{v}[k]\right) + \boldsymbol{c_q} \tag{11.29}$$

$$\Delta\mathbf{v}[k] = \begin{bmatrix} \boldsymbol{u}[k] \\ \boldsymbol{u}[k+1] \\ \vdots \\ \boldsymbol{u}[k_f - 1] \end{bmatrix} - \begin{bmatrix} \boldsymbol{u}[k-1] \\ \boldsymbol{u}[k] \\ \vdots \\ \boldsymbol{u}[k_f - 2] \end{bmatrix} \tag{11.30}$$

$$\mathbf{v}_{\min}[k] \leq \mathbf{v}[k] \leq \mathbf{v}_{\max}[k] \tag{11.31}$$

where q_d is the desired quality (i.e., quality set-point). The first term of the objective function penalizes deviations of the projected quality from the desired quality, and the second term penalizes large input moves. More specifically, $\mathbf{M}$ is a positive-definite (diagonal) matrix where each diagonal element sets the penalty for the corresponding quality variable. $\mathbf{P}$ is also a positive-definite (diagonal) matrix where diagonal elements set the penalty on input moves for the corresponding sampling instant. The first constraint of the problem enforces the dynamic and quality model (arrived at by substituting Eq. (11.25) into (11.27)). The second constraint defines the input movement, and the third constraint applies hard bounds on the input trajectories based on process constraints.

The application of the proposed MPC is carried out as follows: First, in the beginning of the batch, the Luenberger observer is initialized with an arbitrary state value. In the beginning of the batch, a nominal, open-loop input trajectory is applied. Since input movements in the beginning of the batch have less impact on quality than those at the end, applying an open-loop input trajectory early in the batch has little impact on the ability to reach the desired quality. As each measurement becomes available, the Luenberger observer is updated. Once the innovation term of the Luenberger observer falls below a threshold, the state estimate is considered converged and the model predictive control is switched on. At the first MPC sampling instant, the optimization is initialized with the nominal input trajectory. The solution of the first MPC calculation is a complete input trajectory for the remainder of the batch. At the next sampling instant, the first input in the candidate trajectory is implemented and the

MPC is recomputed with the updated state estimate. This procedure is repeated for the remainder of the batch.

Remark 11.8 One of the primary advantages of the proposed MPC is that the required optimization is a convex, quadratic programming problem. This class of problems can be efficiently and quickly solved to a global optimal solution. This is in contrast to many batch MPC approaches which require the solution of a nonlinear programming (NLP) problem at every sampling instant. NLP solvers can be slow to converge on a solution, and global optimality is generally not guaranteed.

Remark 11.9 The proposed MPC incorporates a feedback mechanism through its inclusion of the latest state estimate. As the process progresses and the state estimate is updated, if the state estimate does not follow the projected trajectory (due to plant-model mismatch, noise, or disturbances), the candidate input trajectory will be updated to respond to the latest measurements. Another nice feature of the MPC design is that as the batch progresses, the model accuracy improves. The improvement is resultant of the shortening prediction horizon. At the same time, as the prediction horizon shrinks, larger input moves are necessary to obtain the desired quality. The naturally arising combined effect of increasing model accuracy concurrent with increasingly aggressive input moves yields a stabilizing effect on the resulting quality; as the predictions get better, more aggressive but also more appropriate input moves are prescribed. The improved model available toward the batch termination benefits the accuracy of the final set of control moves that are implemented, which in turn significantly affect the final quality. This can be understood by recognizing that there could be multiple paths from the initial set of states to the final desired (in that they yield the desired quality) states. The initial input moves influence which path is taken, and (as long as this path does not lead to a set of states from where the final quality becomes unreachable), the final set of control moves eventually determine the proximity to the target states.

Remark 11.10 In the literature on batch MPC, it is common to limit MPC calculation to a few predefined decision points [8, 27]. The motivation for these limited decision times is to avoid unnecessary excitation of the process. While the proposed approach could be easily adapted to implement decision points, we choose instead to carry out control calculations at each sampling instant. There are two reasons for this decision. First, it allows the incremental feedback from each new measurement to be immediately acted upon for quick disturbance rejection. Second, since these decision points are triggered by reaching a specified elapsed time from batch initiation, the use of decision times inherently introduces a time dependence on the modeling and control approach. Note that one of the key contributions of the present chapter is the design of a model based on a state-space perspective and avoiding explicit time dependence. One elegant way to incorporate some of the positive aspects of the decision time approach would be to calculate a new input trajectory at every sampling instant (thus incorporating the frequent feedback) but only update the input trajectory if a 'large enough' change in the calculated input trajectory occurs.

Remark 11.11 In addition to hard constraints on the inputs, the projected state trajectories calculated in each sampling instant of the proposed MPC allow the implementation of output constraints as well. Specifically, the output trajectory can be calculated from the projected state trajectory through the identified **C** matrix. In a traditional sense, the ability to add output constraints is important for safety-critical process variables (i.e., reactor temperature). However, applying output constraints is also a viable way to prevent the controller from extrapolating too far from the process operations in the training data.

11.6 Application to the PMMA Process Example

In this section, we compare traditional PI trajectory tracking, a simple implementation of LVMPC, and the proposed SQMPC for quality control. To this end, each of the aforementioned control approaches is implemented on a simulation of the PMMA process previously described. The section is organized as follows: In the first subsection, we discuss the training database to be used in model identification. Next, we use the training data to fit a state-space dynamic model using subspace identification. Subsequently, a second model is identified relating the terminal states of the system and the resulting quality. The following subsection presents the closed loop results from the proposed approach and compares the results to those obtained using a nominal implementation of LVMPC and PI trajectory tracking.

11.6.1 Database Generation

The first step in this simulation study was to generate a database of batch data for use in model identification. This database is used in both the LVMPC approach and the proposed SQMPC and is comprised of complete input and output trajectories as well as the resulting quality measurements. The objective in the database development was to mimic the data that would be available in an industrial setting. To this end, a modest database of 17 historical and identification batches was developed (note that the batches were not of equal duration). Each of these batches was initialized from initial conditions with natural variations (as previously described). White noise signals of appropriate magnitude were added to all simulated output measurements to simulate measurement noise. Table 11.2 summarizes the composition of the training database for reference.

In order to simulate historical data, nine batches operated under PI trajectory tracking control (as described previously) were included. In addition to historical batches, eight identification batches were included in the training database. These were batches with inputs designed to excite the system so that the causality between inputs and outputs could be observed. Of the eight identification batches, three identification batches were generated by adding a pseudo-random binary sequence (PRBS)

Table 11.2 Summary of training database composition

Input policy	Number of batches
Historical batches (PI trajectory tracking)	9
PRBS added to PI set-point trajectory	3
PRBS added to nominal input trajectory	5
Total batches	**17**

signal to the set-point trajectory of the PI controller. This method of introducing an identification signal provides the benefit that, since the batch is still in closed loop, the deviation from normal operations is small. This is beneficial for two reasons: First, the observed behavior is similar to what is expected for a normal batch. Second, the resulting product may still be usable which minimizes the cost of the experiment. The primary disadvantage of this type of identification batch is that the input is still dependent on the output. For this reason, the remaining five identification batches were operated in open loop. In these batches, the input was created by adding a PRBS to the nominal process input. These batches provide the necessary independence between inputs and outputs for subspace identification. However, they may be costly because the open-loop operation may lead to off-spec product.

Remark 11.12 In the current work, we included open-loop identification batches to satisfy the subspace identification requirement for independence between input and output measurements. This approach was chosen in order to simplify the model identification task. However, a large body of the literature exists addressing subspace identification from closed-loop data [15, 28–30] and can be readily used as part of the proposed framework.

Remark 11.13 While white noise was added to simulated output measurements, we assumed the availability of noise-free quality measurement. Of course, in practice there will always be some noise in this type of measurement. The motivations for this assumption are twofold. First, measurement of quality is typically done off-line in a laboratory. The analytical ability in such settings is significantly better than that available from online sensors. Furthermore, repeated measurements can be made to improve the quality measurement. Future work will address the situation where this assumption is relaxed (necessitating a more stochastic approach to quality prediction and control).

Figure 11.4 plots three selected input trajectories representative of the three input policies used in the training database. All three input trajectories follow the same general trend, but each policy provides a different level of information pertaining to the process dynamics. In the next subsection, we discuss building a dynamic model from this training data using subspace identification.

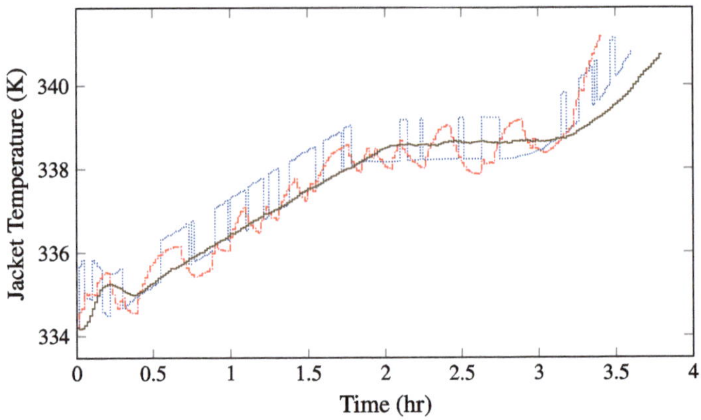

Fig. 11.4 Selected representative examples of the three input policies used in the training batches: nominal input plus PRBS (............), PI with set-point trajectory shifted by PRBS (_._._._.), and traditional PI trajectory tracking (______)

11.6.2 Dynamic Model Fitting

The input and output trajectories from the training database described in the previous subsection were used to identify a dynamic process model. To this end, subspace identification was applied as previously described. First, the data was unfolded into the appropriate Hankel matrices as in Eq. (11.19). As previously mentioned , one of the key modeling decisions in subspace identification is the number of rows to use in the Hankel matrices. In this simulation study, we experimented with the number of Hankel rows to gage the effect on the resulting model. A range of values between 10 and 25 was tested. To evaluate the effect, the resulting models were used to predict the dynamic evolution of a large set of validation batches. In general, we found that the resulting models were insensitive to the number of Hankel rows. From a theoretical perspective, this makes sense because adding 'extra' Hankel rows does not increase the rank of the Hankel matrix since the additional observations are dependent on the prior observations. For the present application, 15 Hankel rows were utilized for building our final model.

The other key model-fitting parameter, as described in the previous section, was the threshold for explained variance. Similarly to the choice of number of Hankel rows, we experimented with different threshold values and used the validation data to determine the best value. First of all, because of the order in which subspace states are fit to the data, each subsequent state is more susceptible to noise. In this regard, choosing to a high threshold, and correspondingly a high system order, results in overfitting and a reduction in the predictive ability for validation data. In contrast, selecting a low threshold results in meaningful dynamic behavior in the data being unmodeled and also results in poorer predictions of the validation data. Note that a parallel result to this logic is that too much noise in the outputs masks meaningful trends in the data and prevents the identification of a model of appropriate order. Thus, a threshold resulting in nine states was selected. With the modeling parameters defined, the dynamic model was identified and validated.

The identified dynamic model was validated by checking its predictive ability for a set of validation batches. Recall that the relationship between the subspace states, $\mathbf{x}$, and the process states, $\boldsymbol{x}$, is unknown. Furthermore, the process states are unmeasured. As a result, there is no point of comparison to validate the evolution of the subspace states directly. Instead, the best practicable validation method is comparison of output predictions. For a new validation batch, however, the initial subspace state and corresponding output predictions are unknown. Therefore, the following validation approach was taken: First, the Luenberger observer was updated with process measurements from the beginning of the validation batch. At a sample instant well after the convergence of the observer (as determined by the innovation), the current state estimate was used as the initial condition to predict the state trajectories corresponding to the input for the remainder of the batch. The resulting state trajectories were then used to predict outputs by multiplication with the identified $\mathbf{C}$ matrix. This predicted output trajectory was then compared with the measured output trajectory to evaluate the fit of the dynamic model.

Figure 11.5 demonstrates the described validation procedure for a validation (i.e., not in the training data) batch. In the beginning of the batch duration, the convergence of the Luenberger observer is apparent. Visually, the outputs corresponding to the state estimates converge after about half an hour of output measurements. To guarantee convergence, we continue updating the observer until it has been updated with two hours worth of measurements. Then, using the current state estimate as an initial condition, the dynamic model is used to predict the output trajectory from the remaining input profile. Clearly, the predicted output trajectory closely follows the true evolution of the process. This demonstrates that the dynamic model effectively captures the dynamic behavior of the process.

Remark 11.14 To attenuate the effect of noise on the identified model, a first-order filter was applied to the output measurements. The smoothing factor for this filter was selected to be high enough to visually reduce noise while not introducing significantly delayed response. This filter was also applied to output measurements before passing them to the proposed SQMPC. To provide an unbiased comparison, the filter was additionally applied to output measurements before they were used in the benchmark LVMPC and PI control.

11.6.3 Quality Model Fitting

Having identified the dynamic model, next a quality model was identified as previously described. Recall that the quality objective for the PMMA process is to reach the desired weight and number average molecular weight. To this end, the quality model should relate the subspace states from the dynamic model identified in the

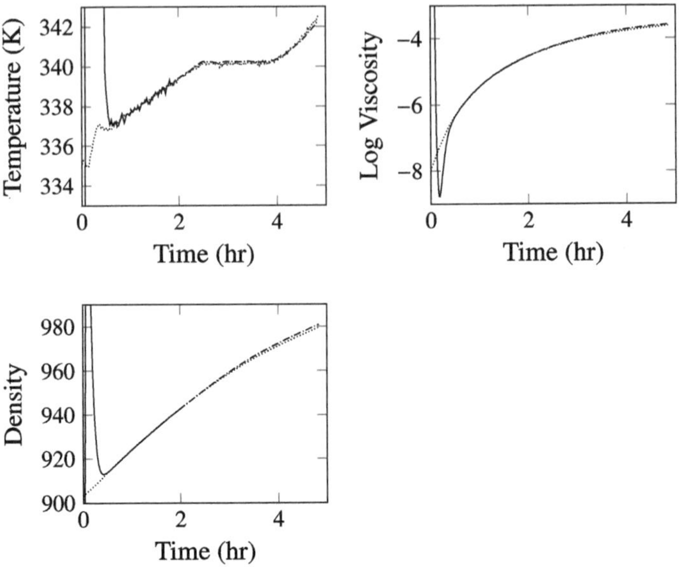

Fig. 11.5 Validation of the dynamic model by comparison of output trajectory prediction for a validation batch: output corresponding to subspace states estimated by updating the Luenberger observer (——), outputs predicted by the dynamic model from the converged state estimate (–·–·–), actual output (............)

previous subsection to these variables. In addition to the molecular weight distribution, another important quality variable for PMMA polymerization is the overall conversion at the end of the reaction. To reflect this, conversion was included as an additional quality variable and was also related to the subspace states. The first step in building the quality model was to design a Luenberger observer.

The performance of a Luenberger observer is parameterized by the location of the observer poles. By placing the poles appropriately, the convergence properties of the observed states can be designed. In this chapter, we choose to place the observer pole for each state on the real axis in the neighborhood of 0.66 (specifically, poles were placed at uniformly distributed random locations between 0.66 and 0.67 because it is not mathematically realizable to place all poles at exactly the same position) . This value was chosen by observing the convergence of the outputs corresponding to state estimates for validation data. Specifically, the value was selected so that the response was stable (poles were within the unit circle) and converged quickly but remained robust to noise.

Next, the Luenberger observer was used to determine the terminal subspace states. As previously described, this was achieved by updating the observer with the output measurements (temperature, density, and log viscosity) for each batch in the training data over its entire duration. The final state estimate for each batch is taken as that batch's terminal subspace state.

Having obtained terminal subspace states, regression was used to relate the state estimates to the quality measurements in the training data. As noted in the previous subsection, the dynamic model we identified had nine states. These states in conjunction with the dynamic model were demonstrated, by validation, to provide a good description of the dynamic evolution of the outputs. However, as previously mentioned, each subsequent state in the identified model is increasingly sensitive to noise. When we examined evolution of the state estimates for the validation data, we observed this effect to be particularly pronounced for the last five states of the identified system. While we were able to verify that these states provided meaningful contributions to modeling the dynamic process, we found that the last five states had a significantly lesser importance in predicting quality. In fact, in some cases, the high level of noise in these states diminished predictive ability of the model. Therefore, we decided to build our quality model only on the first four states of the process.

Remark 11.15 In general, we propose using PCR to relate subspace states and quality. As noted earlier, the use of PCR is particularly important when the terminal subspace states are correlated. In this study, however, the four states that we used to build our quality model were not correlated. In this case, PCR is equivalent to using ordinary least squares regression.

Figure 11.6 is three scatter plots showing the impeccable ability of the model in predicting quality for 70 validation batches. Specifically, the x data for these plots is the actual quality and the y data is the quality predicted. Datum points falling on the plotted 45-degree line represent accurate prediction. This result validates the overall methodology of using subspace states to capture the process condition and explain quality.

Remark 11.16 As previously mentioned, one common characteristic of batch processes is sparsity of available process measurements. To demonstrate the response of the proposed method to sparse measurements, we repeated both model-fitting steps, but instead of using the three process outputs from before, we only considered temperature measurement. In doing this, the number of observable states identified in the subspace identification step fell from 9 to 3. The quality model was built as before. Quality predictions for the same 70 validation batches as the training data are shown in Fig. 11.7. Note that, while the quality predictions are less accurate, the resulting model still has a reasonable predictive ability. This demonstrates the capacity of the proposed method to elegantly use the information identifiable from sparse measurements.

11.6.4 Closed-Loop Results

In the previous two subsections, we utilized the proposed modeling methodology to fit a dynamic model and a quality model for the PMMA process. In this subsection, we will present results obtained by applying the identified model in the proposed

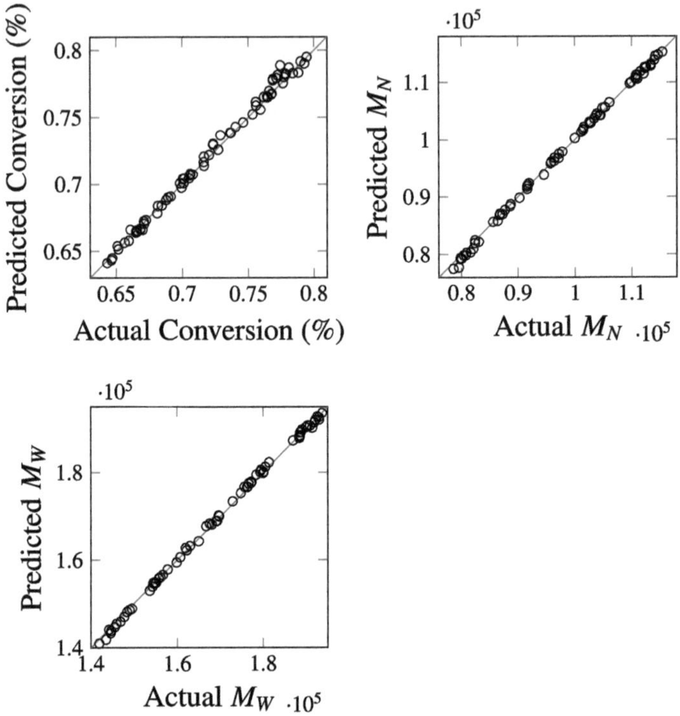

Fig. 11.6 Quality model predictions for 70 validation batches using terminal state estimates obtained by updating a Luenberger observer with output measurements from the duration of the batch

SQMPC. First, we will discuss implementation and tuning aspects of the approach. Our discussion of the closed-loop results will be presented in two parts. In the first part, we will demonstrate the 'regulatory' ability of the control design: in other words, its ability to reject disturbances and drive the process to the nominal condition from the training data. In the second part, we will demonstrate the 'set-point tracking' ability of the control design. Specifically, we will demonstrate the ability of the controller to drive the process to a quality set-point that differs significantly from that of the nominal batch. Note that in both of these cases we will use the same tuning and implementation. The only difference from an implementation perspective is the selection of the quality set-point.

Once the required models are identified, implementation and tuning of the proposed SQMPC are straightforward. Because the models effectively capture and characterize the behavior of the process, tuning the SQMPC is an intuitive procedure. This is particularly true when compared to other batch control methods, such as gain scheduling, where the tuning itself takes on the primary role in accounting for the dynamic behavior of the batch. Recall from its definition that the proposed SQMPC depends on two tuning matrices. These matrices, **M** and **P**, both have intuitive meaning in terms of the process. Specifically, **M** penalizes deviations of the predicted quality from the set-point and **P** penalizes large input moves.

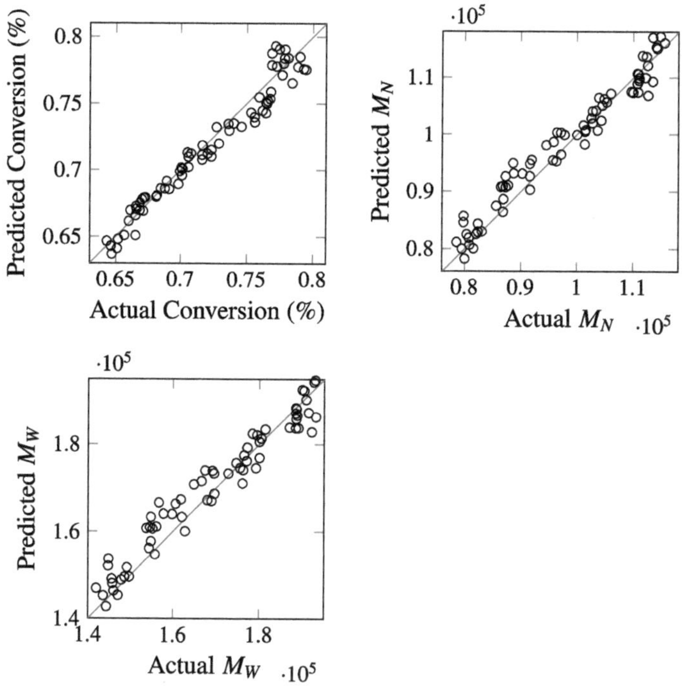

Fig. 11.7 Quality model predictions for the same 70 validation batches as the base case but using only temperature measurement demonstrating the robustness of the approach to sparse measurements

Remark 11.17 When the quality model is fit, the first step in the PCR modeling procedure is to scale both the **X** and **Y** spaces to be centered with unit variance. We refer to the centered and scaled values as being in the scale space. When implementing SQMPC, there are two options: Either the predicted quality variables can be rescaled to measurement space or the set-point can be correspondingly scaled to scale space. Both options are mathematically equivalent in terms of the solution to the optimization problem, but leaving the quality predictions in scale space has the advantage that the penalty matrix, **M**, has a more interpretable meaning. Specifically, because the quality variables are scaled to have unit variance, each of the penalties, m_i, directly represents the relative importance of meeting that quality set-point. For instance, setting all of the diagonals to the same value would apply an equal weighting to every quality. Setting $m_i = 2m_j$ applies twice as much weight on meeting the ith quality as it does on meeting the jth. For this reason, in this study, we choose to keep the quality predictions in the scale space.

In the PMMA process, we have considered three qualities: number average molecular weight, weight average molecular weight, and conversion. Correspondingly, the **M** matrix is a three-by-three diagonal matrix with entries m_1, m_2, and m_3. To tune these values, we started with all three penalties equal to one and then adjusted the values based on our observation of the closed loop results for a small number of batches. Our final tuning was $m_1 = 187$, $m_2 = 58$, and $m_3 = 9.2$ where m_1, m_2, and m_3 correspond to the weights on number average molecular weight, weight average molecular weight, and conversion, respectively.

The other tuning parameter is the penalty on input moves, **P**. The individual diagonal elements of **P** represent the penalty on input moves for each corresponding sampling instant. Due to the shrinking horizon nature of the MPC, the dimension of the **P** matrix is decremented at each sampling instant. From an implementation perspective, this necessitates reforming matrix **P** at every sampling instant as a submatrix of the **P** matrix from the beginning of the batch. Doing so allows a penalty profile to be placed on the input moves of the batch. For instance, higher penalties could be placed in the beginning of the batch to keep input moves smaller during that time. However, there is a strong argument to be made for instead using a single, constant penalty for all sampling instances of the batch. This argument is based on the desire to avoid introducing unnecessary time-indexed elements to the control design. For this reason, we have chosen to use the same penalty, $p_1 = 3$ for all sampling instances. (Note that if the system had multiple inputs a constant input penalty would instead be represented by repeating a vector $\boldsymbol{p_c}$ along the diagonal of the **P** matrix.)

In addition to the penalty on input moves, hard constraints were applied to the input at plus and minus 15°C from the nominal jacket temperature trajectory. These bounds simulate limitations on the process arising from either equipment constraints or operator requirements. These bounds were rarely reached in the presented case studies. However, further experiments designed to test the behavior of the system under significantly tighter input constraints demonstrated that the proposed SQMPC is able to propose input trajectories that respect the input constraints while providing improved quality. These results are omitted from explicit presentation in the chapter for brevity.

Remark 11.18 While the training data sets used in the present chapter are of unequal length, the closed-loop batches are picked to be of the same length. This is done because both LVMPC and the proposed SQMPC required knowledge of the batch duration (which could very well be different for each batch) before the batch starts, so the input trajectories for that duration can be computed. Thus, for fairness of comparison, both controllers are given the same batch length, and this value is kept the same for all closed-loop batches. In general, the batch duration could also be a decision variable in the proposed SQMPC framework. This possibility will be pursued in a future research study.

Implementation of the SQMPC was carried out in MATLAB™[1] using the built-in QP solver to calculate the input move at each sampling instant. The optimization problem in this chapter is relatively efficient to solve, and global optimality is guaranteed (see Remark 11.8 for details). For instance, for the regulation application, the solution of the optimization problem was solved to the global optimal in an

[1] ©2015 The MathWorks, Inc. MATLAB and Simulink are registered trademarks of the MathWorks, Inc. See www.mathworks.com/trademarks for a list of additional trademarks. Other product or brand names may be trademarks or registered trademarks of their respective holders.

average time of 0.0132 seconds (on a 2.40 GHz, Intel, Quad-core, 32-bit machine). The maximum solution time was 0.2207 seconds. Clearly, these values are far shorter than the 60-second sampling frequency meaning that the controller would be readily operable online.

All of the closed-loop results presented in this section were obtained by applying the various control approaches to batches initialized from the same 186 initial conditions. These initial conditions were drawn from the normal distributions used to generate the training data as previously described. The same set of initial conditions were used for each proposed method and over both case studies to enable unbiased performance comparison. The number of batches, 186, was chosen to be sufficiently large to provide statistically relevant results. The following two cases present these results.

Case 1: Regulatory Control

In the first case study, we consider the problem of rejecting disturbances in initial conditions to obtain the quality set-point for each batch. Specifically, the set-point in this case study is the quality from the nominal batch. The temperature trajectory for the nominal batch is the set-point tracked by the existing control. Furthermore, the initial condition for the nominal batch is the mean initial value for all subsequent batches. Therefore, the existing control is most likely to be effective in reaching this quality set-point. This choice of set-point is also a meaningful base case for LVMPC and SQMPC. The base open-loop input and closed-loop set-point trajectory in the training data for the LVMPC and SQMPC are taken from the nominal batch. As a result, driving to the nominal quality is the case that is least likely to cause extrapolation from the training data. For these reasons, this case can be thought of similarly to the regulation problem in continuous processes where the objective is to maintain operations at the original set-point in the face of disturbances.

Figure 11.1 shows the PI, LVMPC, and SQMPC input trajectories for a typical batch. The corresponding output trajectories for this batch are plotted in Fig. 11.2. The relative quality errors for the plotted batch are given in Table 11.3. Some of the notable characteristics of the outputs plotted in these figures are consistent with the general behavior of each of these control methods. First of all, notice that the SQMPC follows a smooth trend over the duration of the batch. This smoothness is partially resultant of the input movement penalty. However, even in cases where the input penalty was dramatically reduced, the input profile tends to remain smooth. Another characteristic of note is that the SQMPC input follows the same general trend as the PI throughout the beginning and middle of the batch up to around 2.4 h. At around 2.4 h, the SQMPC significantly diverges from the trend of both the LVMPC and PI inputs. This divergence occurs as quality predictions improve allowing the SQMPC to take the appropriate corrective action to account for deviations from the set-point.

Remark 11.19 SQMPC is able to obtain especially smooth inputs when compared to other data-driven control approaches. This is a direct result of the way the models are identified. When identifying time-indexed models, coefficients are identified for each sampling instant from observations at that sampling instant. Therefore, the

Table 11.3 Relative error in controlled qualities for characteristic batch plotted in Figs. 11.1 and 11.2

	Number ave. mol. wt. (%)	Weight ave. mol. wt. (%)	Conversion (%)
SQMPC	0.13	2.42	7.90
LVMPC	4.04	4.74	4.99
PI	8.97	8.41	1.76

number of observations available to identify each coefficient is equal to the number of training batches. This small number of observations reduces the ability to average out the impact of noise on the coefficients. Furthermore, the effect of the noise on coefficients is time dependent. When applied in closed loop, the modeled time-dependent noise may be amplified resulting in a jittery input trajectory. In contrast, in the proposed approach, the formation of the Hankel matrices allows a much larger number of observations to be used in identifying each model coefficient. As a result, noise from the training data can be effectively averaged out. More importantly, these coefficients are constant at every sampling instant. Therefore, any modeled noise affects the predictions consistently through time and does not induce a jittery input trajectory.

Figure 11.3 is a table of histograms showing the distributions of each of the quality variables for each of the control approaches under case 1. These results demonstrate the clear gains in control of the molecular weight distribution under the proposed SQMPC. Specifically, the distributions of both number and weight average molecular weight are narrowed when compared to the existing PI and LVMPC approaches. Furthermore, the distributions are centered around the quality set-point. The distribution of conversions suffers slightly under the proposed SQMPC but remains within acceptable bounds of approximately plus and minus ten percent. The results of this case study strongly support the efficacy of the proposed SQMPC for disturbance rejection. Note that we use only the simplest version of LVMPC for benchmarking purposes; improved performance with the LVMPC could very well be possible. The key recognition is the fact that comparable (and in this instance better) performance is achievable using the proposed approach that does not require the existence of any alignment variable as with the LVMPC.

Remark 11.20 While we were tuning the MPC, we found that there was an apparent trade-off obtaining between the desired conversion and the molecular weight distribution. Upon further investigation, we discovered that more than 99.9% of the variance in the quality measurements from the training data can be explained by two dimensions. While this result is not a conclusive proof, it strongly suggests that conversion and the molecular weight properties cannot be independently controlled. In designing the SQMPC, we made the decision to trade-off conversion in favor of reaching the desired molecular weight distribution. This is reflected in the tuning parameters where conversion has a significantly lower weighting. The decision is justifiable from an industrial perspective since the molecular weight distribution is the primary defining property of the product. This explains the reduced performance of the SQMPC in regulating the conversion.

Case 2: Set-point Tracking

In the previous case, we demonstrated the ability of the proposed SQMPC to reject disturbances and drive the process to the nominal quality set-point. In this case, we consider the added complexity of driving the process to a quality set-point that is markedly different from the qualities observed in the training data. This type of problem is of particular relevance industrially for processes where there is a desire to produce multiple grades of product. The set-point tracking ability demonstrated by the results in this case study also provides evidence for the theoretical integrity of the proposed method.

The only change from the previous case was the quality set-point of the controller. All other aspects of both the LVMPC and the SQMPC —including the initial batch conditions, tuning, models, and filtering—remained unchanged. The new set-point was selected from an outlier in a large database of historical operations. The reason for selecting an outlier, instead of inventing a quality set-point, was to ensure that the new quality set-point was physically reachable. The selected quality set-point was well outside the range of qualities observed in the training database.

Figure 11.8 is a scatter plot comparing closed-loop SQMPC qualities for the first 15 batches in case 1 with those achieved by changing the set-point in case 2.

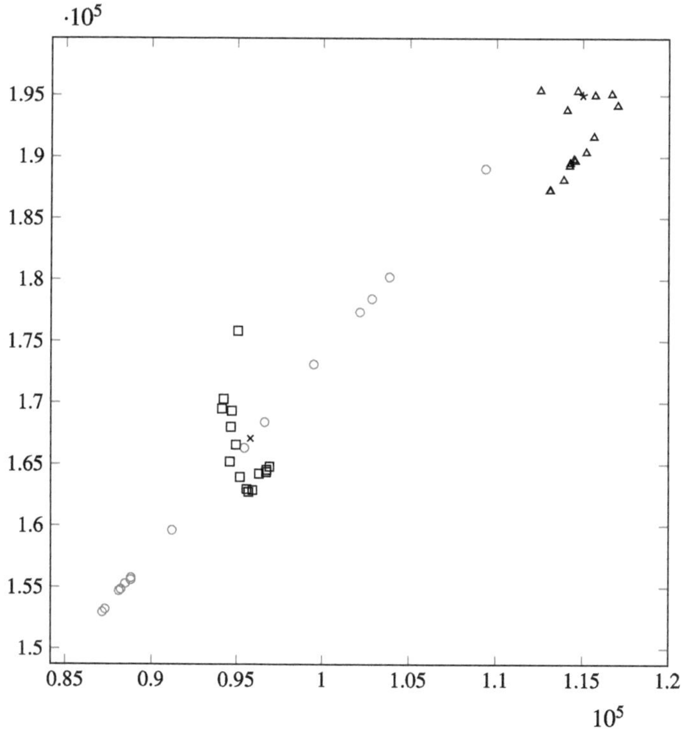

Fig. 11.8 Scatter plot comparing the PI qualities (○), SQMPC qualities for case 1 (□), and SQMPC qualities for case 2 (△) where the set-point for case 1 is the nominal quality (×) and the set-point for case 2 is an outlier from historical operation (⋆)

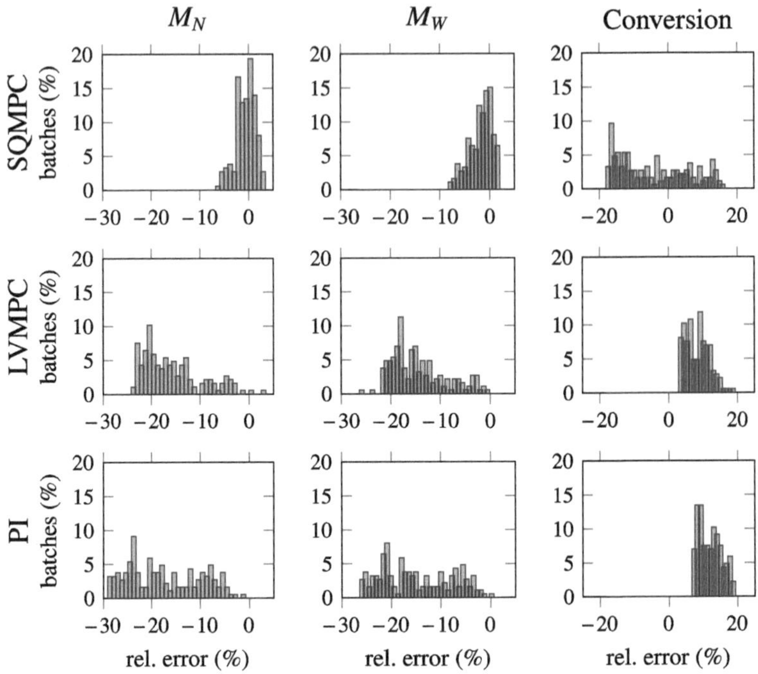

Fig. 11.9 Histograms of case 2 qualities resulting from 186 batches operated under SQMPC, LVMPC, and PI control

For reference, the corresponding qualities using the existing PI are also presented. Figure 11.9 shows the resulting quality distributions under SQMPC, LVMPC, and PI control for case 2. Note that since the existing PI control does not directly control quality, there is no change in the PI results from case 1 to case 2. As with case one, the distributions clearly demonstrate the efficacy of the proposed method. Of the three control designs (PI, LVMPC, and SQMPC), SQMPC is the only one with a distribution centered around the new quality set-point. In fact, none of the LVMPC batches reach the desired weight average molecular weight. Furthermore, while PI and LVMPC produce broad distributions, those of SQMPC are only slightly broader than in case 1. It is also noteworthy that, for this case, SQMPC also produces a conversion distribution that is broad but centered around the set-point.

Remark 11.21 The superior set-point tracking ability of the SQMPC results directly from the fact that the dynamic evolution of the batch is modeled based on state rather than time. When the operation of a batch differs from those in the training data, the process states deviate from the trajectories they took in the database. Fundamentally, the dynamics of the process are dependent on the process state (as described by Eq. (11.1)). For a time-indexed model, therefore, when the process is moved to a new process state trajectory, time-indexed models fail to capture the dynamic behavior.

11.7 Conclusions

In this chapter, we presented a novel data-driven model predictive control scheme for batch processes. The method adapts subspace identification techniques to identify a LTI, dynamic, state-space model from historical and identification batch data. The states of the identified system are then related to quality. The overall model relates process inputs and measurements in a causal manner to batch quality. The identified model is then applied in a model predictive control scheme. Simulation results for a batch PMMA reaction show the efficacy of the approach. Specifically, we are able to demonstrate favorable results when compared with other benchmark batch control schemes. The favorable results obtained by the proposed approach are explicable by the beneficial theoretical properties inherent to the method.

11.8 Appendix A: Identified PMMA Models

11.8.1 Dynamic Model (Subspace Identification)

$$\mathbf{A} = \begin{bmatrix} 0.98516 & -0.041768 & -0.11588 & -0.081012 & 0.14133 & 0.027337 & 0.065944 & -0.008974 & 0.026833 \\ -0.053849 & 0.79861 & 0.063273 & -0.074642 & 0.41775 & 0.21953 & 0.16301 & 0.074074 & 0.0076323 \\ 0.014815 & 0.053163 & 0.98982 & 0.039297 & -0.078398 & 0.084098 & 0.075658 & -0.01062 & -0.015769 \\ -0.0084043 & -0.029059 & 0.010815 & 0.96612 & 0.11911 & -0.033243 & 0.078468 & 0.039254 & 0.0071832 \\ -0.005217 & -0.019198 & 0.0048115 & -0.011641 & 0.8878 & -0.080295 & -0.030649 & -0.13453 & 0.1784 \\ -0.001108 & -0.0040268 & 0.00085731 & -0.004503 & -0.03083 & 0.93209 & -0.038058 & -0.12716 & 0.052801 \\ -0.00041708 & -0.001586 & -0.00040162 & -0.00030684 & 0.1618 & 0.086359 & 0.80687 & -0.077694 & -0.20116 \\ -0.006212 & -0.022836 & 0.004279 & -0.011486 & -0.076859 & -0.10873 & -0.20005 & 0.18359 & 0.014277 \\ 0.0037065 & 0.013566 & -0.002334 & 0.0087541 & -0.13833 & -0.0099258 & 0.39496 & 0.66942 & 0.4205 \end{bmatrix}$$

$$\lambda_A = \begin{bmatrix} 0.2224 \\ 0.56704 + 0.19029i \\ 0.56704 - 0.19029i \\ 0.80866 \\ 0.86491 \\ 0.99958 + 0.0039544i \\ 0.99958 - 0.0039544i \\ 0.98534 \\ 0.95601 \end{bmatrix}$$

$$\mathbf{B} = \begin{bmatrix} -0.058874 \\ -0.20903 \\ 0.057585 \\ -0.032671 \\ -0.020277 \\ -0.004307 \\ -0.0016209 \\ -0.024143 \\ 0.014406 \end{bmatrix}$$

$$\mathbf{C} = \begin{bmatrix} -0.26209 & -0.97571 & 0.49382 & -0.081618 & 0.077041 & 0.15654 & 0.3934 & 0.34012 & -0.064527 \\ 0.003204 & 0.16186 & 0.23888 & -0.63143 & 0.010558 & -0.1874 & -0.030857 & 0.00037325 & -0.01769 \\ -0.72052 & 0.17384 & 0.02821 & 0.024757 & 0.21978 & 0.045767 & 0.094573 & -0.051717 & 0.00075872 \end{bmatrix}$$

11.8.2 Quality Model

$$\mathbf{Q} = \begin{bmatrix} -0.0043689 & -0.016241 & -0.073542 & 0.038459 \\ 1458.7852 & 12206.803 & 21917.8887 & -41581.0424 \\ 2432.0509 & 19726.6489 & 33292.0309 & -72136.2077 \\ -0.00039146 & -0.021493 & -0.056014 & 0.03811 \end{bmatrix}$$

$$c_k = \begin{bmatrix} -5.0845 \\ 1985711.0276 \\ 3323464.3048 \\ 1.3826 \end{bmatrix}$$

11.9 Appendix B: Tuning Parameters for Benchmark Control

11.9.1 PI

$$k_c = -0.1$$

$$\tau_I = 30$$

11.9.2 LVMPC

Number of principal components = 5

$$\mathbf{Q}_1 = \begin{bmatrix} 0.01 & 0 & 0 & 0 \\ 0 & 0.01 & 0 & 0 \\ 0 & 0 & 0.01 & 0 \\ 0 & 0 & 0 & 0 \end{bmatrix}$$

$$\mathbf{Q}_2 = \begin{bmatrix} 10 & 0 & 0 & 0 & 0 \\ 0 & 10 & 0 & 0 & 0 \\ 0 & 0 & 10 & 0 & 0 \\ 0 & 0 & 0 & 10 & 0 \\ 0 & 0 & 0 & 0 & 10 \end{bmatrix}$$

$$\lambda = 100;$$

$$|\Delta t| \leq 0.5.$$

References

1. Corbett, B., Macdonald, B., Mhaskar, P.: Model predictive quality control of polymethyl methacrylate. IEEE Trans. Control Syst. Technol. **23**, 687–692 (2015)
2. VanOverschee, P., DeMoor, B.: N4SID - Subspace algorithms for the identification of combined deterministic stochastic-systems. Automatica **30**, 75–93 (1994)
3. Moonen, M., Demoor, B., Vandenberghe, L., Vandewalle, J.: Online and off-line identification of linear state-space models. Int. J. Control **49**, 219–232 (1989)
4. Shi, R., MacGregor, J.: Modeling of dynamic systems using latent variable and subspace methods. J. Chemom. **14**, 423–439 (2000). 6th Scandinavian Symposium on Chemometrics, PORSGRUNN, NORWAY, 15–19 AUG 1999
5. Overschee, Pv: Subspace Identification for Linear Systems: Theory, Implementation Applications. Kluwer Academic Publishers, Boston (1996)
6. Yao, Y., Gao, F.: Subspace identification for two-dimensional dynamic batch process statistical monitoring. Chem. Eng. Sci. **63**, 3411–3418 (2008)
7. Dorsey, A., Lee, J.: Building inferential prediction models of batch processes using subspace identification. J. Process Control **13**, 397–406 (2003)
8. Flores-Cerrillo, J., MacGregor, J.F.: Control of batch product quality by trajectory manipulation using latent variable models. J. Process Control **14**, 539–553 (2004)
9. Kassidas, A., MacGregor, J., Taylor, P.: Synchronization of batch trajectories using dynamic time warping. AIChE J. **44**, 864–875 (1998)
10. Nomikos, P., Macgregor, J.: Multivariate SPC charts for monitoring batch processes. Technometrics **37**, 41–59 (1995)
11. Kourti, T., Nomikos, P., Macgregor, J.: Analysis, monitoring and fault-diagnosis of batch processes using multiblock and multiway PLS. J. Process Control **5**, 277–284 (1995). IFAC Symposium on Advanced Control of Chemical Processes (ADCHEM 94), KYOTO, JAPAN, 25–27 MAY 1994

12. Kourti, T., Lee, J., MacGregor, J.F.: Experiences with industrial applications of projection methods for multivariate statistical process control. Comput. Chem. Eng. **20**, S745–S750 (1996)
13. Neogi, D., Schlags, C.: Multivariate statistical analysis of an emulsion batch process. Ind. Eng. Chem. Res. **37**, 3971–3979 (1998)
14. Viberg, M.: Subspace-based methods for the identification of linear time-invariant systems. Automatica **31**, 1835–1851 (1995)
15. Qin, S.J.: An overview of subspace identification. Comput. Chem. Eng. **30**, 1502–1513 (2006). 7th International Conference on Chemical Process Control (CPC 7), Lake Louise, Canada, 08–13 Jan 2006
16. VanOverschee, P., DeMoor, B.: A unifying theorem for three subspace system identification algorithms. Automatica **31**, 1853–1864 (1995). 10th IFAC Symposium on System Identification, Copenhagen, Denmark, 04–06 JULY 1994
17. Ekpo, E.E., Mujtaba, I.M.: Evaluation of neural networks-based controllers in batch polymerisation of methyl methacrylate. Neurocomputing **71**, 1401–1412 (2008)
18. Fan, S., Gretton-Watson, S., Steinke, J., Alpay, E.: Polymerisation of methyl methacrylate in a pilot-scale tubular reactor: modelling and experimental studies. Chem. Eng. Sci. **58**, 2479–2490 (2003)
19. Rho, H., Huh, Y., Rhee, H.: Application of adaptive model-predictive control to a batch MMA polymerization reactor. Chem. Eng. Sci. **53**, 3729–3739 (1998)
20. Flores-Cerrillo, J., MacGregor, J.: Latent variable MPC for trajectory tracking in batch processes. J. Process Control **15**, 651–663 (2005)
21. Flores-Cerrillo, J., MacGregor, J.: Iterative learning control for final batch product quality using partial least squares models. Ind. Eng. Chem. Res. **44**, 9146–9155 (2005)
22. Golshan, M., MacGregor, J.F., Bruwer, M.-J., Mhaskar, P.: Latent variable MPC for trajectory tracking in batch processes: role of the model structure. In: Proceedings of the American Control Conference (345 E 47TH ST, New York, NY 10017 USA), vol. 1–6, pp. 4779–4784. IEEE (2009). American Control Conference 2009, St Louis, MO, 10–12 JUNE 2009
23. Golshan, M., MacGregor, J.F., Bruwer, M.-J., Mhaskar, P.: Latent variable model predictive control (LV-MPC) for trajectory tracking in batch processes. J. Process Control **20**, 538–550 (2010)
24. Golshan, M., MacGregor, J.F., Mhaskar, P.: Latent variable model predictive control for trajectory tracking in batch processes: alternative modeling approaches. J. Process Control **21**, 1345–1358 (2011)
25. Nelson, P., Taylor, P., MacGregor, J.: Missing data methods in PCA and PLS: score calculations with incomplete observations. Chemom. Intell. Lab. Syst. **35**, 45–65 (1996)
26. Arteaga, F., Ferrer, A.: Dealing with missing data in MSPC: several methods, different interpretations, some examples. J. Chemom. **16**, 408–418 (2002). 7th Scandinavian Symposium on Chemometrics, Copenhagen, Denmark, 19–23 AUGUST 2001
27. Flores-Cerrillo, J., MacGregor, J.F.: Control of particle size distributions in emulsion semibatch polymerization using mid-course correction policies. Ind. Eng. Chem. Res. **41**(7), 1805–1814 (2002)
28. Verhaegen, M.: Application of a subspace model identification technique to identify LTI systems operating in closed-loop. Automatica **29**, 1027–1040 (1993)
29. Katayama, T.: Subspace identification of closed-loop systems. In: SICE 2002: Proceedings of the 41st SICE Annual Conference (345 E 47TH ST, NEW YORK, NY 10017 USA), vol. 1–5, pp. 1517–1520, Soc Instrument & Control Engineers, IEEE (2002). 41st Annual Conference of the Society-of-Instrument-and-Control-Engineers (SICE 2002), OSAKA, JAPAN, 05–07 AUGUST 2002
30. Veen, G., Wingerden, J.-W., Bergamasco, M., Lovera, M., Verhaegen, M.: Closed-loop subspace identification methods: an overview IET. Control Theory Appl. **7**, 1339–1358 (2013)

Chapter 12
Handling Multi-rate and Missing Data in Variable Duration Economic Model Predictive Control of Batch Processes with Application to Electric Arc Furnace Operation

12.1 Introduction

In the previous chapters, the subspace model identification approach was tailored for batch process model identification and control. The fundamental operation in subspace model identification is projection, which may emanate from prudent numerical techniques like singular value decomposition (SVD) or even QR factorization [1–6]. However, an adverse consequence of their convenience is that the algorithms are incapable of directly handling the multi-rate nature of the measurement availability, missing data, and incongruous batch process data. Although recent work has studied the problem of system identification of linear time-invariant models under nonideal sampling conditions with missing input and output data, the resulting approaches typically involve solving complicated optimization problems such as nuclear norm-based structural rank minimization [6, 7] and maximum likelihood estimation (MLE) through expectation–maximization algorithms [8]. The area of batch process control thus stands to gain from a computationally tractable alternative that retains the strengths of the subspace identification modeling approach.

Predictive control algorithms that make explicit use of a (deterministic first-principles-based or empirical data-driven) process model in the optimization of a cost function to obtain the optimal control signal have been widely proposed and successfully implemented in the process industries. One excellent approach for the application of MPC in regulating batch processes utilizes time-indexed latent variable methods for trajectory tracking, where the correlation between subsequent measurement samples is employed to predict the dynamic evolution of the process through the use of a missing data algorithm [9, 10]. As stated earlier, the time-indexed model necessitates that all batches are of equal duration, which is often enforced through data alignment algorithms. While possible for monitoring, alignment of variables for an ongoing batch remains an unaddressed problem. Another method is based on tracking the necessary conditions of optimality, where an optimality criteria-based parameterization of the input profiles is used to design a multivariable feedback scheme along with model adaptation to track the first-order optimality conditions,

P. Mhaskar et al., *Modeling and Control of Batch Processes*,
Advances in Industrial Control, https://doi.org/10.1007/978-3-030-04140-3_12

thereby driving the system toward optimality [11, 12]. More recently, the concept of reachability regions is used to implement model predictive control strategies where the controller, rather than driving the process to the desired end-point at all computation times, guides the process through a set of precomputed states that ensure the end-point is satisfied at batch termination [13, 14]. Existing batch control approaches, especially those based on data-driven models, have thus far not utilized batch duration as a decision variable.

An energy-intensive batch process where both the challenge of multi-rate and missing data, and the opportunity of variable duration control are exemplified is the electric arc furnace (EAF) process that is widely employed in steelmaking, using electricity to melt postconsumer scrap metal to produce new steel. The process provides significant reductions in the labor, energy, and environmental costs of steelmaking over conventional blast furnaces that forge steel out of iron ore, coal, and natural gas. Despite these efficiency gains, the EAF process requires generating tremendous amounts of heat to recycle the scrap steel, while the industrial sector struggles with electricity costs that are exorbitantly high [15]. The energy efficiency and economics of the EAF process can hence be improved by leveraging new technologies in optimization and advanced control systems. However, the harsh environment and high corrosiveness of molten steel mean that online measurements of the molten steel temperature and chemical composition are often disrupted with unmeasured process variables and missing data, thus making many prior existing modeling and control approaches inapplicable. Note that the control objective of the arc furnace, analogous to other batch processes, deals with the effective allocation of a set of limited resources over a finite time duration [16, 17]. When considering competing criteria and from an industrial perspective where the prevailing incentive is of an economic nature (often stated in terms such as cost, revenue, profit or investment payback time or rate of return), the multi-objective optimization problem for the EAF process is to economically reach the desired product end-point target at the termination of the batch. While recent contributions have demonstrated the advantages of economic model predictive control of the EAF process under the assumption that all batches are of equal durations [18], using a data-driven model that is capable of handling variable batch lengths opens new and beneficial possibilities of optimizing the operation of batches with variable durations.

Motivated by the above considerations, in this chapter, we develop a system identification method that handles multi-rate and missing data for variable duration economic model predictive control (EMPC) and apply it to the EAF process. The remainder of the article is organized as follows. In the next section, we briefly define the class of multi-rate sampled batch processes considered in this chapter and review the conventional subspace-based system identification technique and representative MPC approach for completeness. Following the preliminaries, we present the novel subspace-based system identification approach for identifying a dynamic model with incomplete measurements and missing data. The system identification method is based on a subspace formulation and uses the incremental singular value decomposition approach, as opposed to conventional SVD, to compute a state realization and identify a dynamic model of the process from measurements complicated by missing and multi-rate data. In the subsequent section, the resulting dynamic model is integrated into a tiered economic model predictive control formulation for optimizing process economics with varying batch durations, where solutions to computationally tractable mixed-integer quadratic programming problems achieve the desired final product end-point specifications by batch termination while minimizing the operating costs. Following the proposed multi-rate modeling and variable duration EMPC formulation, we describe the electric arc furnace process and present the model development and validation simulation results. Furthermore, we present the corresponding closed-loop simulation results that illustrate the use and effectiveness of the proposed multi-rate subspace-based system identification technique and EMPC framework through implementation on the EAF process, subject to the limited availability of process measurements, missing data, measurement noise, and constraints. Finally, we conclude this article with some remarks in the last section.

12.2 Preliminaries

In this section, we first provide a brief description of the general class of batch processes that are considered in the study. Then, we succinctly review the conventional subspace-based system identification and economic model predictive control approach for batch processes.

Problem Statement

In this chapter, we consider a class of multi-rate sampled batch processes, such as the EAF process (Fig. 12.1), that have measurements available at different sampling frequencies. For notational convenience, consider that the variables $y^{\phi_1} \in \mathbb{R}^{p^{\phi_1}}$ and $y^{\phi_2} \in \mathbb{R}^{p^{\phi_2}}$ denote the vector of frequent and infrequent measurements, respectively, $u \in \mathbb{R}^m$ denotes the vector of constrained input variables, taking values in a nonempty convex set $\mathcal{U} \subseteq \mathbb{R}^m$ with $\mathcal{U} := \{u \in \mathbb{R}^m : u_{\min} \leq u \leq u_{\max}\}$, and $u_{\min} \in \mathbb{R}^m$ and $u_{\max} \in \mathbb{R}^m$ denote the lower and upper bounds on the manipulated input, respectively. The sets $\mathbb{K}^{\phi_1}$ and $\mathbb{K}^{\phi_2}$ are the sampling instances where either the frequent or infrequent measurements are available. Note that the infrequent measurements typically also include all the output variables available at the frequent sampling instances. The specific problem we address in the manuscript is the subspace-based model identification for batch processes that handles multi-rate and missing data and variable duration batch control.

Subspace Identification

In this subsection, we briefly review the conventional subspace-based state-space system identification [2, 4] methods used to determine the system matrices of a discrete time state-space model using synchronous data, where a model of the following form is identified:

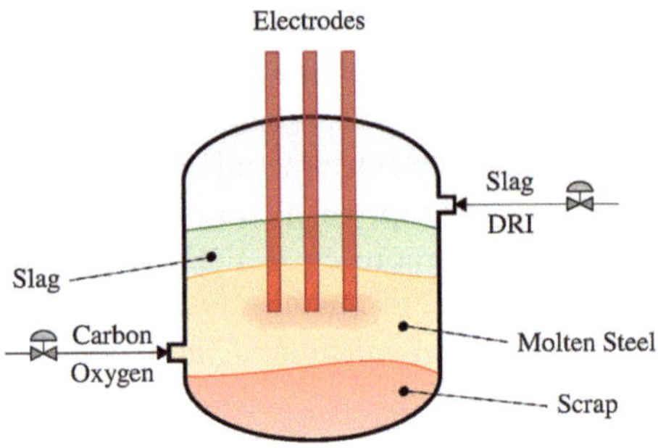

Fig. 12.1 An illustrative diagram of the EAF process

$$\begin{aligned} x_{k+1} &= Ax_k + Bu_k \\ y_k &= Cx_k + Du_k \end{aligned} \tag{12.1}$$

The subspace-based system identification techniques utilize Hankel matrices constructed from the process measurements and manipulated inputs. To establish these Hankel matrices, define a $pi \times 1$ vector of stacked output measurements as

$$y_{k|i} = \left[y_k^T \; y_{k+1}^T \; \cdots \; y_{k+i-1}^T \right]^T \tag{12.2}$$

where i is a user-specified parameter greater than the observability index or, for simplicity, the system order n. Similarly, define a vector of stacked input variables as $u_{k|i}$. Through the repeated iterative application of the state equations (Eq. (12.1)), it is straightforward to verify the expression for the stacked quantities:

$$y_{k|i} = \Gamma_i x_k + \Phi_i u_{k|i} \tag{12.3}$$

where:

$$\Gamma_i = \begin{bmatrix} C \\ CA \\ \vdots \\ CA^{i-1} \end{bmatrix}$$

$$\Phi_i = \begin{bmatrix} D & 0 & \cdots & 0 \\ CB & D & \cdots & 0 \\ \vdots & \vdots & \ddots & \vdots \\ CA^{i-2}B & CA^{i-3}B & \cdots & D \end{bmatrix}$$

Consider the block Hankel matrix for the outputs

$$Y_i = \left[y_{1|i} \; y_{2|i} \; \cdots \; y_{n_s|i} \right]$$

and similarly define U_i as a block Hankel matrix of inputs [3–6]. Then, it is clearly evident that Y_i is given by

$$Y_i = \Gamma_i X_i + \Phi_i U_i \tag{12.4}$$

where

$$X_i = \begin{bmatrix} x_1 \; x_2 \; \cdots \; x_{n_s} \end{bmatrix}$$

The next step in the system identification methods is to estimate either the extended observability matrix or the state sequence, followed by computing the system matrices [2, 19]. We first review the conventional approach where the extended observability matrix is identified to estimate the system matrices, and subsequently, we provide a brief overview of the later approach where the state sequence is identified to estimate the model.

Extended Observability Matrix Approach

The basic underlying idea of some common system identification methods is the orthogonal projection matrix on the null space of U_i as

$$\Pi^{\perp}_{U_i^T} = I - U_i^T \left(U_i U_i^T\right)^{-1} U_i \tag{12.5}$$

Multiplying Eq. (12.4) by the projection matrix $\Pi^{\perp}_{U_i^T}$ yields

$$Y_i \Pi^{\perp}_{U_i^T} = \Gamma_i X_i \Pi^{\perp}_{U_i^T} \tag{12.6}$$

where $Y_i \Pi^{\perp}_{U_i^T}$ can be readily computed using prudent numeral algorithms such as LQ factorization. An efficient implementation of this scheme is the multivariable output error state-space (MOESP) algorithm, where an estimate of Γ_i is obtained through the dominant left singular vectors of $Y_i \Pi^{\perp}_{U_i^T}$. Moreover, numerous variations of this approach (for instance, multiplying the matrix $Y_i \Pi^{\perp}_{U_i^T}$ with instrumental variables and/or nonsingular weight matrices before computing the SVD) are proposed to improve the consistency of the estimate. Four major variants of this method are past outputs MOESP (PO-MOESP), numerical algorithms for subspace state-space system identification (N4SID), instrumental variable method (IVM), and canonical variate analysis (CVA) approach, which differ by the choice of weight matrices that pre- and post-multiply the matrix $Y_i \Pi^{\perp}_{U_i^T} \Psi^T$, where $\Psi = \begin{bmatrix} U_j^T \; Y_j^T \end{bmatrix}^T \in \mathbb{R}^{2j \times n_s}$ is the instrumental variable matrix constructed by combining the Hankel matrices of 'past' inputs and outputs [3–6]. Once an estimate of Γ_i has been determined from the dominant left singular vectors of $W_1 Y_i \Pi^{\perp}_{U_i^T} \Psi^T W_2$, where W_1 and W_2 denote the weight matrices, a system realization can be calculated in a multitude of ways. One approach for computing the system realization involves retrieving estimates of system matrices A and C from Γ_i, while estimates of B and D can be computed by solving a least squares problem [4, 6].

The key step in the classical subspace methods described above is the SVD of the matrix $W_1 Y_i \Pi^{\perp}_{U_i^T} \Psi^T W_2$ to estimate the extended observability matrix (Γ_i). However, the rank of the matrix Γ_i may not be equal to the order of the system, a phenomenon known as rank cancelation. Furthermore, the classical system identification approaches are limited to measurements sampled at the same frequency and cannot readily handle missing or unmeasured data [6, 7]. Consequently, in this chapter, we use an alternative subspace approach that first estimates the state sequence and then computes the system matrices in order to model the multi-rate sampled batch processes.

State Sequence Approach

We now review one particular example of a system identification approach that first estimates the matrix of states [1]. In this approach, the input and output data are divided into 'past' and 'future' measurements as follows

$$U_{2i} = \begin{bmatrix} U_p \\ U_f \end{bmatrix}, \quad Y_{2i} = \begin{bmatrix} Y_p \\ Y_f \end{bmatrix}$$

where the matrices $U_p \in \mathbb{R}^{mi \times n_s}$ and $U_f \in \mathbb{R}^{mi \times n_s}$ are referred to as the past and future input Hankel matrices, and $Y_p \in \mathbb{R}^{pi \times n_s}$ and $Y_f \in \mathbb{R}^{pi \times n_s}$ are referred to as the past and future output Hankel matrices, respectively. It is readily shown that Y_f and Y_p are given by the relations

$$\begin{aligned} Y_f &= \Gamma_i X_f + \Phi_i U_f \\ Y_p &= \Gamma_i X_p + \Phi_i U_p \end{aligned}$$

where X_p and X_f are the past and future states that are yet to be identified. One approach to estimating the state sequence involves computing the intersection between the past and future data given by

$$\text{span}\left(X_f\right) :- \text{row-space}\left(\begin{bmatrix} Y_f \\ U_f \end{bmatrix}\right) \cap \text{row-space}\left(\begin{bmatrix} Y_p \\ U_p \end{bmatrix}\right)$$

which provides an estimate of the future states X_f [1]. Once an estimate of X_f is known, it is straightforward to calculate a system realization through solving a least squares problem since X_f, Y_f, and U_f are all known. More details on the derivation of this result and the application of this system identification approach to batch processes are provided in the next section.

Beyond the noniterative subspace-based state-space system identification techniques that are readily implemented using efficient algorithms such as SVD and QR decomposition, other system identification techniques for modeling uniformly sampled processes include the maximum likelihood estimation approach and the closely related prediction error methods (PEMs). These MLE and PEM approaches solve a (possibly) nonconvex optimization problem to identify a system realization. Furthermore, they have well-established theoretical properties including asymptotically achieving the Cramér–Rao lower bound and extensively studied theoretical underpinnings for practical considerations such as variance and bias distributions [8, 20, 21]. Their theoretical advantages in addressing important practical issues such as error analysis and performance trade-offs notwithstanding, the practical implementation of MLE methods is not always straightforward due to the embedded nonconvex optimization problem (which may be poorly conditioned as a result of the chosen model parametrization) [4, 8, 20]. Moreover, the identification approach involving MLE is more meaningful for continuous processes because sufficient data collected around a nominal operating point enables the estimators to achieve consistency and asymptotic efficiency [8, 20]. In contrast, batch processes are typically not operated around an equilibrium operating point and a sufficient quantity of data may not be available for the finite duration of batch processes, making consistency and efficiency difficult to achieve under the limited data conditions of batch processes.

The multi-rate model structure for batch processes that we identify in this chapter is similar to the form considered Raghavan et al. [20] in the context of continuous processes through maximum likelihood estimation. In this chapter, we extend the multi-rate modeling problem to the application of batch processes through subspace-based system identification techniques that avoid solving complex and possibly nonconvex optimization problems. Although the estimate of the subspace in these system identification algorithms, as with the proposed multi-rate system identification approach, is correct and consistent as the number of data samples tends to infinity, for a finite number of data samples, as is the case in batch data where a sufficient quantity of data may not be available, there is no guarantee that the estimated subspace will be consistent [6, 7]. Furthermore, the multi-rate state-space realization obtained through the subspace-based system identification techniques can be employed as an initialization for a batch-specific MLE formulation to obtain a solution with possibly strong theoretical properties including consistency and asymptotic efficiency.

Model Predictive Control

In this subsection, we provide a standard formulation of a representative MPC to illustrate the paradigm of end-point control for batch processes. In a typical batch MPC, the control action at each sampling instance is computed by solving an optimization problem of the following form:

$$\min_{u \in \mathcal{U}} \quad \mathcal{J} = \delta \cdot \sum_{i=k}^{n_t - 1} c^T u_i \tag{12.7}$$

$$\text{s.t.} \quad \begin{cases} x_{k+1} = A x_k + B u_k \\ \hat{y}_k = C x_k + D u_k \\ \hat{y} \in \mathcal{Y} \\ \hat{y}_{n_t} = y_{\text{des}} \end{cases}$$

where $u \in \mathbb{R}^m$ denotes the vector of constrained input variables, taking values in a nonempty convex set $\mathcal{U} \subseteq \mathbb{R}^m$. The objective function minimizes the costs of the manipulated inputs, thus penalizing resource usage. While the end-point-based economic model predictive controllers have demonstrated improved economic operation in various applications, the control paradigm typically does not consider batches of varying durations. This is due to the fact that the model utilized in the above predictive control framework often originates from system identification techniques that cannot handle multi-rate sampled measurements and missing data.

12.3 Multi-rate System Identification with Missing Data

In this section, we propose a novel multi-rate subspace-based system identification algorithm for batch processes that uses the incremental singular value decomposition (iSVD) to compute a realization of the state sequence since regular SVD is inapplicable in subspace-based system identification from Hankel matrices composed of multi-rate measurements with missing data. Recognizing the multi-rate nature of the data, we pose the problem as one of identifying a discrete time, linear time-invariant state-space model of the following form:

$$\begin{aligned} x_{k+1} &= Ax_k + Bu_k \\ \hat{y}_k^{\phi_1} &= C^{\phi_1} x_k + D^{\phi_1} u_k, \ \forall k \in \mathbb{K}^{\phi_1} \\ \hat{y}_k^{\phi_2} &= C^{\phi_2} x_k + D^{\phi_2} u_k, \ \forall k \in \mathbb{K}^{\phi_2} \end{aligned} \tag{12.8}$$

where $x \in \mathbb{R}^n$ denotes the state vector of the identified model. Consider that there are n_b batches of varying durations and define a vector of stacked multi-rate sampled output measurements as

$$y_{k|i} = \begin{bmatrix} {y_k^{\phi_2}}^T & {y_{k+1}^{\phi_1}}^T & \cdots & {y_{k+i-1}^{\phi_1}}^T \end{bmatrix}^T \tag{12.9}$$

Then for an arbitrary batch b of the n_b total batches, with $n_s^b + 2i - 1$ input and output data samples, we develop the output Hankel matrices for the bth batch as follows

$$Y_f^b = \begin{bmatrix} y_{i+1|i} \ y_{i+2|i} & \cdots & y_{i+n_s^b|i} \end{bmatrix} \tag{12.10}$$

$$Y_p^b = \begin{bmatrix} y_{1|i} \ y_{2|i} & \cdots & y_{n_s^b|i} \end{bmatrix} \tag{12.11}$$

where i is a user-specified parameter that is greater than the observability index or the system order. It should be noted that $y_{k|i}$ in this case is composed of vectors of stacked output measurements that include both the infrequent and frequent measurements, as shown in Eq. (12.9). Each output Hankel matrix for an arbitrary batch (batch no. b of n_b total batches) is of size $pi \times n_s^b$ where the entries in the block Hankel matrices corresponding to unmeasured variables at the frequent sampling instances are left empty. The partitioning of the data into Y_p^b and Y_f^b is sometimes referred to as past and future. Define U_f^b and U_p^b as input block Hankel matrices similar to Eqs. (12.10) and (12.11), respectively. The individual block Hankel matrices of the various batches are then assembled together as

$$Y_f = \begin{bmatrix} Y_f^1 \; Y_f^2 \; \cdots \; Y_f^{n_b} \end{bmatrix} \tag{12.12}$$

and similarly to create Y_p, U_f, and U_p. It was proposed by Moonen et al. [1], who showed that a realization of the unknown system states can be obtained through computing the intersection of the past input–output and the future input–output spaces via the application of singular value decomposition, as we will show below.

From Moonen et al. [1], it is readily shown that Y_f and Y_p are given by the relations

$$Y_f = \Gamma_i X_f + \Phi_i U_f \tag{12.13}$$

$$Y_p = \Gamma_i X_p + \Phi_i U_p \tag{12.14}$$

where

$$X_f = \begin{bmatrix} x_{i+1}^{(1)} \; \cdots \; x_{i+n_s^{(1)}}^{(1)} \; x_{i+1}^{(2)} \; \cdots \; x_{i+n_s^{(2)}}^{(2)} \; x_{i+1}^{(n_b)} \; \cdots \; x_{i+n_s^{n_b}}^{(n_b)} \end{bmatrix}$$

$$X_p = \begin{bmatrix} x_1^{(1)} \; \cdots \; x_{n_s^{(1)}}^{(1)} \; x_1^{(2)} \; \cdots \; x_{n_s^{(2)}}^{(2)} \; x_1^{n_b} \; \cdots \; x_{n_s^{n_b}}^{(n_b)} \end{bmatrix}$$

Note that Γ_i and Φ_i are defined in terms of the full output measurement vector available at the infrequent sampling times using the multi-rate state-space model of Eq. (12.8) as

$$\Gamma_i = \begin{bmatrix} C^{\phi_1} \\ C^{\phi_1} A \\ \vdots \\ C^{\phi_1} A^{i-1} \end{bmatrix} \quad \Phi_i = \begin{bmatrix} D^{\phi_1} & 0 & \cdots & 0 \\ C^{\phi_1} B & D^{\phi_1} & \cdots & 0 \\ \vdots & \vdots & \ddots & \vdots \\ C^{\phi_1} A^{i-2} B & C^{\phi_1} A^{i-3} B & \cdots & D^{\phi_1} \end{bmatrix}$$

while keeping the missing data at the frequent sampling instances as empty entries in the block Hankel matrices. This ensures that the equations are properly specified and that the resulting state sequence is uninterrupted and congruous, a necessary requirement for estimating the system realization. Further, X_f can be related to X_p as follows

$$X_f = A^i X_p + \Delta_i U_p \tag{12.15}$$

where Δ_i is the reversed extended controllability matrix given by

$$\Delta_i = \begin{bmatrix} A^{i-1} B \; A^{i-2} B \; \cdots \; AB \; B \end{bmatrix}$$

Solving X_f in Eq. (12.13) yields

$$X_f = \left[\Gamma_i^{\dagger} \; -\Gamma_i^{\dagger}\Phi_i\right] \begin{bmatrix} Y_f \\ U_f \end{bmatrix} \tag{12.16}$$

where $\Gamma_i^{\dagger}$ denotes the Moore–Penrose matrix inverse of Γ_i. Equation (12.16) implies that the row-space of X_f is contained within the row-space of $\left[Y_f^T \; U_f^T\right]^T$. Similarly, solving X_p in Eq. (12.14) and substituting into Eq. (12.15) give

$$X_f = \left[A^i\Gamma_i^{\dagger} \; \Delta_i - A^i\Gamma_i^{\dagger}\Phi_i\right] \begin{bmatrix} Y_p \\ U_p \end{bmatrix} \tag{12.17}$$

which implies that the row-space of X_f is equally contained within the row-space of $\left[Y_p^T \; U_p^T\right]^T$. Therefore, the intersection between the past and future data is given by

$$\text{span}\left(X_f\right) :- \text{row-space}\left(\begin{bmatrix} Y_f \\ U_f \end{bmatrix}\right) \cap \text{row-space}\left(\begin{bmatrix} Y_p \\ U_p \end{bmatrix}\right) \tag{12.18}$$

For the present case of multi-rate sampled measurements as well as missing data, however, traditional SVD of the Hankel matrices cannot be determined because the Hankel matrices are composed of multi-rate sampled output measurements, and thus, this intersection cannot be computed using singular value decomposition of the two spaces as is done in traditional subspace identification to compute an estimate of the states X_f [1]. Therefore, in this chapter, we propose the use of incremental singular value decomposition algorithm to compute the intersection between Hankel matrices, and thus the estimate of the states X_f, where the Hankel matrices are composed of multi-rate sampled measurements with missing data.

Remark 12.1 The conventional subspace model, based on the underlying assumption that each measurement or data vector is a linear combination of a small number of principal component vectors or singular vectors, is widely used to build prediction models. In contrast to the conventional subspace models, the subspace-based state-space system identification approach considered in this chapter identifies a dynamic model from a subspace realization that also constitutes the state sequence underlying the observed data. Consequently, a dynamic state-space model is identified in this chapter for the purposes of predictive control.

Remark 12.2 Alternative approaches for handling multi-rate sampled data as missing values include interpolation techniques that impute values for the missing variables. The prediction capability of such missing data interpolation techniques is inherently predicated on the efficiency of the underlying model used to infer the unknown values. Moreover, the selection of a specific type of model to infer the values of the unknown data is not trivial and is the whole point of the modeling exercise. Thus, filling these unobserved data and then determining a subspace model are inconsistent with the notion of estimating a linear time-invariant subspace model. Furthermore, to simply down-sample the measurements to a common sampling frequency means that the frequent measurement samples available between infrequent sampling instances are discarded. Thus, the relevant process information that could be used to identify a model and to implement frequent feedback control anachronisms is neglected. In contrast, the proposed approach uses all available information as best as possible.

In this chapter, in order to calculate the intersection between two spaces that contain many missing variables, we employ the use of an incremental singular value decomposition [22, 23]. We will briefly review the key equations involved in the incremental SVD algorithm for completeness. Although the iSVD algorithm has been used to compute the principal components or singular values by identifying a subspace of low dimension given a data set consisting of missing values, the approach has not been utilized to identify a dynamic state-space model from the data. In the proposed approach, after the iSVD is used to compute the intersection between the subspaces of the past and future Hankel matrices, as shown in Eq. (12.18), the regular singular value decomposition is applied to the resulting intersection space to compute the state sequence.

To deal with the missing data in an arbitrary vector $v_t \in \mathbb{R}^{n_v}$ of a matrix $[v_1\ v_2\ \ldots\ v_t \ldots v_T]$ to be factorized using SVD in which only the components indicated by the set $\Omega_t \subset \{1, \ldots, n_v\}$ are measured or known, we observe the following subvector at iteration t:

$$(v_t)_{\Omega_t} = \left(\bar{U} w_t\right)_{\Omega_t} \tag{12.19}$$

where w_t is a known weight vector, $\bar{U}$ is an orthonormal matrix, and the subscript Ω_t on a matrix or vector is used to indicate restriction to the rows indicated by Ω_t. Given initial SVD matrices U_t, Σ_t, and V_t at some sampling instant t, we can calculate the SVD of a vector v_t with missing data by computing w_t and r_t as the least squares weight and residual vector, respectively, defined with respect to only the set of observed indices Ω_t as follows

$$w_t = \arg\min_{w_t} \| (U_t w_t)_{\Omega_t} - (v_t)_{\Omega_t} \|_2^2$$

$$r_t = v_t - U_t w_t$$

Noting that

$$\left[U_t \Sigma_t V_t\ v_t\right] = \left[U_t\ \frac{r_t}{\| r_t \|}\right] \begin{bmatrix} \Sigma_t & w_t \\ 0 & \| r_t \| \end{bmatrix} \begin{bmatrix} V_t & 0 \\ 0 & 1 \end{bmatrix}^T \tag{12.20}$$

and computing the SVD of the updated matrix:

$$\begin{bmatrix} \Sigma_t & w_t \\ 0 & \| r_t \| \end{bmatrix} = \hat{U} \hat{\Sigma} \hat{V}^T \tag{12.21}$$

and setting

$$U_{t+1} = \begin{bmatrix} U_t & \frac{r_t}{\parallel r_t \parallel} \end{bmatrix} \hat{U} \tag{12.22}$$

$$\Sigma_{t+1} = \hat{\Sigma} \tag{12.23}$$

$$V_{t+1} = \begin{bmatrix} V_t & 0 \\ 0 & 1 \end{bmatrix} \hat{V} \tag{12.24}$$

while retaining only the $\tilde{n}$ largest singular values yields an update of the initial SVD, taking into account the new vector of measurements v_t, which had missing data within it. The procedure is then repeated in order to incorporate the next vector v_{t+1} in the incremental SVD factorization. Therefore, the missing data incremental SVD algorithm can be applied to compute an estimate of the states $\tilde{X}_i$ and ultimately the model matrices.

By definition, only the observable part of the system can be identified from the manipulated inputs and measured outputs; therefore, the multi-rate system $\left(A, \left[C^{\phi_1}, C^{\phi_2}\right]\right)$ is always observable. The order of the identified system $\tilde{n}$ is selected based on how well the validation data is predicted, and the system order is always chosen so that the system (A, B) is controllable [1, 3, 4].

The system identification procedure requires that the manipulated inputs are persistently exciting signals, which corresponds to the fact that the projection matrix $\Pi^{\perp}_{U_i^T}$ can be computed and that the matrix inverse in Eq. (12.5) exists. Nevertheless, the criteria for ensuring that the inverse exists amount to the input block Hankel matrix being full rank [1, 3, 4]. In this chapter, we consider this rank criterion to determine whether the inputs are persistently exciting, similar to the procedure in conventional system identification.

Remark 12.3 Note that the conventional subspace-based system identification procedure cannot handle missing values, and the standard model is built using the correct corresponding values of the inputs. In contrast, the proposed multi-rate subspace-based system identification techniques utilizes the incremental singular value decomposition algorithm to identify a state sequence and thus the state-space model parameters, while avoiding the use of missing data imputation. Since the incremental SVD algorithm computes the decomposition of matrix with arbitrarily missing values, the approach is not limited to just measurements sampled at different frequencies. The approach thus readily accommodates missing input values in the present application.

12.4 Variable Duration Economic Model Predictive Control

In this section, we propose a novel two-tiered economic model predictive control algorithm capable of handling batches of varying durations. The proposed economic model predictive control formulation employs a tiered framework to achieve product that satisfies the end-point target while optimizing an economics-based cost function. To this end, the first tier of the EMPC computes the optimal achievable end-point that satisfies the desired product specifications at each sampling instance by solving the following mixed-integer quadratic programming problem

$$\min_{u \in \mathcal{U},\, n_t} \quad \mathcal{J}_1 = \| \hat{y}_e - y_{\text{des}} \|_{\Xi}^2 \tag{12.25}$$

$$\text{s.t.} \begin{cases} x_{k+1} = Ax_k + Bu_k \\ \hat{y}_k^{\phi_1} = C^{\phi_1} x_k + D^{\phi_1} u_k, \ \forall k \in \mathbb{K}^{\phi_1} \\ \hat{y}_k^{\phi_2} = C^{\phi_2} x_k + D^{\phi_2} u_k, \ \forall k \in \mathbb{K}^{\phi_2} \\ \hat{y}^{\phi_1} \in \mathcal{Y}^{\phi_1}, \quad \hat{y}^{\phi_2} \in \mathcal{Y}^{\phi_2} \\ \hat{y}_e = g\left(\hat{y}_{n_t}^{\phi_1}\right) \end{cases}$$

where $\hat{y}_e = g\left(\hat{y}_{n_t}^{\phi_1}\right)$ defines the batch end-point characteristics through the estimates of the output measurement variables at batch termination (n_t), $u \in \mathbb{R}^m$ denotes the vector of constrained input variables, taking values in a nonempty convex set $\mathcal{U} \subseteq \mathbb{R}^m$, Ξ is a positive-definite matrix used to penalize the squared deviation of the predicted end-point variables with respect to the desired end-point values, $\mathcal{Y}^{\phi_1} \subseteq \mathbb{R}^{\phi_1}$ and $\mathcal{Y}^{\phi_2} \subseteq \mathbb{R}^{\phi_2}$ denote the constraints on the infrequent and frequent measurement variables, respectively, y_{des} denotes the desired end-point target specified by plant operators, and n_t is the batch duration.

Subsequent to obtaining the optimal achievable end-point $\hat{y}_e^*$ that best satisfies the desired product specifications y_{des}, the next tier computes the economically optimal trajectory of manipulated input variables, albeit constricted with an additional constraint to satisfy the optimal achievable end-point product specifications. The second tier of the EMPC solves a mixed-integer quadratic programming problem to determine the optimal manipulated input trajectory by minimizing an economic objective function at every sampling time as follows

$$\min_{u \in \mathcal{U},\, n_t} \quad \mathcal{J}_2 = \delta \cdot \sum_{i=k}^{n_t - 1} c^T u_i + c_{\Delta u}^T \Delta u_i^2 \tag{12.26}$$

$$\text{s.t.} \begin{cases} x_{k+1} = Ax_k + Bu_k \\ \hat{y}_k^{\phi_1} = C^{\phi_1} x_k + D^{\phi_1} u_k, \ \forall k \in \mathbb{K}^{\phi_1} \\ \hat{y}_k^{\phi_2} = C^{\phi_2} x_k + D^{\phi_2} u_k, \ \forall k \in \mathbb{K}^{\phi_2} \\ \hat{y}^{\phi_1} \in \mathcal{Y}^{\phi_1}, \quad \hat{y}^{\phi_2} \in \mathcal{Y}^{\phi_2} \\ h\left(\hat{y}_{n_t}^{\phi_1}, \hat{y}_e^*\right) \leq 0 \end{cases}$$

where $h\left(\hat{y}_{n_t}^{\phi_1}, \hat{y}_e^*\right) \leq 0$ denotes end-point constraints that are appropriately relaxed to an interval with $\hat{y}_e^*$ as a end-point restricting bound, δ is the sampling time or hold time for the control action, and $c \in \mathbb{R}^m$ and $c_{\Delta u} \in \mathbb{R}^m$ denote the operating costs associated with the manipulated variables and the rate of change of the manipulated input variables given by Δu.

In this chapter, we penalize the rate of change of the input variables through the term $c_{\Delta u}^T \Delta u_i^2$ in the objective function. Another approach to minimize the difference between successive inputs involves penalizing the absolute value of the input change as $c_{\Delta u}^T |\Delta u_i|$, thus reformulating the optimization of the second tier into a linear programming problem. Regardless of the specific formulation used, the rate of change of the input variables may need to be penalized in the proposed economic MPC to mitigate the abrupt input moves that arise as a result of the shorter batch durations preferred from a cost perspective.

Remark 12.4 There have been several recent results focusing on EMPC including those that allow a nonquadratic cost function where the assumption of positive definiteness of the cost is not satisfied [24, 25]. In order to ensure that the closed-loop system resulting from the application of an economic MPC algorithm does in fact converge to the optimal equilibrium operating point, dissipativity conditions have been employed that are shown to be sufficient for optimal steady-state operation of a system as well as for convergence and stability analysis [24–28]. The key feature that differentiates continuous processes from batch and semi-batch processes, such as the EAF process, is that continuous processes have an equilibrium operating point whereas batch processes are typically not run long enough to reach a steady state (and the steady state is typically not the desired target point anyway). This precludes the direct implementation of these EMPC techniques on batch processes and necessitates the development of batch-specific control algorithms.

Remark 12.5 Considerable effort has been devoted in recent years to improve the computational tractability of model-based optimization problems through the derivation of convex formulations and by employing algorithms that exploit the underlying structure of the problem [29, 30]. Other attempts to avoid the challenges of online computing have involved moving the heavy computations off-line, where time and computational power are much more abundant, thus limiting the online operations to expedited decisions and simpler calculations. One such approach is multi-parametric programming, which generates off-line a register of explicit control laws to be employed online based on the estimated states of the system [31]. Furthermore, advanced-step MPC strategies are proposed that solve the detailed optimization problem in the background between sampling instances, assuming the computations can be completed within one sampling time, and apply sensitivity-based updates online when measurements become available [32, 33]. Another approach is the real-time nonlinear iteration scheme, which uses a continuation Newton-type framework and solves one quadratic programming problem at each iteration, thus allowing for multiple active set changes and ensuring that the nonlinear MPC algorithm performs at least as well as a linear MPC approach [34]. Another method is based on tracking the necessary conditions of optimality, where an optimality criteria-based parameterization of the input profiles is used to design a multivariable feedback scheme along with model adaptation to track the first-order optimality conditions, thereby driving the system toward optimality [11, 12]. More recently, the concept of reachability regions is used to implement model predictive control strategies where the controller, rather than driving the process to the desired end-point at all computation times, guides the process through a set of precomputed states that ensure the end-point is satisfied at batch termination [13, 14]. Note that the proposed approach poses minimal online computational issues due to the fact that the underlying model is linear.

12.5 Application to the Electric Arc Furnace

In this section, we first describe the electric arc furnace process, followed by simulation results that demonstrate the efficiency of the proposed approach to identify models from multi-rate sampled data and missing measurements using the test bed EAF process. Finally, the identified multi-rate subspace-based model is integrated into the proposed economic model predictive control framework, and the closed-loop simulation results demonstrate the improvement in economic performance through the minimization of operating costs and optimization of batch durations.

Electric Arc Furnace Process

Integrated steel mills typically use the electric arc furnace process to produce steel predominantly from recycling postconsumer scrap and occasionally using supplementary iron sources such as direct reduced iron (DRI). The EAF process is a batch process, and a batch is referred to as a heat, with a duration time of about one to two hours. A heat of an electric arc furnace begins with scrap metal being placed inside the furnace. The furnace will generally have some molten steel in place from the preceding heat in order to aid in the melting of the scrap metal. Once the furnace is charged with the scrap metal, a high-intensity electric arc, which is typically the largest energy contributor in the melting operation, is discharged from electrodes into the furnace in order to melt the scrap metal. After a significant amount of scrap has been melted, oxygen gas and raw carbon are injected into the molten steel that react within the molten steel to create iron oxide and carbon monoxide. The injected oxygen and carbon also serve to foam slag on the surface of the molten steel through the diffusion of carbon monoxide, which aids in the removal of oxides and impurities from the molten steel. Furthermore, the foamed slag also serves as an insulator retaining heat, thus improving the efficiency of the steel production [15, 35]. Once the desired steel composition and temperature are obtained, the heat is tapped and the molten steel is transported to downstream operations for further processing.

In this study, we utilize the electric arc furnace process as an application example to demonstrate the efficacy of modeling and control of multi-rate sampled process measurements. The electric arc furnace operational challenges that we address in this study are the subspace-based identification of state-space models and the economically optimal control of the EAF process, subject to multi-rate and missing measurements. To this end, we utilize a first-principles mechanistic model as a test bed to implement and evaluate the efficiency of the proposed modeling and control approach. Specific details regarding the structure and configuration of the test bed mechanistic model can be found in previous studies [18] and are omitted here for brevity. Although the electric arc furnace produces significantly fewer infrequent measurements than what we consider in this case study, we utilize the electric arc furnace process mainly as an application example to demonstrate the efficacy of multi-rate subspace-based system identification algorithm. The more extreme infrequent sampling scenarios encountered in industrial practice are not in the scope of the multi-rate modeling problem considered in this chapter and will be studied in future research studies.

Recognizing the limited availability of process measurements in practice, in this chapter, the measurements available to build the data-driven model and to implement the proposed control approach include infrequent and frequent measurement variables, where the infrequent measurements related to the slag and molten steel are sampled twice as slowly as the frequent measurements corresponding to the off-gas composition that are available at each sampling instance. A list of the infrequent and frequent process measurement variables are given in Tables 12.1 and 12.2, respectively, and the manipulated inputs along with the associated costs are listed in Table 12.3. Furthermore, each historical heat has a duration of 70 min and a sampling time of $\delta = 1$ min for the frequent measurements. Additionally, the variability of the feed, measurement noise, and conventional decentralized trajectory tracking control schemes of the EAF process are kept consistent with previous studies [18]. In this chapter, the desired end-point attributes of the heat at batch termination are explicitly characterized through the melt temperature ($\mathrm{T}_{t_f} \geq 1890$ K), mass fraction of carbon in the molten steel ($x_{\mathrm{C},t_f} \leq 0.005$ kg/kg), and the mass fraction of iron oxide in the slag ($x_{\mathrm{FeO},t_f} \leq 0.378$ kg/kg). Moreover, path constraints that limit the maximum achievable temperature of the molten steel ($\mathrm{T}_t \leq 2000$ K, $t \in [0, t_f]$) and the relative pressure within EAFs ($\mathrm{P}_t \leq 0$ Pa, $t \in [0, t_f]$) for safety considerations are considered as well.

Electric Arc Furnace Modeling Results

A historical database of past heats is generated to evaluate the efficacy of the multi-rate system identification procedure. To this end, a deterministic EAF process model is simulated to generate 50 normal operation batches of varying durations starting from diverse initial conditions. The heats are terminated between 62 min and 70 min, depending on whether the aforementioned end-point target criteria are satis-

Table 12.1 List of infrequent measurement variables of the EAF process

Variable	Description	Units
T	Temperature of molten steel	K
x_{Fe}	Mass fraction iron in molten steel	kg/kg
x_{C}	Mass fraction carbon in molten steel	kg/kg
x_{Slag}	Mass fraction lime/dolime in slag	kg/kg
x_{FeO}	Mass fraction iron oxide in slag	kg/kg
$x_{\mathrm{SiO_2}}$	Mass fraction silicon dioxide in slag	kg/kg

Table 12.2 List of frequent measurement variables of the EAF process

Variable	Description	Units
P	Relative pressure	Pa
x_{CO}	Mass fraction carbon monoxide in gas	kg/kg
x_{CO_2}	Mass fraction carbon dioxide in gas	kg/kg
x_{N_2}	Mass fraction nitrogen in gas	kg/kg

Table 12.3 List of manipulated variables and corresponding input costs for the EAF process

Variable	Description	Units	Costs	Units of cost
m_{gas}	Off-gas turbine flow	kg/s	4.6296×10^{-5}	$\$ \cdot kg^{-1}$
m_{O_2}	Oxygen lanced	kg/s	0.01749	$\$ \cdot kg^{-1}$
m_{DRI}	DRI additions	kg/s	0.0900	$\$ \cdot kg^{-1}$
m_{Slag}	Slag additions	kg/s	0.0600	$\$ \cdot kg^{-1}$
E	Electric arc power	kW	0.0625	$\$ \cdot kWh^{-1}$
m_C	Carbon injected	kg/s	0.2500	$\$ \cdot kg^{-1}$

fied. To represent practical measurement issues, 10% of the manipulated input variables have a missing input value, where the input value transmitted to the actuators was not recorded. In this chapter, each batch of the electric arc furnace process has on average approximately 19.4% of the overall batch process data as missing values, while almost 30% of the output measurements are missing because of the multi-rate sampling of the process measurements. In addition to the normal operation heats, five identification batches are used to augment the database, where a low-amplitude pseudo-random binary sequence (PRBS) signal is added to the input values computed by the proportional–integral (PI) controllers. The input Hankel matrix for the electric arc furnace (EAF) was confirmed to be of full rank, thus ensuring that the assumption for persistently exciting inputs holds for the EAF process.

In addition to the batches used for modeling purposes, a total of 50 additional heats are simulated for model validation purposes. For an objective evaluation of the multi-rate system identification procedure, we compare our proposed approach to a standard system identification method that predicts all available process variables at only the infrequent sampling instances. The number of states to incorporate in the standard infrequent-only model and proposed multi-rate model is determined through evaluating the root-mean-square error (RMSE) in predicting back the validation samples, which yielded 17 states for the proposed multi-rate modeling approach and 15 for the standard infrequent-only model. The model validation results for the measurement variables are given in Table 12.4, and the corresponding output predictions of the multi-rate model, standard infrequent-only model, and actual variable profile trajectories from the test bed for a set of new initial conditions are shown in Figs. 12.2 and 12.3. For training batches, the subspace identification approach generates state trajectories for all batches, thus determining the initial subspace state for each batch. For a new batch, however, the underlying subspace state is unknown, requiring a

Table 12.4 Model validation results for the measurement variables of the EAF process

Measurement type	Variable	RMSE		Units
		Infrequent-only model	Multi-rate model	
Infrequent measurements	T	22.4722	8.8419	K
	x_{Fe}	0.0025	0.0012	kg/kg
	x_C	0.0013	0.0004	kg/kg
	x_{Slag}	0.0111	0.0090	kg/kg
	x_{FeO}	0.0139	0.0124	kg/kg
	x_{SiO_2}	0.0026	0.0032	kg/kg
Frequent measurements	P	0.3863	0.1333	Pa
	x_{CO}	0.0289	0.0208	kg/kg
	x_{CO_2}	0.0133	0.0094	kg/kg
	x_{N_2}	0.0158	0.0112	kg/kg

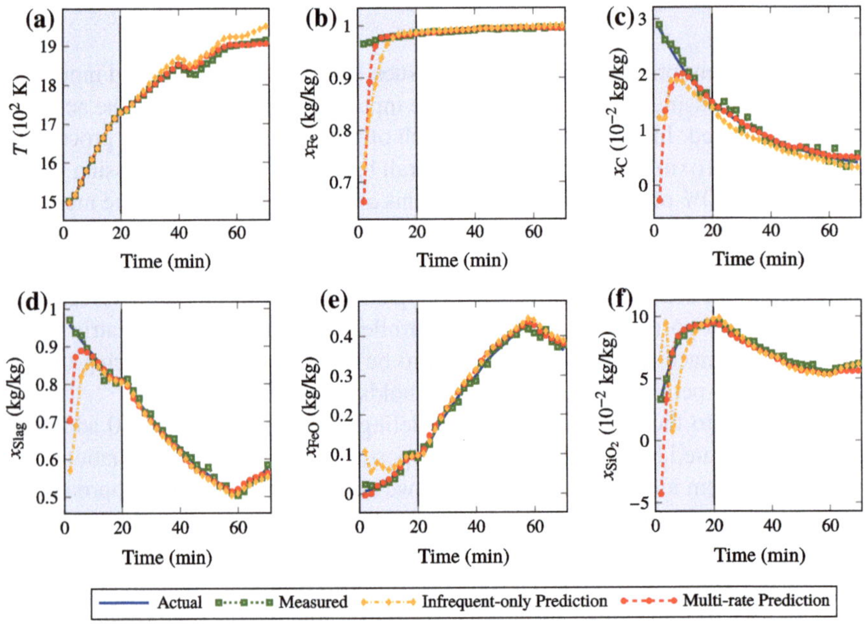

Fig. 12.2 Model validation results for the infrequent measurement variables of the EAF process

state observer to first estimate the process state to in turn be able to predict the states and measured outputs forward in time and validate the model [36]. To this end, we utilize a multi-rate Kalman filter, which incorporates an online real-time adaptive learning algorithm to provide updated state estimates. In some sense, estimating the underlying state of the model can be thought of as a 'learning' of the new batch

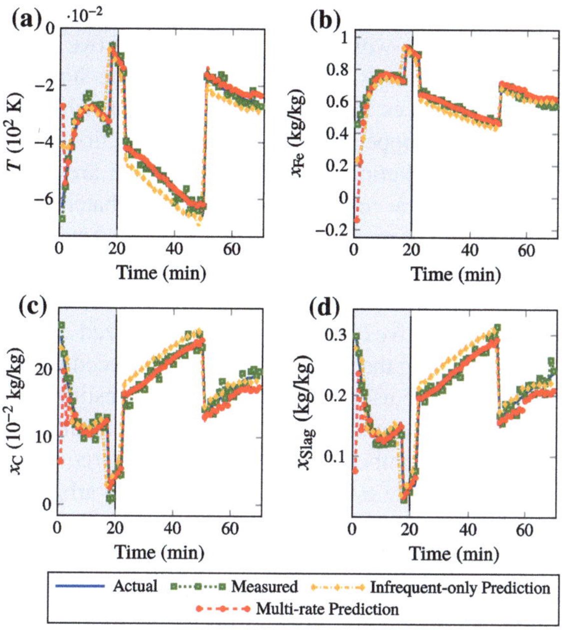

Fig. 12.3 Model validation results for the frequent measurement variables of the EAF process

conditions. The shaded area until 20 min on the plots represents the initial phase of the heat where feedback from the measurements is used to correct the state estimates until convergence. For the proposed multi-rate models, a multi-rate Kalman filter is implemented as

$$\hat{x}_k^- = A\hat{x}_{k-1} + Bu_k$$
$$P_k^- = AP_{k-1}A^T + Q$$

$$\left.\begin{aligned} L_k^{\phi_1} &= P_k^- C^{\phi_1 T}\left(C^{\phi_1}P_k^- C^{\phi_1 T} + R^{\phi_1}\right)^{-1} \\ \hat{x}_k &= \hat{x}_k^- + L_k^{\phi_1}\left(y_k^{\phi_1} - C^{\phi_1}\hat{x}_k^-\right) \\ P_k &= \left(I - L_k^{\phi_1}C^{\phi_1}\right)P_k^- \end{aligned}\right\} \quad \forall k \in \mathbb{K}^{\phi_1}$$

or

$$\left.\begin{aligned} L_k^{\phi_2} &= P_k^- C^{\phi_2 T}\left(C^{\phi_2}P_k^- C^{\phi_2 T} + R^{\phi_2}\right)^{-1} \\ \hat{x}_k &= \hat{x}_k^- + L_k^{\phi_2}\left(y_k^{\phi_2} - C^{\phi_2}\hat{x}_k^-\right) \\ P_k &= \left(I - L_k^{\phi_2}C^{\phi_2}\right)P_k^- \end{aligned}\right\} \quad \forall k \in \mathbb{K}^{\phi_2}$$

where L^{ϕ_1} and L^{ϕ_2} denote the Kalman gains for the infrequent and frequent sampling instances, respectively, and $\hat{x}^-$ and $\hat{x}$ denote the prior and posterior state estimates with error covariances P^- and P, respectively. Furthermore, the positive-definite matrices Q and R denote the covariances of the process disturbances and measurement noise, which can be tuned to improve the Kalman filter performance. For validation purposes, the convergence of the state estimates is evaluated using the error covariances to determine the point at which the Kalman filtering update of the states is stopped and subsequent predictions are made in an open-loop fashion where the predictions throughout the batch are not corrected and the model errors are allowed to accumulate over the entire batch duration. This is to test the ability of the model to predict reasonable future behavior for a candidate input profile and thus establish its utility for feedback control purposes.

Note that the state sequence for the model validation results, and in the development of predictive controllers, is not initialized arbitrarily. Rather, the state estimates obtained during the model development procedure (Eq. (12.18)) for one of the training batches are used to initialize the state estimates (instead of initializing them at zero), which results in faster convergence. Although the multi-rate sampling can increase the number of sampling instances required for the states to converge, the fact that the state estimator is not initialized arbitrarily assists in ensuring relatively fast convergence of the state estimates. Moreover, the lack of measurement output variables at the sampling instances where the infrequent measurements are not available means that the covariance of state estimate error will be larger at those incomplete sampling times. However, this higher uncertainty in the state estimates when a few output measurement variables are available is common to many state estimation algorithms. In this chapter, we determined the amount of sampling instances required for the states to converge in the application example by observing the norm of the covariance matrix of the state estimation error (P), which suggested that 20 sampling instances are sufficient for the state estimates to converge. Further, we determined the effects of the multi-rate sampling on the state estimates by observing the closed-loop results, which demonstrated good predictions of the output variables.

It is readily observed that the standard system identification approach that neglects the frequent measurement variables does not predict the process variables well, resulting in higher RMSE values. Note that the only variable predicted well by the standard approach is the silicon dioxide in the slag, which is not an end-point target controlled variable. Accordingly, we do not consider the standard infrequent-only model to be sufficiently accurate to drive a predictive controller, and is hence not utilized further in this study.

Economic Model Predictive Control Results

The closed-loop simulation of 50 new initial conditions is performed using the proposed economic model predictive controller, and the performance of the proposed approach is compared against standard EAF operating policies involving PI control. The conventional PI control approach used as a comparison in this chapter involves closely tracking reference variable trajectories based on a historical batch that produced steel of the desired end-point. Details on the pairing of the controlled and manipulated variables, and the associated PI tuning parameters are available in Rashid et al. [18] Furthermore, batches controlled using the conventional PI trajectory tracking approach were allowed to be terminated earlier than the duration of the reference batch if the end-point targets are satisfied in a shorter batch duration. All of the new initial conditions for evaluating the closed-loop performance of the proposed approach are obtained from the same distribution as the initial conditions in the training data. The batch termination time is allowed to vary in a predefined set $n_t = \{62\ , \ldots , 70\ \text{min}\}$, and the two tiers of the proposed EMPC are executed consecutively at each sampling instance through explicit enumeration. The first tier of the economic MPC determines the optimal end-point for a given batch duration, and if a feasible solution exists, the second tier computes the economically optimal inputs. The most economical input variable trajectory with the lowest cost is implemented on the EAF process, and the tiered EMPC problem is solved at subsequent sampling instances with a shrinking horizon and with updated state estimates.

A representative set of closed-loop simulation results (batch no. 7) is presented in Figs. 12.4 and 12.5 for the infrequent and frequent process measurements, respectively. The closed-loop trajectories of the end-point target variables and the corresponding intervals of the end-point constraints are shown in Fig. 12.4 as well. Although both the proposed EMPC and the standard PI control approach are able to meet the desired product quality, the proposed economic MPC offered a significant cost advantage of approximately $\$5.5794 \times 10^3$ compared to $\$6.3632 \times 10^3$ for the standard approach, an average savings of $783.80 per batch or approximately 12.32% improvement per batch. The proportional–integral (PI) controllers were designed and tuned to closely track the trajectories of a historical batch that resulted in on-spec product. As the simulation results confirm, the PI controllers are not conservatively designed because they did indeed enable tracking of the prescribed trajectory (and in this case, also the product satisfying the end-point target specifications). Furthermore, the average heat duration obtained through the proposed EMPC is 64.96 min, a 5.86% shorter batch duration compared to the conventional PI control approach.

The proposed economic MPC design was efficiently solvable with the average CPU time required to solve both the mixed-integer quadratic programming problems involved in the tiered approach with the longest prediction horizon as 4.42 s (maximum 4.97 s) using MATLAB on an Intel Core i7 machine with 8 GB RAM. Although the proposed EMPC formulation uses the subspace-based model and thus is computationally tractable, the efficiency of the optimization problem can be improved by parallelization [37] where the EMPC problems for varying batch termination times are solved in parallel using multiple CPU cores. The closed-loop input profiles are shown in Fig. 12.6, and it is readily observed that the EMPC approach recognizes that in certain instances, the end-point constraints are capable of being satisfied while maintaining the manipulated variables close to their lower bounds, due to which the operating costs of the EMPC approach are significantly lower. It is readily observed in Fig. 12.6 that the manipulated variables may move abruptly in the proposed EMPC formulation because the batch duration is also a decision variable, which may result in shorter batches being preferred from an economic perspective. A potential undesired consequence of the preferred shorter batch durations is that the manipulated inputs move impulsively. A remedy for the abrupt input moves brought upon by the inclusion of batch duration as a decision variable is to add an input move suppression term in the economic model predictive control algorithm, as in Eq. (12.26). If the rate

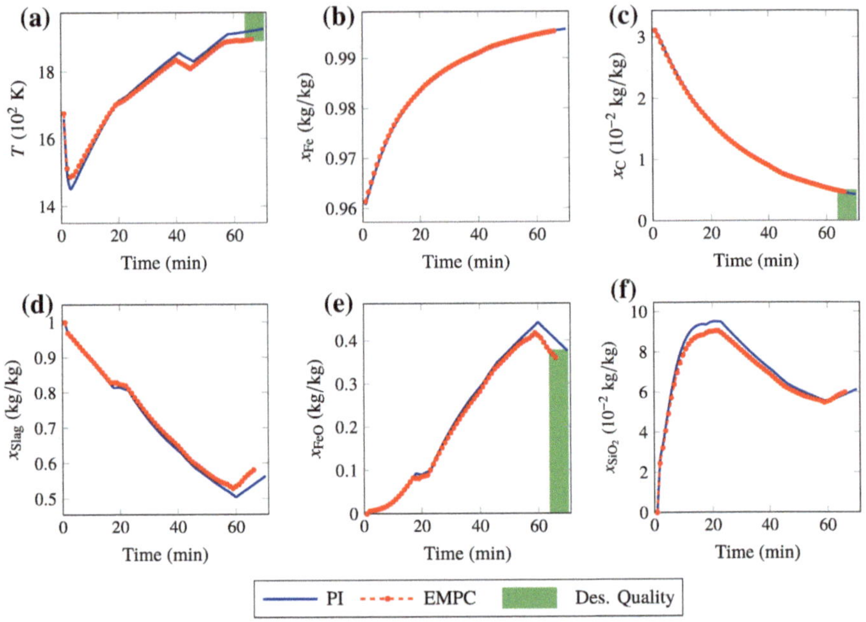

Fig. 12.4 Comparison of the trajectories for the infrequent measurement variables obtained from the proposed economic MPC and conventional method

Fig. 12.5 Comparison of the trajectories for the frequent measurement variables obtained from the proposed economic MPC and conventional method

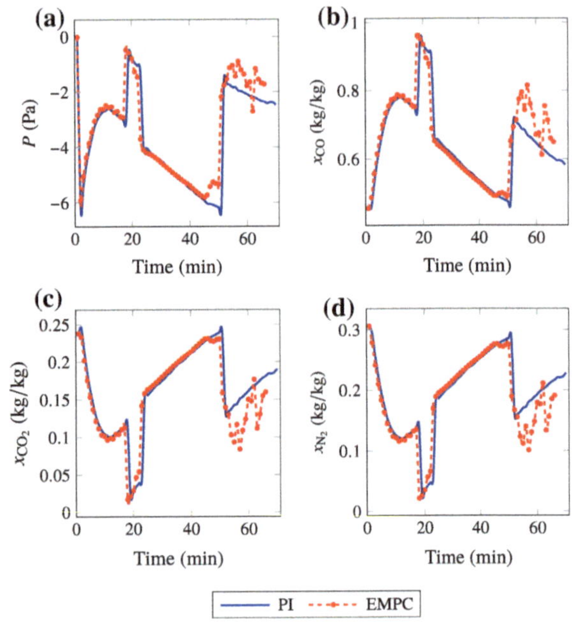

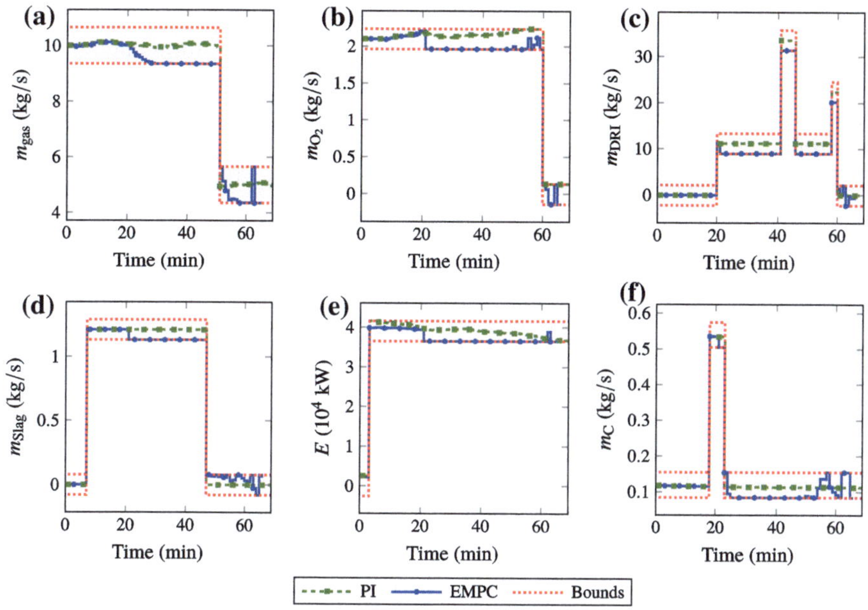

Fig. 12.6 Closed-loop profiles of the manipulated variables obtained from the proposed economic MPC and conventional method for a selected batch of the EAF process

of change of the manipulated inputs is deemed to be excessive by plant operators, the penalty cost associated with the rate of change of the manipulated inputs can be increased to mitigate the abrupt input moves.

Note that if the proposed data-driven model was utilized to generate optimal tracking trajectories online for each batch, and then these were tracked by a low-level PI controller, the closed-loop profiles and performance will likely be similar. The key contribution of the present chapter is the development of the data-driven model that enables computation of appropriate trajectories for the batch in a way that accounts for the specific variations in the current batch, and leads to performance improvement. Overall, the simulation results demonstrated the advantages of implementing the proposed economic predictive controller over the standard operating policies.

12.6 Conclusion

In this chapter, we develop a novel multi-rate subspace-based system identification and variable duration economic model predictive control framework for batch systems and apply it to the electric arc furnace process. The system identification method uses the incremental singular value decomposition to identify a dynamic model from a finite number of noisy data samples disrupted with unmeasured process variables and asynchronous data. Furthermore, the proposed identification approach is capable of handling inconsistent batch lengths to optimize batches with variable durations. The identified dynamic model is integrated into a tiered EMPC formulation, where solutions to computationally tractable problems achieve the desired final product end-point specification by batch termination, while minimizing the operating costs. The proposed multi-rate modeling and EMPC framework are implemented on the EAF process, and simulation case studies demonstrate the efficacy of the proposed approach subject to the limited availability of process measurements, missing data, measurement noise, and constraints.

References

1. Moonen, M., DeMoor, B., Vandenberghe, L., Vandewalle, J.: On- and off-line identification of linear state-space models. Int. J. Control **49**, 219–232 (1989)
2. VanOverschee, P., DeMoor, B.: Subspace Identification for Linear Systems: Theory, Implementation, Applications. Kluwer Academic Publishers, USA (1996)
3. Jansson, M., Wahlberg, B.: On consistency of subspace methods for system identification. Automatica. **34**, 1507–1519 (1998)
4. Ljung, L.: System Identification: Theory for the User. Prentice-Hall, Inc., New Jersey (1999)
5. Qin, S.J.: An overview of subspace identification. Comput. Chem. Eng. **30**, 1502–1513 (2006)
6. Liu, Z., Hansson, A., Vandenberghe, L.: Nuclear norm system identification with missing inputs and outputs. Syst. Control Lett. **62**, 605–612 (2013)
7. Liu, Z., Vandenberghe, L.: Interior-point method for nuclear norm approximation with application to system identification. SIAM J. Math. Anal. Appl. **31**, 1235–1256 (2009)
8. Gibson, S., Ninness, B.: Robust maximum-likelihood estimation of multivariable dynamic systems. Automatica **41**, 1667–1682 (2005)
9. Flores-Cerrillo, J., MacGregor, J.F.: Latent variable MPC for trajectory tracking in batch processes. J. Process Control **15**, 651–663 (2005)
10. Golshan, M., MacGregor, J.F., Bruwer, M.J., Mhaskar, P.: Latent variable model predictive control (LV-MPC) for trajectory tracking in batch processes. J. Process Control **20**, 538–550 (2010)
11. Srinivasan, B., Bonvin, D.: Real-time optimization of batch processes by tracking the necessary conditions of optimality. Ind. Eng. Chem. Res. **46**, 492–504 (2007)
12. Bonvin, D., Srinivasan, B.: On the role of the necessary conditions of optimality in structuring dynamic real-time optimization schemes. Comput. Chem. Eng. **51**, 172–180 (2013)
13. Aumi, S., Mhaskar, P.: Safe-steering of batch process systems. AIChE J. **55**, 2861–2872 (2009)
14. Aumi, S., Mhaskar, P.: Robust model predictive control and fault handling of batch processes. AIChE J. **57**, 1796–1808 (2011)
15. MacRosty, R.D.M., Swartz, C.L.E.: Dynamic modeling of an industrial electric arc furnace. Ind. Eng. Chem. Res. **44**, 8067–8083 (2005)
16. Srinivasan, B., Palanki, S., Bonvin, D.: Dynamic optimization of batch processes: I. characterization of the nominal solution. Comput. Chem. Eng. **27**, 1–26 (2003)
17. Srinivasan, B., Bonvin, D., Visser, E., Palanki, S.: Dynamic optimization of batch processes: II. role of measurements in handling uncertainty. Comput. Chem. Eng. **27**, 27–44 (2003)
18. Rashid, M.M., Mhaskar, P., Swartz, C.L.E.: Multi-rate modeling and economic model predictive control of the electric arc furnace. J. Process Control **40**, 50–61 (2016)
19. VanOverschee, P., DeMoor, B.: A unifying theorem for three subspace system identification algorithms. Automatica **31**, 1853–1864 (1995)
20. Raghavan, H., Tangirala, A.K., Gopaluni, R.B., Shah, S.L.: Identification of chemical processes with irregular output sampling. Control Eng. Practice **14**, 467–480 (2006)

21. Gopaluni, R.B.: Nonlinear system identification under missing observations: the case of unknown model structure. J. Process Control **20**, 314–324 (2010)
22. Balzano, L., On, Wright S.J.: On GROUSE and incremental SVD. In: IEEE 5th International Workshop on Computational Advances in Multi-Sensor Adaptive Processing (CAMSAP), pp. 1–4 (2013)
23. Balzano, L., Wright, S.J.: Local convergence of an algorithm for subspace identification from partial data. Found. Comput. Math. **15**, 1279–1314 (2015)
24. Heidarinejad, M., Liu, J., Christofides, P.D.: Economic model predictive control of nonlinear process systems using Lyapunov techniques. AIChE J. **58**, 855–870 (2012)
25. Amrit, R., Rawlings, J.B., Biegler, L.T.: Optimizing process economics online using model predictive control. Comput. Chem. Eng. **58**, 334–343 (2013)
26. Diehl, M., Amrit, R., Rawlings, J.B.: A Lyapunov function for economic optimizing model predictive control. IEEE Trans. Autom. Control **56**, 703–707 (2011)
27. Ellis, M., Durand, H., Christofides, P.D.: A tutorial review of economic model predictive control methods. J. Process Control **24**, 1156–1178 (2014)
28. Ellis, M., Christofides, P.D.: On finite-time and infinite-time cost improvement of economic model predictive control for nonlinear systems. Automatica **50**, 2561–2569 (2014)
29. Chachuat, B., Marchetti, A., Bonvin, D.: Process optimization via constraints adaptation. J. Process Control **18**, 244–257 (2008)
30. Chachuat, B., Srinivasan, B., Bonvin, D.: Adaptation strategies for real-time optimization. Comput. Chem. Eng. **33**, 1557–1567 (2009)
31. Bemporad, A., Morari, M., Dua, V., Pistikopoulos, E.N.: The explicit linear quadratic regulator for constrained systems. Automatica **38**, 3–20 (2002)
32. Yang, X., Biegler, L.T.: Advanced-multi-step nonlinear model predictive control. J. Process Control **23**, 1116–1128 (2013)
33. Jäschke, J., Yang, X., Biegler, L.T.: Fast economic model predictive control based on NLP-sensitivities. J. Process Control **24**, 1260–1272 (2014)
34. Diehl, M., Bock, H.G., Schlöder, J.P.: A real-time iteration scheme for nonlinear optimization in optimal feedback control. SIAM J. Control Optim. **43**, 1714–1736 (2005)
35. Bekker, J.G., Craig, I.K., Pistorius, P.C.: Modelling and simulation of an electric arc furnace process. ISIJ Int. **39**, 23–32 (1999)
36. Kozub, D.J., MacGregor, J.F.: State estimation for semi-batch polymerization reactors. Chem. Eng. Sci. **47**, 1047–1062 (1992)
37. Washington, I.D., Swartz, C.L.E.: Design under uncertainty using parallel multiperiod dynamic optimization. AIChE J. **60**, 3151–3168 (2014)

Chapter 13
Data-Driven Modeling and Quality Control of Variable Duration Batch Processes with Discrete Inputs

13.1 Introduction

In the previous chapters, a subspace-based modeling and control approach for batch quality [1] control was presented comprising a state-space dynamic model and a quality model, that addressed the problem of model building form batches of unequal duration. However, the control designs did not include batch duration as a decision variable. Furthermore, the control designs in the previous chapters considered a relatively simplistic batch recipe which did not include mid-batch additions which often play a critical role in batch processes.

The current chapter generalizes the method proposed in [1] to address both mid-batch additions and batch duration. Specifically, novel identification method is introduced that accounts for and identifies the effects of mid-batch additions. Also, the ability to use the identified models in batch quality control where batch duration is a decision variable is demonstrated. The remainder of the paper is structured as follows: First, in Sect. 13.2, background and notation needed to support the contributions in subsequent sections is provided. Section 13.3 outlines the model structure used for identification. Next, Sect. 13.4, goes step by step through the proposed identification method. Section 13.5 proposes a shrinking horizon model predictive control scheme to use the identified models for direct quality control. To demonstrate the efficacy of the proposed approach, Sect. 13.6 demonstrates the proposed identification and control methods applied to a simulation of a polymethyl methacrylate (PMMA) batch reactor. Finally, Sect. 13.7 draws a few brief conclusions.

13.2 Preliminaries

The first subsection of this section explicitly defines the class of processes considered by this study. Then the well-researched field of subspace identification is briefly reviewed.

P. Mhaskar et al., *Modeling and Control of Batch Processes*,
Advances in Industrial Control, https://doi.org/10.1007/978-3-030-04140-3_13

13.2.1 Class of Processes

The key contributions of this chapter are motivated by the Markov assumption which states that future behavior of the process is dependent solely on its current state and future input excitation.

For chemical systems, the state is captured by chemical composition of the reaction media and thermodynamic variables (temperatures and pressures). In this chapter, states in this physically motivated variable space are referred to as fundamental states and are denoted as $\boldsymbol{x}$.

The current chapter considers processes with two types of inputs: continuous and discrete. Continuous inputs are defined as independently manipulatable variables whose values impact process behavior through dynamics. A cooling jacket which continually removes heat from the reactor is a good example of a continuous input. Continuous inputs are denoted with the standard $\boldsymbol{u}$ notation. Discrete inputs are defined as factors that affect the process state instantaneously (or sufficiently fast so its easiest to treat it as instantaneous). For example, addition of an ingredient causes concentration of the added species to instantly increase. Notationally, discrete inputs are represented as $\boldsymbol{u}_d$.

To implement digital control, some narrowing of the input space is necessary. Specifically, this chapter makes two assumptions: First, that continuous inputs occur at a regular sampling frequency, equal to the measurement sampling frequency for convenience of feedback, with a zero-order hold. Second, that discrete inputs can only occur at the end of sampling instances, immediately prior to the measurement of the subsequent sampling instant. Note that the proposed approach would readily work with any other policy, for instance a first-order hold policy. Also, the timing of the addition of the discrete inputs could be easily changed.

In general, measurements are nonlinear functions of state. For instance, for a polymerization reactor, the viscosity of the reaction media is a nonlinear function of the temperature and composition. Further, process measurements are classified into two types: online measurements and off-line quality measurements. As the labels imply, online measurements are taken from process sensors and are sampled frequently in (nearly) real time. This chapter denotes online measurements with the standard $\boldsymbol{y}$ notation. Quality measurements, denoted by the vector $\boldsymbol{q}$, provide a more direct and detailed evaluation of the reaction product. These measurements capture the useful properties of the product and, as such, define the production objectives. However, quality measurements are generally only available infrequently (often at the end of a batch) and with substantial delay.

The system described above can be represented mathematically as follows:

$$\boldsymbol{x}[k+1] = \boldsymbol{f}\left(\boldsymbol{x}[k], \boldsymbol{u}[k]\right) \quad \forall \quad k \notin \mathbb{K}_{\mathrm{d}} \tag{13.1}$$

$$\boldsymbol{x}[k+1] = \boldsymbol{f_d}\left(\boldsymbol{x}^{-}[k+1], \boldsymbol{u}_d[k]\right) \quad \forall \quad k \in \mathbb{K}_{\mathrm{d}} \tag{13.2}$$

$$\boldsymbol{x}^{-}[k+1] = \boldsymbol{f}\left(\boldsymbol{x}[k], \boldsymbol{u}[k]\right) \quad \forall \quad k \in \mathbb{K}_{\mathrm{d}} \tag{13.3}$$

$$\boldsymbol{y}[k] = \boldsymbol{g}\left(\boldsymbol{x}[k]\right) \tag{13.4}$$

$$q[k] = h(x[k]) \tag{13.5}$$

where $f(x[k], u[k])$ is the nonlinear function governing the dynamic evolution of the fundamental states. f_d is the function capturing the effect of the discrete addition at the end of a sampling instant. $x^-[k+1]$ is the state after the kth sampling instant but before a discrete addition. g describes the relationship between states and frequently sampled measurements. h is the relationship between fundamental states and quality. Finally, $\mathbb{K}_d$ is the set of all sampling instances at the end of which a discrete input occurs. Note that the subsequent results do not require knowledge of the above functions; they have only been introduced for notational purposes.

Remark 13.1 By definition, every batch has at least one discrete input at $t = 0$ corresponding to the initial charge of the reactor and another at $t = T_f$ when the reactor is discharged. This conceptualization could extend the approaches presented in this paper to cases where there is a statistical dependency between subsequent batches. This type of dependency is the theoretical basis for batch-to-batch control approaches. However for simplicity, this chapter focuses on the within-batch control problem.

13.2.2 Subspace Identification

This subsection provides a brief overview of subspace identification methods. These methods were developed to be stable, noniterative approaches for identifying state-space, dynamic models directly from input and output data. Traditionally, these methods are used to identify models around steady-state operating conditions, as is the case for continuously fed chemical processes. Standard practice is to excite the system using an identification signal (for instance, pseudo-random binary sequence signals (PRBS)). This excitation is necessary to correctly isolate the influence of each input on each of the states.

Consider the linear, time-invariant, discrete, state-space system:

$$x[k+1] = \mathbf{A}x[k] + \mathbf{B}u[k] \tag{13.6}$$

$$y[k] = \mathbf{C}x[k] + \mathbf{D}u[k] \tag{13.7}$$

To begin the identification procedure, a contiguous sequence of $i + j - 1$ output measurements (i.e., $y[1], y[2], \ldots, y[i + j - 1]$) and the corresponding input values from this system are arranged into a Hankel matrix as follows:

$$\mathbf{Y}_{1|i} = \begin{bmatrix} y[1] & y[2] & \cdots & y[j] \\ \vdots & \vdots & & \vdots \\ y[i] & y[i+1] & \cdots & y[i+j-1] \end{bmatrix}$$

Note that the notation $\mathbf{Y}_{1|i}$ in this context refers to the Hankel matrix in which the first column contains outputs from 1 to i. A corresponding unfolding of the state vectors is defined as:

$$\mathbf{X}_i = \left[\boldsymbol{x}[i]\ \boldsymbol{x}[i+1]\ \cdots\ \boldsymbol{x}[i+j-1]\right]$$

Having defined these matrix arrangements of the available observations, the update and output state-space equations (Eqs. (13.6) and (13.7)) can be re-expressed as:

$$\mathbf{X}_{i+1} = \mathbf{A}^i\mathbf{X}_1 + \Delta_i\mathbf{U}_{1|i} \tag{13.8}$$

$$\mathbf{Y}_{1|i} = \Gamma_i\mathbf{X}_1 + \mathbf{H}_i\mathbf{U}_{1|i} \tag{13.9}$$

where $\mathbf{A}^i$ is the dynamic A matrix to the ith power. Δ_i, Γ_i, and $\mathbf{H}_i$ can be derived by appropriate iterative substitutions of Eqs. (13.6) and (13.7) (see [2] for the definition and derivation of these matrices). Next, using the pseudo-inverse (denoted $\bullet^*$), Eq. (13.9) is solved for $\mathbf{X}_1$ which can be plugged into Eq. (13.8) and re-arranged to get:

$$\mathbf{X}_{i+1} = \left[\Delta_i - \mathbf{A}^i\Gamma_i^*\mathbf{H}_i\ \ \mathbf{A}^i\Gamma_i^*\right]\begin{bmatrix}\mathbf{U}_{1|i}\\ \mathbf{Y}_{1|i}\end{bmatrix} \tag{13.10}$$

Next, rewrite Eq. (13.9) for observations from $i+1$ to $2i$ to get:

$$\mathbf{Y}_{i+1|2i} = \Gamma_i\mathbf{X}_{i+1} + \mathbf{H}_i\mathbf{U}_{i+1|2i} \tag{13.11}$$

Equations (13.10) and (13.11) together demonstrate the fundamental principle behind subspace identification methods. Particularly, note that the matrix $\mathbf{X}_{i+1}$ appears in both of these equations and forms a partial basis for the output measurements in both cases. Assuming that inputs are independent from outputs (i.e., not purely closed loop), it is possible to identify this basis through a series of appropriate projections (perpendicular to 'future' inputs, $\mathbf{U}_{i+1|2i}$) (see Remark 13.12 for a discussion of open-loop data requirements in batch identification). In practice, care must be taken to exploit this theoretical result in a numerically stable fashion. To this end, a number of algorithms have been proposed.

This chapter does not present a new subspace identification algorithm; accordingly, readers are referred to the existing literature for algorithmic information. For specific details of the subspace identification principles used in this chapter, including numerically stable algorithms, see [2]. For a good review and analysis of different subspace identification methods, see [3]. In previous work, [1], it was demonstrated that by appropriately modifying Hankel matrices, one could apply the methods from [2] to identify dynamic models for batch processes. This work generalizes these results for batches with discrete inputs.

13.3 Proposed Model Structure

The idea motivating the selected model structure is to break the overall model into multiple distinct components. This component based structure facilitates identification by allowing each component of the model to be identified individually. The selection of the components, and structure of each model component, is motivated by our fundamental understanding of these processes as presented in Sect. 13.2.1. Specifically, the selected model components are: a dynamic model to capture the effects of continuous inputs, another for capturing the effect of the discrete inputs, and a quality model relating final states to final quality. All three model component types are related through the state, $\mathring{\boldsymbol{x}}$, defined by the dynamic model. Mathematically, the model is represented as:

$$\mathring{\boldsymbol{x}}[k+1] = \begin{cases} \mathbf{A}\mathring{\boldsymbol{x}}[k] + \mathbf{B}\boldsymbol{u}[k] & \forall k \notin \mathbb{K}_\mathrm{d} \\ \mathbf{A}_d\left(\mathbf{A}\mathring{\boldsymbol{x}}[k] + \mathbf{B}\boldsymbol{u}[k]\right) + \mathbf{B}_d\mathbf{u}_\mathbf{d}[k] + \boldsymbol{c}_d & \forall k \in \mathbb{K}_\mathrm{d} \end{cases}$$

$$\boldsymbol{y}[k] = \mathbf{C}\mathring{\boldsymbol{x}}[k] + \mathbf{D}\boldsymbol{u}[k]$$

$$\boldsymbol{q}[T_f] = \mathbf{M_q}\mathring{\boldsymbol{x}}[T_f] + \mu_q$$

where **A**, **B**, **C**, and **D** as well as the selected length (order) of $\mathring{\boldsymbol{x}}$ comprise the dynamic state model; the matrices $\mathbf{A}_d$, $\mathbf{B}_d$, and the vector $\boldsymbol{c}_d$ comprise the dth discrete addition model; and matrix $\mathbf{M_q}$ and vector μ_q comprise the quality model relating the terminal state, $\mathring{\boldsymbol{x}}_{T_f}$, to the terminal quality, $\boldsymbol{q}_{T_f}$. Specific details of each of these models are discussed in subsequent sections in conjunction with details regarding model identification.

Note that this model structure is based on a latent state, $\mathring{\boldsymbol{x}}$. Section 13.4.1, where identification of the dynamic model using subspace methods is discussed, briefly describes the advantages afforded by this latent state including the ability to adequately describe the evolution of the process dynamics using the proposed linear structure. For a deeper examination of the way that LTI models capture behavior that is nonlinear in fundamental states space, readers are referred to the detailed treatment of the topic in [1]. Other identification sections discuss the linear representations of the quality and discrete input models as well.

13.3.1 Observer

The fact that the proposed model structure is state-space based introduces the state estimation problem. Before presenting the proposed approach, this subsection introduces a receding horizon observer design based on one presented in [4]. This optimization-based observer is used to estimate the states of the proposed model structure for a series of inputs and the corresponding measured response. Importantly, the observer intuitively incorporates the full model structure proposed in this section, including the discrete-input component of the model structure. To estimate a state trajectory at sampling instant k, $\hat{\mathring{x}}_i \forall i \leq k$, from measurements, y_i, continuous inputs, u_i, and discrete inputs, $u_d[i]$, the optimization problem is formulated as:

$$\min_{\hat{\mathring{x}}[1],\ldots,\hat{\mathring{x}}[k]} \left(\hat{\mathring{x}}[1|k-1] - \hat{\mathring{x}}[1|k]\right)' \mathbf{Q}_0 \left(\hat{\mathring{x}}[1|k-1] - \hat{\mathring{x}}[1|k]\right) + \cdots$$

$$\sum_{i=1}^{k} \left(y[i] - \hat{y}[i]\right)' \mathbf{R} \left(y[i] - \hat{y}[i]\right) + \cdots$$

$$\sum_{i=1}^{k-1} \omega_i' \mathbf{Q} \omega_i$$

$$s.t.\ \omega_i = \begin{cases} \mathring{x}[i+1] - \left(\mathbf{A}\mathring{x}[i] + \mathbf{B}u[i]\right) & \forall i \notin \mathbb{K}_\mathrm{d} \\ \mathring{x}[i+1] - \left(\mathbf{A}_d\left(\mathbf{A}\mathring{x}[i] + \mathbf{B}u[i]\right) + \mathbf{B}_d\mathbf{u}_\mathbf{d}[i] + c_d\right) & \forall i \in \mathbb{K}_\mathrm{d} \end{cases}$$

$$\hat{y}[i] = \mathbf{C}\hat{\mathring{x}}[i] + \mathbf{D}u[i]$$

where $\mathbf{Q}_0$, $\mathbf{R}$, and $\mathbf{Q}$ are symmetric, positive-definite tuning matrices. $\hat{\mathring{x}}[1|k-1]$ is the state estimate at the first sampling instant of the process calculated during the last update of the observer. Similarly, $\hat{\mathring{x}}[1|k]$ is equivalent in our notation to $\hat{\mathring{x}}[1]$ and is the new estimate of the initial state. With these two definitions in mind, the first term and the corresponding symmetric, positive-definite, tuning matrix, $\mathbf{Q}_0$ provides an iterative dampening effect. The purpose of this term is to stabilize the state estimate trajectories when the observer is updated online at each sampling instant with new observations. The second term is similar to the innovation term in a Luenberger or Kalman estimator and penalizes deviations between the observed output and that predicted by the model. The final term enforces model validity. While it would be equally feasible to include this term as a constraint (i.e., $\omega_i = 0$), doing so would not enable accounting for plant-model mismatch.

Remark 13.2 An alternative approach would be to use a Kalman filter. In fact, for a linear system with Gaussian process and measurement noise, a Kalman filter is provably optimal. There are, however, a number of reasons not to use a Kalman filter for the present framework. First and foremost, the underlying assumptions for Kalman filters are not necessarily met. Specifically, as discussed later, in this chapter nonlinear dynamics are approximated using a linear dynamic model. As a result, using a Kalman filter would result in a biased estimate because the state update error would be autocorrelated and would not be Gaussian centered on the true state.

Furthermore, the presented batch estimator is provably equivalent to a Kalman filter if the process is well described by the linear system and process noise is Gaussian [4]. Therefore, there is no possible benefit to using a Kalman filter in place of the proposed estimator. Another advantage of using a batch estimator is the ability to 'even out' the effect of the measurement noise over the entire batch without unduly detuning the estimator.

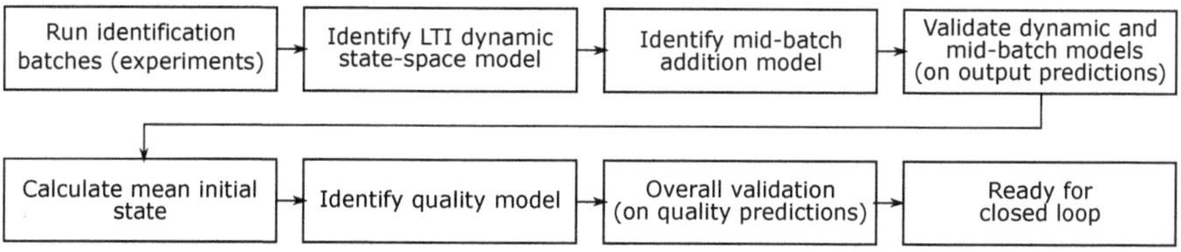

Fig. 13.1 Flow diagram summarizing the proposed model identification procedure discussed in Sect. 13.4 and applied in the simulation study in Sect. 13.6

13.4 Model Identification

The previous section proposed a model structure for batch processes motivated by facilitating identification from process data. This section provides methodology for identifying the corresponding model parameters. The proposed method is simple enough to be practical for industrial applications, yet general enough to describe batches with a wide variety of recipes.

Figure 13.1 is flow diagram that summarizes the proposed identification and validation steps. Model identification is broken into four distinct parts: First, subspace identification is used to find a dynamic model. Next, discrete input models are identified. To improve the early-batch performance of the proposed observer, a prototypical initial state is determined. Finally, a model is developed relating terminal state to terminal quality. Each of these steps is dependent on the preceding steps. Together, these steps provide a method for capturing the complex behavior of batch processes from historical operation data and a limited amount of identification data.

13.4.1 Dynamic Model Using Subspace Identification

Reference [1] illustrated that subspace identification could be applied to capture the dynamic behavior of nonlinear systems over a finite duration of time. The key to the success of this method was use of a LTI model of (potentially) higher order than the nonlinear system, in conjunction with the fact that the model is required to approximate the nonlinear dynamics over a finite duration. Another aspect critical to success was appropriately using data from multiple training batches in the identification step.

In [1], observations from a number of batches were combined in the pseudo-Hankel matrix construction step of subspace identification. To achieve this, first Hankel matrices were constructed from each individual batch using observations from the entire batch duration as a single block of contiguous data. The resulting Hankel submatrices were then concatenated horizontally to form a pseudo-Hankel matrix containing information from multiple batches. After concatenation, the completed pseudo-Hankel matrices possess the same row-space relationship as a training matrix constructed from a single block of contiguous data. Accordingly, these pseudo-Hankel matrices can be used in unaltered subspace identification methods. Moreover, by including a number of different batches in the training data, a sufficient range of the fundamental state space was exposed to identify a model capable of accurately capturing the dynamic behavior of the system.

Remark 13.3 Note that these results incorporated batches of varying durations in the training data. This was possible because the duration of the training data altered only the number of columns in each submatrix of the pseudo-Hankel matrix. For horizontal concatenation, only the number of rows must be consistent. The number of Hankel rows is a user parameter (defining a size of a history of observations around the identified state) and was therefore consistent regardless of batch duration.

The subspace identification approach adopted in the current chapter is a generalization of the framework presented in [1] to handle discrete inputs. To include batches with discrete inputs in the training data, the pseudo-Hankel matrix construction must be adequately modified. Subspace identification methods assume that all state changes occur as a result of process dynamics or as a dynamic response to input variables. By definition, these dynamic response state trajectories are smooth. However, discrete inputs cause a discontinuity in state trajectories. Such discontinuities violate the assumptions made by subspace methods about the evolution of state variables. As a result, including observations around discrete inputs in subspace training data would be inconsistent with assumptions and could substantially skew identified model dynamics. This is the motivation for using a separate model component to capture discrete addition effects.

To address the effects of discrete inputs, the Hankel submatrix for each batch is further divided into submatrices. In some sense, one can think of the period of time between each discrete input as an individual batch; for the duration between each addition, the state of the process is transformed from an initial state (the result of the last discrete input) to a final state (the state before the next discrete input). Note of course that this assumption would fail if multiple discrete additions are made one after the other and would necessitate appropriate recognition of this fact in (a) choosing the sampling time for building the dynamic model and (b) approximating the entire set of discrete additions as one event over this larger sampling time. For the present case, the transformation of the process between two discrete events is governed by process dynamics and continuous inputs made during that time. Hence, it is appropriate to build Hankel submatrices for each of these intervals. It is assumed that discrete inputs can only modify the system state and that process dynamics remain unaltered by discrete inputs. Therefore, instead of modifying system dynamics, the effect of each discrete input is accounted for by allowing a discontinuity in the identified state. With this in mind, it makes sense to identify one dynamic model for the entire duration of the batch (including all intervals between discrete events). Mathematically, these ideas can be represented as follows.

For the set of L discrete input times:

$$\mathbb{K}_d = \{d_1, d_2, \dots, d_L\}$$

Define the set of $l + 1$ corresponding continuous intervals

$$\begin{aligned}\mathbb{I}_1 &= \{k \,|1 \le k \le d_1\} \\ \vdots &= \vdots \\ \mathbb{I}_l &= \{k \,|d_{l-1} < k \le d_l\} \\ \vdots &= \vdots \\ \mathbb{I}_{L+1} &= \left\{k \,\middle|d_L < k \le T_f\right\}\end{aligned}$$

Then define the following sub-Hankel matrices for each batch, b, and interval $\mathbb{I}_l$:

$$\mathbf{Y}_{1|i}^{b,\mathbb{I}_l} = \begin{bmatrix} y[d_{l-1}+1] & y[d_{l-1}+2] & \cdots & y[d_l - 2i] \\ y[d_{l-1}+2] & y[d_{l-1}+3] & \cdots & y[d_l - (2i-1)] \\ \vdots & \vdots & & \vdots \\ y[d_{l-1}+i] & y[d_{l-1}+i+1] & \cdots & y[d_l - i] \end{bmatrix}^{(b)}$$

Likewise for $\mathbf{Y}_{i+1|2i}$ get:

$$\mathbf{Y}_{i+1|2i}^{b,\mathbb{I}_l} = \begin{bmatrix} y[d_{l-1}+i+1] & y[d_{l-1}+2] & \cdots & y[d_l - i] \\ y[d_{l-1}+i+2] & y[d_{l-1}+3] & \cdots & y[d_l - (i-1)] \\ \vdots & \vdots & & \vdots \\ y[d_{l-1}+2i] & y[d_{l-1}+i+1] & \cdots & y[d_l] \end{bmatrix}^{(b)}$$

Then combine the data from all intervals and all sampling instances to form:

$$\mathbf{Y}_{1|i} = \left[\mathbf{Y}_{1|i}^{1,1}\ \mathbf{Y}_{1|i}^{1,2} \cdots \mathbf{Y}_{1|i}^{1,L}\ \mathbf{Y}_{1|i}^{2,1}\ \mathbf{Y}_{1|i}^{2,2} \cdots \mathbf{Y}_{1|i}^{2,L} \cdots \mathbf{Y}_{1|i}^{B,L}\right]$$

The corresponding concatenation of sub-Hankel matrices $\mathbf{Y}_{i+1|2i}^{b,\mathbb{I}_l}$ gives the complete pseudo-Hankel matrix $\mathbf{Y}_{i+1|2i}$. Similarly, the same procedure can be used go generate $\mathbf{U}_{1|i}$ and $\mathbf{U}_{i+1|2i}$. Note that by partitioning the sub-Hankel matrices around discrete inputs, the Hankel matrices of inputs only contain continuous inputs. Critically, by partitioning data into segments omitting sample instances with discrete inputs, the row-space relationship of the resulting Hankel matrix is consistent with that described in Sect. 13.2.2. Therefore, from this point forward, it is permissible to apply subspace identification algorithms without further alteration.

The result of the subspace identification algorithm's state identification step is a state vector for each training batch, b, of the following form:

$$\mathbf{X}_i^{(b)} = \left[x[i]^{(b)}\ x[i+1]^{(b)} \cdots x[d_1 - i]^{(b)}\ x[d_1 + i]^{(b)}\ x[d_1 + i]^{(b)} \cdots x\left[T_f - i\right]^{(b)}\right]$$

From this state progression and the corresponding input and output progressions, it is trivial to identify the state-space matrices, **A**, **B**, **C**, and **D**. In this chapter, this step was done using linear regression (see [2] for details). However, as discussed in the literature it is also possible to back calculate these values directly from the intermediate calculation matrices. Linear regression was chosen for simplicity. The state matrices identified in this section are used extensively in the remaining identification steps to inform the state observer. These remaining identification steps are discussed next.

Remark 13.4 One side effect of partitioning the Hankel submatrices around discrete inputs is that only states up to i sampling instances before and after the discrete input can be identified. This is, however, not a restriction of the proposed approach. As long as there is sufficient time between discrete inputs (for batch processes, there always is because of implementation practicalities), and the sampling frequency is sufficiently fast compared to the process dynamics (necessary from a process identification standpoint), the approach enables capturing the discrete and continuous components of the process dynamics.

13.4.2 Identification of Discrete Input Model

With continuous dynamics of the system identified, the next step in the proposed method is to identify the effect of discrete inputs. The objective is to identify a model relating the state immediately prior to the discrete input to that immediately following it. One approach would be to use the state estimates resulting directly from the subspace identification step as training data. However, recall from the previous section that the state estimates provided by subspace identification leave a gap around each discrete addition. This is undesirable since it is necessary to isolate the effects of discrete inputs from the effects of continuous inputs implemented during the gap. Therefore, another approach is required.

In order to identify the effect of the discrete input in isolation, first the state immediately before and after the input takes place must be estimated. To achieve this, a pared-down version of the observer proposed in Sect. 13.3.1 is used on the continuous observations from each of the $L + 1$ segments of each batch. After completing the dynamic identification, the only model parameters available to use from the overall model proposed in Sect. 13.3 are the continuous dynamic matrices $\mathbf{A}$, $\mathbf{B}$, $\mathbf{C}$, and $\mathbf{D}$. With these matrices, enough information is available to apply the observer independently on each of the $L + 1$ segments of continuous observations in the training data. Note that since this is not an iterative application, and a good initial state estimate is not available, $\mathbf{Q}_0$ is set to zero, $\mathbf{Q}_0 = \mathbf{0}$. The resulting observer is given by:

$$\min_{\hat{\mathring{x}}_i, i \in \mathbb{I}_l} \sum_{i \in \mathbb{I}_l} \left(y[i] - \hat{y}[i]\right)' \mathbf{R} \left(y[i] - \hat{y}[i]\right) + \cdots \tag{13.12}$$

$$\sum_{i \in \mathbb{I}_l} \boldsymbol{\omega}_i' \mathbf{Q} \boldsymbol{\omega}_i \tag{13.13}$$

$$\boldsymbol{\omega}_i = \mathring{\boldsymbol{x}}[i+1] - \left(\mathbf{A}\mathring{\boldsymbol{x}}[i] + \mathbf{B}\boldsymbol{u}[i]\right) \tag{13.14}$$

$$\hat{y}[i] = \mathbf{C}\hat{\mathring{\boldsymbol{x}}}[i] + \mathbf{D}\boldsymbol{u}[i] \tag{13.15}$$

The solution of the above optimization problem is an estimated state trajectory, $\hat{\mathring{\boldsymbol{x}}}[i], i \in (d_{l-1}, d_l]$. By solving for all $L+1$ intervals, a complete state estimate trajectory is obtained for each training batch. Of specific interest, for each discrete input, l, now available are the states one sampling instant before the discrete input, $\hat{\mathring{\boldsymbol{x}}}[d_l]$ and the state immediately after the discrete input, $\hat{\mathring{\boldsymbol{x}}}[d_l+1]$. Recall, however, that discrete inputs are made at the end of sampling intervals, immediately prior to the next measurement. Accordingly, the state at the end of the sampling instant, directly preceding the discrete input, must be determined before a model from this data can be built. Since no measurement is available after the sampling instant but before the discrete addition, the best method to accomplish this is to propagate the state estimate at $\hat{xs}[d_l]$ forward using the continuous model. This is represented mathematically as:

$$\hat{\mathring{\boldsymbol{x}}}[d_l]^- = \mathbf{A}\hat{\mathring{\boldsymbol{x}}}[d_l] + \mathbf{B}\boldsymbol{u}[d_l]$$

where $\hat{\mathring{\boldsymbol{x}}}[d_l]^-$ is our best estimate of the state immediately before the discrete addition given our knowledge of the continuous process dynamics, measurements, and continuous inputs.

With $\hat{\mathring{\boldsymbol{x}}}[d_l]^-$ and $\hat{\mathring{\boldsymbol{x}}}[d_l+1]$ identified, it is possible to construct regression matrices to identify $\mathbf{A}_d$, $\mathbf{B}_d$, and $\boldsymbol{c}_d$. These regression matrices are given by:

$$\mathbf{Y}_{reg} = \left[\hat{\mathring{\boldsymbol{x}}}[d_l+1]^{(1)}\ \hat{\mathring{\boldsymbol{x}}}[d_l+1]^{(2)} \cdots \hat{\mathring{\boldsymbol{x}}}[d_l+1]^{(B)}\right] \tag{13.16}$$

$$\mathbf{X}_{reg} = \begin{bmatrix} \hat{\mathring{\boldsymbol{x}}}[d_l]^{-(1)} & \hat{\mathring{\boldsymbol{x}}}[d_l]^{-(2)} & \cdots & \hat{\mathring{\boldsymbol{x}}}[d_l]^{-(B)} \\ \boldsymbol{u}_d[d_l]^{(1)} & \boldsymbol{u}_d[d_l]^{(2)} & \cdots & \boldsymbol{u}_d[d_l]^{(B)} \end{bmatrix} \tag{13.17}$$

Then solving the regression problem yields:

$$\mathbf{Y}_{reg} = \left[\mathbf{A}_d\ \mathbf{B}_d\right]\mathbf{X}_{reg} + \boldsymbol{c}_d \tag{13.18}$$

where $\mathbf{A}_d$ captures the influence of preaddition states on post-addition states, $\mathbf{B}_d$ captures the direct effects of the addition, and $\boldsymbol{c}_d$ is a constant.

Remark 13.5 Note that the proposed approach is designed to handle situations where multiple discrete additions are part of the batch recipe. If the discrete additions are at fixed instances during the batch, or involve the addition of separate ingredients, they could be accounted for using separate models. On the other hand, the approach could be readily generalized to handle variable discrete addition time by building a single model for the discrete addition events.

Remark 13.6 An alternative to the form of the mid-batch model proposed above is to instead model the change in state $\Delta\hat{\mathring{x}}$. This alternative form is motivated by the assertion that a discrete input of zero magnitude should have no effect on the state of the system. To use this alternative form, the steps to identify $\hat{\mathring{x}}[d_l]^-$ and $\hat{\mathring{x}}[d_l+1]$ are unaltered. The new regression matrices are:

$$\mathbf{Y}_{reg} = \left[\hat{\mathring{x}}[d_l+1]^{(1)} - \hat{\mathring{x}}[d_l]^{-\,(1)}\ \hat{\mathring{x}}[d_l+1]^{(2)} - \hat{\mathring{x}}[d_l]^{-\,(2)}\ \cdots\ \hat{\mathring{x}}[d_l+1]^{(B)} - \hat{\mathring{x}}[d_l]^{-\,(B)}\right] \tag{13.19}$$

$$\mathbf{X}_{reg} = \left[\boldsymbol{u}_d[d_l]^{(1)}\ \boldsymbol{u}_d[d_l]^{(2)}\ \cdots\ \boldsymbol{u}_d[d_l]^{(B)}\right] \tag{13.20}$$

For regression matrices in this form, it makes sense to force the regression solution to pass through the origin. The direct result of the regression is $\mathbf{B}_d$ while $\mathbf{A}_d$ for this case is the identity matrix and $\boldsymbol{c}_d$ is the zero vector.

13.4.3 Identification of Initial State Estimate

In state estimation problems where a first-principles, state-space model is used, it may be possible to provide an initial state estimate (at least for some of the states) based on process knowledge. For instance, initial concentration of product in a batch reactor would likely be close to zero. However, in the proposed approach, the subspace states of the system have no direct physical meaning. As a result, it is not possible to draw a reasonable initial system state from process knowledge. Instead, to help with convergence, a reasonable initial state must be obtained from training data. One way to accomplish this is by using an average of the initial subspace states from the training data. To determine this average initial state, the observer outlined in Sect. 13.3.1 was applied to each training batch. Note that, with the discrete input model(s) identified, the observer can be applied to complete output, input, and discrete input trajectories from each batch. Once again, the $\mathbf{Q}_0$ matrix is set to zero because the observer is not being used iteratively. From the resulting subspace state trajectories, the average of each initial subspace state is taken across all training batches. The resulting average initial state vector is a good starting state estimate guess for new batches, which is subsequently refined with measurements from the batch in question.

Remark 13.7 For batch processes where measurements in excess of the first online measurement are available, rather than using the average value for the initial state, a model relating these additional measurements to initial state could also be built. Doing so would leverage available initial measurements to make appropriate early control decisions.

13.4.4 Identification of Quality Model

The last step to the proposed identification process is to identify the quality model. Recall from Sect. 13.2.1 that quality is a function of state. Because the identified models will be used to implement quality control, it is expected that the controller will drive the quality to a narrow range. Therefore, our quality prediction model only needs to be accurate for a small range of qualities near the set-point. As such, it is reasonable to use a simple linear model to relate the quality to the terminal state. This has the added advantage that it simplifies the identification process.

To identify a linear model relating terminal state and quality, the procedure is similar to that from the previous subsection for identifying the initial state. First, an observer of the form from Sect. 13.3.1 is constructed using the dynamic model and discrete input model(s) identified in the previous sections. Note that once again $\mathbf{Q}_0$ is set to $\mathbf{0}$. Next, the observer is solved with the measurements, continuous inputs, and discrete inputs from each batch in the training data. The result, as in the previous subsection, is a state trajectory for each batch in the training data.

Once a state trajectory has been identified for each batch, quality can be regressed onto the terminal state of each batch to yield a linear model of the desired form. Specifically, the matrices in this regression are:

$$\mathbf{X}_{reg} = \left[\hat{\mathring{\boldsymbol{x}}}\left[T_f\right]^{(1)} \; \hat{\mathring{\boldsymbol{x}}}\left[T_f\right]^{(2)} \cdots \hat{\mathring{\boldsymbol{x}}}\left[T_f\right]^{(B)}\right] \tag{13.21}$$

where $\hat{\mathring{\boldsymbol{x}}}\left[T_f\right]^{(b)}$ denotes the state estimate after the final sampling instant in the bth batch. Note that there is no requirement for the training batches to have equivalent length (i.e., T_f may be different for each batch).

$$\mathbf{Y}_{reg} = \left[\boldsymbol{q}\left[T_f\right]^{(1)} \; \boldsymbol{q}\left[T_f\right]^{(2)} \cdots \boldsymbol{q}\left[T_f\right]^{(B)}\right]$$

where $\boldsymbol{q}\left[T_f\right]^{(b)}$ is the terminal quality measurement in the database for the bth batch. The result of this regression is a model of the form:

$$\boldsymbol{q}\left[T_f\right] = \mathbf{M}_q \mathring{\boldsymbol{x}}\left[T_f\right] + \mu_q$$

where $\mathbf{M}_q$ is the coefficient matrix and μ_q is the intercept term that results from appropriately centering the variables before the regression.

13.5 VD-SQMPC

This section presents a model predictive control scheme that applies the model, defined in Sect. 13.3 and identified in Sect. 13.4, for direct closed-loop quality control. The decision space of this optimization problem includes future continuous inputs and future discrete inputs. In addition, the proposed controller has the ability to select the optimal duration of the batch. The proposed approach is one of the first purely data-driven, within-batch control designs to effectively incorporate batch duration as a decision variable. Furthermore, this result is achieved in a computationally tractable manner that is suitable for online control.

To maintain a modest computational load, the proposed control problem is divided into a two-level optimization problem. The lower level controller is responsible for determining the best input trajectory for a fixed batch duration. The upper level selects between the solutions of the lower level to find the optimal duration. In addition to the two optimization levels, a decision level is included in each calculation step. This decision step is responsible for determining whether or not to terminate the batch at the current sampling instant.

13.5.1 Inner MPC: Fixed Batch Duration

This subsection discusses the lower level MPC for the proposed overall control approach. The objective of this level of MPC is to calculate the best input policy for a fixed duration.

The optimization problem for this level of the control problem is given by:

$$\min_{\boldsymbol{u}[k+1],\boldsymbol{u}[k+2],\cdots,\boldsymbol{u}[T_f],\boldsymbol{u}_{d_1},\cdots,\boldsymbol{u}_{d_L}} \left(\hat{\boldsymbol{q}} - \boldsymbol{q}_{set}\right)' \mathbf{M} \left(\hat{\boldsymbol{q}} - \boldsymbol{q}_{set}\right) + \sum_{i=k+2}^{Tf} \left(\boldsymbol{u}_i - \boldsymbol{u}_{i-1}\right)' \mathbf{P} \left(\boldsymbol{u}_i - \boldsymbol{u}_{i-1}\right) + \cdots \tag{13.22}$$

$$\sum_{l=1}^{L} \left(\boldsymbol{u}_{d_l} - \boldsymbol{u}_{d_l,set}\right)' \mathbf{P}_d \left(\boldsymbol{u}_{d_l} - \boldsymbol{u}_{d_l,set}\right)$$

$$s.t. \quad \hat{\boldsymbol{q}} = \mathbf{M_q} \mathring{\boldsymbol{x}}[T_f] + \mu_q \tag{13.23}$$

$$\mathring{\boldsymbol{x}}[i+1] = \begin{cases} \mathbf{A}\mathring{\boldsymbol{x}}[i] + \mathbf{B}\boldsymbol{u}[i] & \forall k \notin \mathbb{K}_\mathbf{d} \\ \mathbf{A}_d \left(\mathbf{A}\mathring{\boldsymbol{x}}[i] + \mathbf{B}\boldsymbol{u}[i]\right) + \mathbf{B}_d \mathbf{u_d}[i] + \boldsymbol{c}_d & \forall k \in \mathbb{K}_\mathbf{d} \end{cases} \quad \forall\, k+1 \le i \le T_f - 1 \tag{13.24}$$

$$\mathring{\boldsymbol{x}}[k+1] = \hat{\mathring{\boldsymbol{x}}}[k+1] \tag{13.25}$$

$$\boldsymbol{u}_{min} \le \boldsymbol{u}_i \le \boldsymbol{u}_{max} \tag{13.26}$$

$$\boldsymbol{u}_{d_l,min} \le \boldsymbol{u}_{d_l} \le \boldsymbol{u}_{d_l,max} \tag{13.27}$$

where **M** is a positive-definite, symmetric, tuning matrix penalizing deviations of the predicted quality for the candidate input from the target quality. In practice, it is a natural choice to make this matrix a diagonal matrix so that the relative weight on deviations from each of the quality targets can be set explicitly to reflect the importance of each product quality to the end use of the product. **P** is also a positive-definite, symmetric, tuning matrix which penalizes large input movements. This term of the objective function provides a means of dampening input movements. $\mathbf{P_d}$ is the final positive-definite, tuning matrix. The third term of the objective requires a bit more analysis. Similar to the second term, $\mathbf{P}_d$ acts as an input dampening matrix. However, since discrete inputs are made infrequently, the objective is not to prevent movement as it is for the continuous inputs. Instead, the goal of this term is to maintain discrete inputs close to a desired set-point. The physical justification of this is twofold. First, from an industrial operation perspective, it is desirable to follow prior recipes as closely as possible. Second, and more importantly, this term in the objective prevents undue extrapolation from the identified model. This is particularly important for the discrete addition. While historical operation data likely contains variations in continuous input policies (i.e., introduced by trajectory tracking controllers rejecting disturbances), it is far less likely that a database rich in discrete inputs is available. Naturally, in order to identify a discrete input model, some variation in discrete inputs must be introduced in training batches. However, even with these identification batches, it is likely that the training data for these discrete input models will be less rich and it is therefore prudent to take extra steps to avoid extrapolation.

Remark 13.8 Note that in the proposed method, the models are identified using the original (unscaled) variables. For good closed-loop control, however, care must be taken when setting tuning penalty matrices on unscaled engineering units used in measurements and input levels. In this chapter, first tuning matrices were set in a scaled space such that equal penalty values would give equal influence in the objective function. Next, these penalty values were rescaled, using the average input and quality measurement levels, to account for the original engineering units of the variables they penalized. This allowed the control optimization problem to be formulated in the original variables while providing consistent, intuitive tuning parameters. An alternative and equivalent approach would be to rescale the measurements before identification and control calculations.

The constraints in this level of the optimization problem enforce the identified models, maintain operations between physical bounds on the system, and provide the feedback mechanism for the control calculations. Specifically, Eqs. (13.23) and (13.24), respectively, enforce the quality and state models presented in Sect. 13.3 and identified in Sect. 13.4. Equations (13.26) and (13.27) restrict continuous and discrete inputs, respectively, between their operational bounds. Equation (13.25) implements the current estimate of the state into the control calculation as the starting point for the predicted state and quality predictions. This step is critical to the proposed method and provides the feedback mechanism that allows the proposed control approach to reject disturbances and drive the process to the desired set-point.

Note that all terms of the proposed optimization problem are quadratic and all constraints are linear. Furthermore, all decision variables on this level of the MPC are continuous bounded variables. As a result, the proposed optimization problem is a relatively modest QP. Furthermore, the problem is convex. Therefore, exact solutions to this lower level of the proposed control scheme can be calculated in a computationally efficient manner.

Observe that, for a given sampling instant during a batch, the solution to the QP proposed in this subsection is dependent only on the current subspace state estimate, $\hat{\boldsymbol{x}}[k+1]$, the current input level, $\boldsymbol{u}[k]$, and a candidate batch duration T_f. Since the problem is a convex QP with an exact solution, this defines a mapping which can be denoted as follows:

$$\{\boldsymbol{u}^*, \boldsymbol{u}_d^*, \hat{\boldsymbol{q}}^*\} = \mathbb{C}_1\left(T_f, \hat{\mathring{\boldsymbol{x}}}[k+1], \boldsymbol{u}[k]\right)$$

Where $\boldsymbol{u}^*$ denotes the optimal continuous input trajectory and $\boldsymbol{u}_d^*$ denotes the optimal discrete input policy. By applying the identified models, the predicted quality resulting from implementing the calculated input trajectories, $\hat{\boldsymbol{q}}^*$, is also obtained. This abbreviation of the QP solution will be used in the remainder of this paper for notational simplicity. Note however, that for the constrained problem, this mapping does not have an explicit analytical solution and instead must be solved numerically for each new set of conditions.

The formulation proposed in this subsection provides a means to efficiently calculate optimal input trajectories for a batch of fixed duration. However, as noted earlier, batch duration is a key factor in determining batch quality. With this in mind, the next subsection describes how the formulation from this section can be used to implement an overall controller that considers batch duration as a decision variable.

13.5.2 Outer MPC: Selecting Duration

The previous subsection described a control scheme to find the optimal input policy for a batch process (including discrete and continuous inputs) given a fixed batch duration, T_f. The control calculation depended on the current state estimate, $\hat{\mathring{\boldsymbol{x}}}[k+1]$, and input, $\boldsymbol{u}[k]$. The next step is to use this relationship to determine the optimal batch duration.

Using the mapping described in the previous subsection, a policy for selecting the optimal duration can be defined. The optimization problem is as follows:

$$\min_{T_f \in \mathbb{Z}} \left(\hat{\boldsymbol{q}}^* - \boldsymbol{q}_{set}\right)' \mathbf{M} \left(\hat{\boldsymbol{q}}^* - \boldsymbol{q}_{set}\right) \tag{13.28}$$

$$s.t. \quad \{\boldsymbol{u}^*, \boldsymbol{u}_d^*, \hat{\boldsymbol{q}}^*\} = \mathbb{C}_1\left(T_f, \hat{\mathring{\boldsymbol{x}}}[k+1], \boldsymbol{u}[k]\right) \tag{13.29}$$

$$T_{f,min} \leq T_f \leq T_{f,max} \tag{13.30}$$

where $T_f \in \mathbb{Z}$ denotes that T_f is restricted to integer values. Note that this restriction is necessary because the discretization of our dynamic models prevents their application for partial sampling instances. The objective of this optimization problem is a pared-down version of the optimization problem at the lower level. In this chapter, the design decision was made to eliminate the input movement suppression term and the discrete input set-point term in the upper level MPC. This decision is motivated by the fact that at the upper level, the input vector is of different lengths for different batch durations. As a result, in extreme cases, a longer duration batch providing better predicted quality could be overtaken in the objective function by a shorter batch with fewer input movements to penalize. There is, however, an argument for including input movement in both levels of the MPC. That argument runs that if smooth input trajectories are desirable, jumpy inputs should be penalized at all levels and doing otherwise is inconsistent. One way to rectify these opposing views would be to incorporate an input penalty weighting so that all input trajectory lengths would be penalized equally.

Remark 13.9 In this chapter, focus is on processes where the primary economic benefits come from tight quality control and therefore quality control is more important than production rate. That said, for batch processes, batch duration fixes production rate. As a result, in many industrial examples one objective of batch operations is to minimize batch durations. In order to address these production goals, an alternative viable configuration of the outer control level would be to select the minimum duration for which the optimal quality, $\boldsymbol{q}^*[T_f]$ falls within an acceptable range. Note that, in this situation, it is better to make quality control a constraint: otherwise the optimal solution could select short batch durations at the direct detriment of product quality. However, note that with quality as a constraint, quality control to an exact quality set-point is not feasible, instead one must settle for driving quality to an acceptable range.

Implementation of the optimization problem specified by Eqs. (13.28)–(13.30) includes more subtleties compared to the lower level. The first subtlety arises from the fact that the decision variable is an integer variable making the optimization an integer programming problem. Furthermore, the constraint governing $\boldsymbol{q}^*$ is itself a constrained quadratic problem. Fortunately, the decision space for this optimization problem is limited to a tightly bounded one-dimensional range. This small decision space and the relatively quick solution of the lower level QP permit solution of this problem by total enumeration. In Sect. 13.5.4, where the practical implementation of the proposed control design is discussed, further steps to solve this outer MPC problem will be addressed.

Further improvement in the ability of our controller to reach the desired quality can be achieved using the quality model to make an informed decision about when to terminate a batch. In the next subsection, logic is proposed to make this batch termination decision.

13.5.3 Batch Termination

Recall that the quality model permits quality predictions from any state estimate. With this in mind, one logical way to determine whether to terminate a batch would be as follows: First calculate the quality corresponding to the current state estimate. Apply the quality deviation term from the MPC objective function to this predicted current quality. Next carry out the MPC calculations. If the objective of the quality at the current sampling instant is less then the quality objective predicted by the

MPC, terminate the batch. This policy makes intuitive sense since it terminates the batch if and only if the current predicted quality is better than any future quality. Unfortunately, this approach will inevitably fail. To understand why, consider that it is extremely improbable for any current state to produce exactly the correct quality. However, the MPC will usually be able to invert the model in a way to drive the quality to the set-point at some time in the future. Therefore, the termination policy proposed above would lead to a controller continually 'chasing' the set-point. The factor that is unaccounted for in this termination method is that there is always plant-model mismatch which is not accounted for in the predicted future qualities. To overcome these difficulties, a different termination strategy is required.

The flawed termination logic proposed above has two logical features that make it beneficial in theory. These are that the batch is terminated if the quality is very close to the desired set-point and that the batch is terminated if no better quality is expected in the future. Ideally, both of these features should be kept in the proposed, effective termination logic. To achieve this, in the proposed approach two possible termination criteria are considered. If either is true, the batch is terminated. The first termination criteria are designed to address the current state. As before, the current state estimate is used to predict the current quality. Then, this predicted current quality is checked to see whether it is within a set of tight terminal quality bounds. If it is, the batch is immediately terminated. This provides the first feature of termination logic that the batch should terminate if it is very close to the desired quality target. However, it does so while avoiding the chasing behavior that is set up by comparing the current quality to potential future quality.

The second termination condition deals with the possibility that the current quality is better then any future qualities will be. This is achieved by asserting that a batch should be terminated if the current time is equal to the optimal duration selected by the controller at the previous sampling instant. Recall that the outer layer of the MPC makes duration decision based on selecting the best quality from the set of predicted qualities for all possible durations. Therefore, if the batch duration selected in the last sampling instant is equal to the current time, it means that at the last control calculation it was determined that any longer batch durations would result in poorer quality objectives. Terminating based on duration selected in the previous sampling instant effectively solves the set-point chasing problem since all of the qualities considered by the outer MPC are pure prediction and therefore incorporate the same plant-model mismatch. In a sense, the comparison made by the outer MPC when selecting duration is a comparison of apples to apples so the plant-model mismatch is not a problem.

The previous three subsections have presented the fundamentals for a data-driven, model predictive control scheme capable of direct quality control considering batch duration as a decision variable. While these fundamentals provide a basis for the method proposed in this chapter, practical implementation considerations and a few heuristic methods are needed to achieve the most effective control. The next subsection outlines these practical considerations and provides a complete overview of the proposed method.

13.5.4 Additional Features and Implementation

This subsection presents the so-called bells and whistles that make the methods proposed in the previous subsection practicable. Figure 13.2 shows the timing of the proposed MPC steps. The process begins when the batch is charged. Immediately, the first measurement is taken and the first continuous input is applied. Note that since there is no time between the first measurement becoming available and the first input being applied, it is not possible to apply the MPC until the subsequent sampling instant. Instead, for the first sampling instant of the process, a nominal input (picked from past successful batches) is used. In the present study, this was chosen from the input profile of one of the training batches. After a new measurement becomes available at sampling instant k, the observer is updated with the new measurement. The observer update provides the state $\hat{\mathring{\boldsymbol{x}}}[k]$. Since the input $\boldsymbol{u}[k]$ has already been implemented, the next input to decide is $\boldsymbol{u}[k+1]$. As such, the state $\hat{\mathring{\boldsymbol{x}}}[k+1]$ needs to be the basis of control calculations. To get $\hat{\mathring{\boldsymbol{x}}}[k+1]$, the dynamic model (and discrete input model if needed) is applied to the last available state estimate $\hat{\mathring{\boldsymbol{x}}}[k]$. Next, the prediction of $\hat{\mathring{\boldsymbol{x}}}[k+1]$ is used in the termination calculations to determine whether to end the batch. If the decision is made to terminate, the controller waits until the end of the current (kth) sampling instant, discharges the batch, and takes the final quality and output measurements. Otherwise, the MPC calculations are carried out to determine the optimal input policy. When the sampling instant is complete, the discrete input and first input of the candidate input profile are implemented and the process repeats.

To address the practical implementation of the proposed method, a few heuristic tweaks are applied in various phases throughout the batch. Figure 13.3 shows the temporal regions for which these heuristics are applied. The details of each region are discussed in detail in the remainder of this subsection.

In the beginning of the batch, the relatively small number of output measurements available means that the state estimate is relatively inaccurate. This inaccuracy arises despite using our initial state estimate because the initial state estimate cannot account for batch-to-batch variations in feedstock. Importantly, the state estimate is the only source of feedback informing the decision of input trajectory by the MPC. As a result, using input trajectories calculated by the MPC based on an inaccurate state estimate is, at best, unhelpful and at worst, could drive the process into a region of state space for which our models are inaccurate or for which the quality is far off-

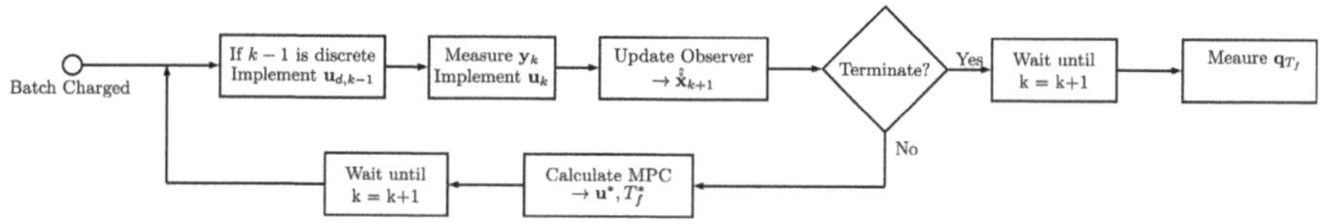

Fig. 13.2 Flow diagram of timing of the proposed MPC scheme

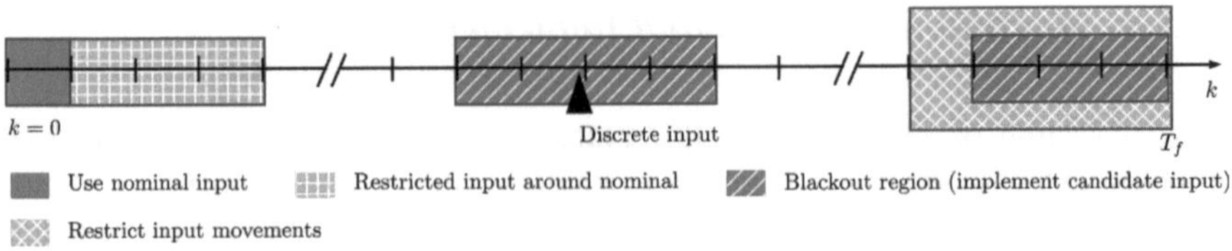

Fig. 13.3 Time line of implementation features that improve the ability of the proposed controller to obtain desired quality

spec. To avoid this, in practice tight bounds on the input trajectory (around a nominal input trajectory) are applied for the first several sampling instances of a new batch. This provides time for the observer to converge to a correct state estimate before large input moves are implemented.

Like the beginning of the batch, special care must be taken toward the end of the batch. However, unlike the beginning of the batch, the issues at the end of the batch do not arise from inaccurate state estimates. Instead, the issue that needs to be avoided is over-excitation of the process. Because the quality term in the objective function only penalizes the last sampling instant, there is a large incentive (from an optimization perspective) to make large input moves in the last few sampling instances of the batch to exploit the linear dynamic model by exciting the process enough to hit the desired quality. Of course, since this kind of excitation does not occur in any of the training data, this input behavior takes the models well outside their range of validity. Therefore, it is unlikely that this kind of policy will actually result in the predicted behavior. To avoid this effect, tight bounds on input movements are placed on a period toward the end of the batch.

Remark 13.10 In principle, there is an argument that tight input movement constraints like the ones discussed above should be applied to all sampling instances throughout the batch. It is true that large input movements cause model validity problems in all time regions of the process. However, for every sampling instant where an input movement constraint is included, two new inequality constraints must be added to the optimization problem. In our experience, the tendency for the controller to proscribe this kind of large input only occurs toward the end of the candidate input (likely because the input movement penalty prevents it in other parts of the batch). Therefore, in order to reduce the number of constraints and increase the computational speed, we elected to only include input movement constraints for a window at the end of the batch.

The final heuristic tweaks used in this chapter are so-called blackout regions. Practically, these are user-specified regions around critical batch events (discrete inputs, batch termination, etc.) where rather then calculating new input trajectories at every sampling instant, the candidate input trajectory from the last sampling instant before the blackout region is implemented. In this chapter, these regions are used around discrete inputs and before batch termination. The justification for using blackout regions after discrete additions is similar to that for using a restricted input range at the beginning of a batch. Discrete inputs generally have a large, instantaneous impact on the system state. As a result, directly after a discrete addition, it is likely that the state estimate will have higher then average error. Applying a blackout region after the discrete addition provides the observer some time to converge and prevents input decisions from being made with poor state estimates.

The justification for applying a blackout phase before discrete additions is more nuanced. The benefit of this policy stems from the identification of the dynamic model. Recall that for a period of i sampling instances before and after the addition, it is impossible to identify a state trajectory during subspace identification. This inability is because a window of i contiguous past and future output measurements is needed. As a result, our dynamic models have a built-in blind spot around the discrete addition sampling instant. This slightly reduces the accuracy of the dynamic model around those instances. From experimental observations, this inaccuracy appears to particularly negatively impact the state observer around those sampling instances. Once again, to avoid making detrimental input decisions based on inaccurate state estimates, a blackout period is applied.

The same justification as the above about blackout regions before discrete inputs can be made about the end of the batch. In the subspace identification step, state values of any given training batch cannot be determined for i sampling instances before batch termination. However, there is a second, stronger justification. It is that applying a blackout period at the end of the batch further avoids the set-point 'chasing' behavior described in Sect. 13.5.3. Applying a blackout period at the end of the batch ensures that the duration decision is made several sampling instances before termination which effectively prevents the controller from chasing the set-point trying to eliminate plant-model mismatch.

Remark 13.11 The most extreme extension of the blackout region concept is to use a small number of fixed decision points throughout the batch and blackout all other sampling instances. This decision point strategy has been widely proposed and accepted in the batch literature [5]. The justification for this policy in general is to prevent undue excitation of the process. It also has the benefit of allowing the operator to approve candidate input trajectories after every decision point. For the proposed approach, using decision points has some theoretical justification as well in that it potentially reduces the amount of excitation of the process which allows the observer to converge to a more accurate state estimate. However, the downside is that by limiting the number of times the trajectory is updated, the speed at which the controller can reject new disturbances is also reduced.

13.6 Simulation Study

This section demonstrates the efficacy of the proposed approach through an industrially relevant simulation example. Specifically, this simulation study considers the free-radical polymerization reaction of polymethyl methacrylate (PMMA). In this section, first a brief overview of the process, its objectives, and its simulation are presented. Next the model building approach is demonstrated and validation results are presented. Finally, the ability of the proposed controller to eliminate quality error introduced through feedstock disturbance is demonstrated.

13.6.1 Overview of PMMA Process

In this study, PMMA quality is characterized by its number and weight average molecular weight as well as the percent conversion of monomer to polymer. Therefore, the control objective is to reach the desired values of these three variables by batch termination.

The reaction takes place in a continuously stirred jacketed reactor where jacket temperature is manipulated following a zero-order hold throughout the batch. Note that it is assumed that jacket temperature dynamics are much faster then the reactor dynamics so that, for modeling purposes, jacket dynamics are considered negligible. At the beginning of each batch, the vessel is charged with the monomer, the initiator (AIBN), and the solvent (toluene). To simulate feedstock disturbances, normally distributed errors are introduced to the initial charge recipe. The nominal recipe for this initial charge, as well as the standard deviation for each ingredient, is presented in Table 13.1.

To achieve the desired number and weight average molecular weight, the recipe for this process calls for a mid-batch addition of monomer. In this chapter, it is assumed that this addition is time indexed (i.e., made a specified amount of time after the initial charge of the reactor). Specifically, the recipe calls for an addition of 1.3 kg of monomer two hours after the start of the batch.

Throughout the batch, online measurements of reaction media temperature, density (calculated using volume measurements), and viscosity (calculated from stirrer torque) are considered available with a sampling frequency of one minute. To simulate measurement noise, white noise was added to each of the measurements. To serve as the simulation test bed, a nonlinear first-principles model from [6] was adapted with appropriate changes from [7, 8]. The final model consisted of nine states dictated by nonlinear ODEs. These states were the concentration of monomer, concentration of initiator, six states for the moments of the molecular weight distribution, and the temperature. Throughout the study, this model was treated as a purely black-box model. The model was used to generate a database of historical and training data and as a test bed for the proposed control.

Table 13.1 Initial conditions for PMMA simulations

Ingredient	Nominal initial concentration	Standard deviation
Methyl methacrylate monomer	$3.0\,kg/m^3$	0.5
AIBN initiator	$1.03 \times 10^{-2}\,kg/m^3$	1×10^{-3}
Temperature	61 °C	2.5

13.6.2 Existing PI Control

It is assumed that the batch has been historically operated under a PI controller tracking an optimal temperature trajectory taken from [6]. The overall rate of reaction for each batch varies because of variations in the initial concentrations of initiator and monomer. To account for this variation, alignment techniques were applied to the set-point trajectory. Specifically, the set-point trajectory for a nominal batch was indexed to density, which increases monotonically and captures overall reaction progress. For new batches, the temperature set-point at any given sampling instant was selected from the set-point from the nominal batch at the corresponding density.

With this alignment in place, the tracking PI controller was tightly tuned for a set of batches to minimize the root-mean-squared tracking error. The resulting tuning parameters for the PI are shown in Table 13.2. Despite the nonlinear dynamic response of the reactor, the trajectory tracking performance of the PI controller was very good as demonstrated in Fig. 13.4.

To address the effects of batch duration on the quality for the PI trajectory tracking case, density measurements were used. Because density reflects the average polymer chain length, it also reflects, on some level, closeness to the desired molecular weight distribution. For the PI case, this was exploited by using density measurements to determine when to terminate a batch. Specifically, a density threshold was taken from the nominal batch. For new batches, the reaction was carried out until either the density threshold from the nominal batch was reached or the maximum amount of time (five hours) had elapsed.

Table 13.2 Tuning for benchmark PI trajectory tracking controller

Parameter	Value
k_c	−0.1
τ_i	30

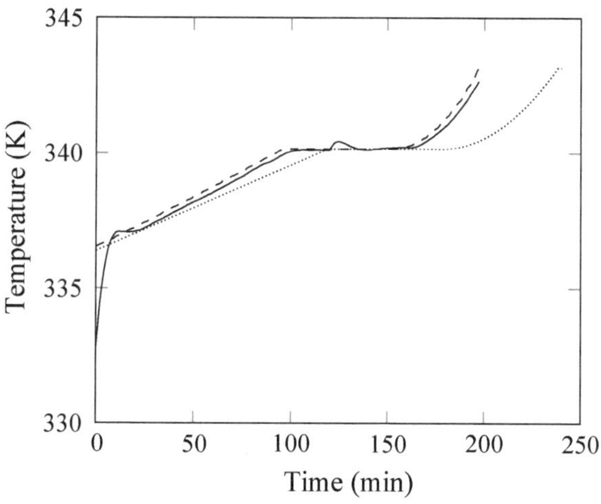

Fig. 13.4 Representative PI trajectory tracking response. Nominal set-point (dotted line), set-point aligned on density (dashed line), response (solid line)

The PI trajectory tracking controller described above served two roles in this work. First, it provided a reasonable way to simulate data available from historical operation of an industrial process. This historical data was used in training and validating models as discussed in the following subsections. The PI controller was also used as a benchmark for comparison of the proposed model predictive control. This PI scheme is a reasonable benchmark both because it is representative of the industrial state of the art and because it provides a rudimentary implementation of variable batch duration.

13.6.3 Training Data Generation

The first step in this simulation study was to generate an appropriate database of historical and identification batches to use as training data for the proposed model identification technique. To this end, a database of 37 batches was constructed. A summary of the input policies for these batches is given in Table 13.3. Of these batches, 20 batches were generated under the PI trajectory tracking control as described in the previous subsection. These batches were intended to simulate data that would be available from historical process operation.

Ideally, all of the models could be identified from historical data which has the distinct advantage of being cost free. In contrast, running identification batches may yield off-spec product which is either less profitable or completely unsalable. Unfortunately, historical data alone is unlikely to provide a rich enough database to adequately identify dynamic behavior. To improve the richness of the training data, seven open-loop training batches were generated (shown in Fig. 13.5). Of these batches, three were generated by implementing the nominal input trajectory shifted by a pseudo-random binary sequence signal (PRBS). The remaining four identification batches were generated by shifting the nominal input by a constant amount (two shifted higher and two lower). These shifted nominal trajectories provide valuable information because they illuminate the behavior of the process when inputs are operated near their limits, expanding the range in which identified models can be applied without extrapolation.

In addition to dynamic identification batches, batches designed to identify effects of monomer addition were also generated. Recall that in the existing control scheme, the amount of monomer added in each batch was constant. In order to quantify the

Table 13.3 Composition of batch database for model identification

Batch input policy	Number of batches
Historical PI operation	20
PI with perturbations of monomer addition	10
PRBS added to nominal continuous input	3
Nominal input shifted up or down by a constant	4

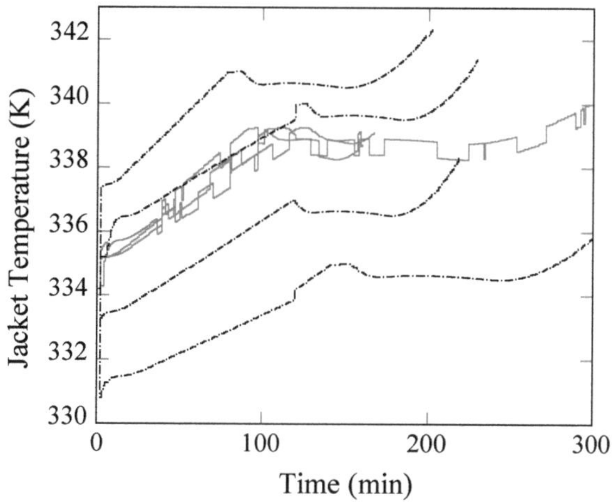

Fig. 13.5 Seven identification input signals (three PRBS (solid line) and four shifted nominal (dash dotted line)) added to the training database

effect of monomer addition, the amount added was varied linearly between 0 and 2.6 kg for the seven identification batches and 10 PI batches. The large number of discrete input identification batches required is reflective of the fact that each of these batches only provides one new datum point. This is in contrast to dynamic identification batches which each provide hundreds of new dynamic observations.

Remark 13.12 For continuous processes, when closed-loop controllers are applied, the inputs are strongly dependent on the previous outputs because of the proportional mode of the controller. Mathematically, this is because of the constant set-point in this kind of control design. Fortunately, for batch processes under trajectory tracking control, this problem is alleviated by the constantly changing set-point trajectory. By implementing a time-varying set-point, there is a component, independent from the output measurements, that is introduced to the error at each sampling instant. Theoretically, this breaks the necessity to have purely open-loop data for subspace identification. However, from a practical standpoint, open-loop data still provides the richest dynamic information for identification purposes. It is also worth noting that alternative subspace algorithms that can handle closed-loop data have been presented and may be applicable [9].

13.6.4 Model Training and Validation

This subsection describes how the database identified in the previous subsection was used to apply the methods discussed in Sect. 13.4 to identify a dynamic, discrete input, initial state, and quality model for the PMMA process.

Dynamic Model

The first step of the proposed identification approach is to identify a dynamic state-space model from training data outputs and inputs using subspace identification.

Recall from Sect. 13.4.1 that the number of Hankel rows i and number of identified states are user-specified parameters. Generally speaking, the number of Hankel rows should be selected to be slightly higher then the number of states. With this in mind, focus can be solely on selecting an appropriate number of states.

In the literature, subspace algorithms are shown to be capable of selecting system order from training data based on the amount of explained variance. However, for the transient systems in this work, selecting system order from training data alone is fraught as it is impossible to tell if undue over-fitting has occurred. The solution is to use a set of validation batches omitted from the training data. To do this, first the appropriate pseudo-Hankel matrix is constructed using only the training data (omitting the validation data). Next, subspace identification is repeatedly carried out using a range of potential system orders. Finally, the identified model of each order is used to make predictions for the validation batches. If the identified model adequately predicts behavior of the validation batches, it is strong evidence that an appropriate system order has been selected.

Remark 13.13 Even if the user has very good process knowledge, it is not straightforward to estimate the model order. This is because process knowledge only informs the nonlinear evolution of fundamental states. A process which can be described well with a small number of nonlinear fundamental states may require substantially more LTI states to be described effectively. Similarly, it is possible that some fundamental states are redundant and the process can be represented with substantially fewer LTI subspace states.

The metric for validation is a comparison of predicted outputs and measured outputs from validation batches. In practice, validation is complicated by discrete inputs. Consider the validation steps: First, an observer is constructed using the dynamic model identified with a given number of states. Next, for each validation batch, the observer is updated with observations from the beginning of the batch. This provides a state estimate for sampling instant $k_{predstart}$. Next the dynamic model is applied to make output predictions for the rest of the batch using the estimated state as an initial condition and the input trajectory in the database. Unfortunately, however, without first identifying a discrete input model, the effects of discrete additions cannot be accounted for. Consequently, outputs occurring after discrete inputs cannot be used for the validation. The solution is to identify the discrete input model prior to validation. Doing so has the added advantage that validation evaluates accuracy of the combined dynamic and discrete input models. As a result, the system order that provides the best overall dynamic description of the process is selected. With this in mind, in the next subsection identification of the monomer addition model is discussed and validation results are presented for the combined models.

Mid-Batch Addition Model

As described in Sect. 13.6.1, the recipe for this process involves a single discrete input, in the form of a shot of monomer, which takes place 120 min into the batch. To identify the discrete input model, the observer based on the dynamic model identified in the previous subsection was updated, for each batch in the training data, first with observations up to 119 min and then again with observations from 120 min on. The resulting state estimates were used to fit a regression model as discussed in Sect. 13.4.2.

As mentioned in the previous subsection, the discrete input model and the dynamic model were validated together. For each potential system order, the root-mean-squared error (RMSE) of output predictions for validation batches is calculated. Each number of states between four and sixteen was tested. Based on these results, it was determined that using six subspace states gave the best output predictions for the validation batches. On this basis, the dynamic and discrete input models with six states were used for the rest of this chapter.

Figure 13.6 shows the output results for the validation of the dynamic model. To generate this figure, for each validation batch, first the observer was updated with observations between 1 and 50 min. This provided the state estimate at $k = 50$. Next the dynamic model was applied utilizing the actual inputs to the process. The identified mid-batch model was applied as appropriate between the 119 and 120th min. Finally, the complete state trajectories were used with the identified C and D matrices to make output predictions for each validation batch. These results demonstrate the predictive ability of both the dynamic model and the monomer addition model.

As discussed earlier, for subsequent validation and closed-loop results, the initial values of the states were the mean value of the identified states from the training data set and are reported in Table 13.4.

Quality Model

The next step in the identification process was to identify the quality model. This was done as discussed in Sect. 13.4.4. In the simplest implementation of the method proposed, there are no user-defined model parameters involved in this identification step. However, in practice, it was found that better results (in terms of overall validation) can be obtained by limiting the number of states used in this step. To explain this finding, recall that the states in subspace identification are identified in order of decreasing explained output variance. As a result, the last states in the identified dynamic model are substantially more difficult to observe and are also more susceptible to noise. While these states demonstrably improve dynamic behavior prediction (see previous section discussing validation of dynamic models), error in model predictions for these states negatively impacts quality predictions. For the PMMA process, it was found that a model built on the first four subspace states gave the best performance. This number was arrived at by comparing validation results. Therefore, the final identified quality model related four terminal subspace states to M_N, M_W, and conversion.

To validate the identified quality model, the observer was applied to obtain terminal state estimates for each of the validation batches. These terminal state estimates were then applied in the quality model and the results compared to the actual measured qualities. This comparison is plotted in the bottom row plots in Fig. 13.7. The results clearly demonstrate the predictive ability of the model.

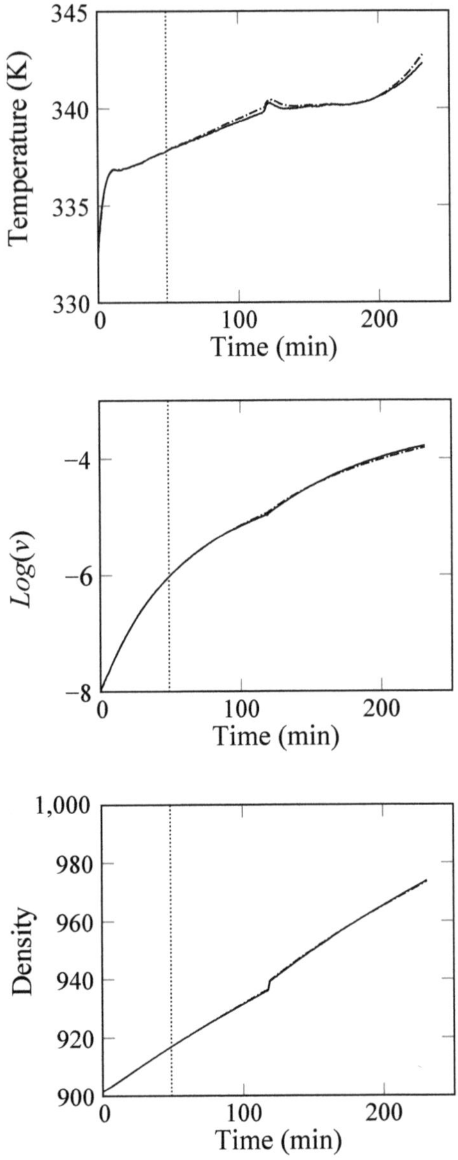

Fig. 13.6 Validation of dynamic and mid-batch addition model for prediction beginning after updating the observer for 50 min of outputs. Dot-dashed line is the true output values, solid line is the prediction

Overall Validation

The last model identification step was a final, overall model validation. In this step, the dynamic, discrete input, and quality models were combined to verify the predictive ability of the coupled models. Similar to the dynamic validation procedure, first the observer was applied for each batch with observations up to a specified prediction start time. Next, these state estimates were applied in the dynamic and discrete input

Table 13.4 Initial state estimate

State	Value
$\mathring{x}_1$	1132.2219
$\mathring{x}_2$	−10.2826
$\mathring{x}_3$	−0.6371
$\mathring{x}_4$	1.9188
$\mathring{x}_5$	−0.2167
$\mathring{x}_6$	0.1616

models (as appropriate), using the input implemented for each validation batch, to predict the terminal subspace state of each batch. Finally, the predicted terminal subspace states were used to make quality predictions. This procedure provides valuable validation insights as it closely resembles the way these models are applied in model predictive control. The results of this procedure for predictions starting at 50 and 120 min are shown in Fig. 13.7.

Remark 13.14 Note that quality predictions improve as the length of the prediction horizon shrinks and more observations are used in the estimator. In the beginning of the batch, using the first 50 min of measurements, quality predictions are not accurate but are good enough to inform input moves of the correct direction. By the end of the batch, predictions improve allowing for tight quality control. This creates a stabilizing effect where, as the batch gets closer to termination, better predictions allow for appropriate inputs to fine-tune the final quality.

13.6.5 Tuning

With the models identified, the next step was to construct the model predictive control for this process. This was done following the framework laid out in Sect. 13.5. In practice, determining reasonable values for each of the tuning parameters can be challenging. For the base case, the tuning parameters needing to be identified are: the quality penalty, the input movement penalty, the observer initial state penalty, the observer output penalty, and the observer state validity penalty. When one also includes the features discussed in Sect. 13.5.4, additional tuning parameters are introduced. These are the duration of restricted input movements at the start of the batch, the duration of the blackout regions at discrete inputs and at the end of the batch, and the duration of the restricted input movement at the end of the batch. In principle, this large number of tuning parameters could be a distinct downside to the proposed method. Fortunately, it was found that the method is fairly insensitive to the values of the selected tuning parameters.

To demonstrate that the proposed method is robust to tuning parameter selection, simulations of batches, starting from a representative set of 10 initial conditions, were repeated for 20 randomly selected tunings. The results, in terms of mean, maximum,

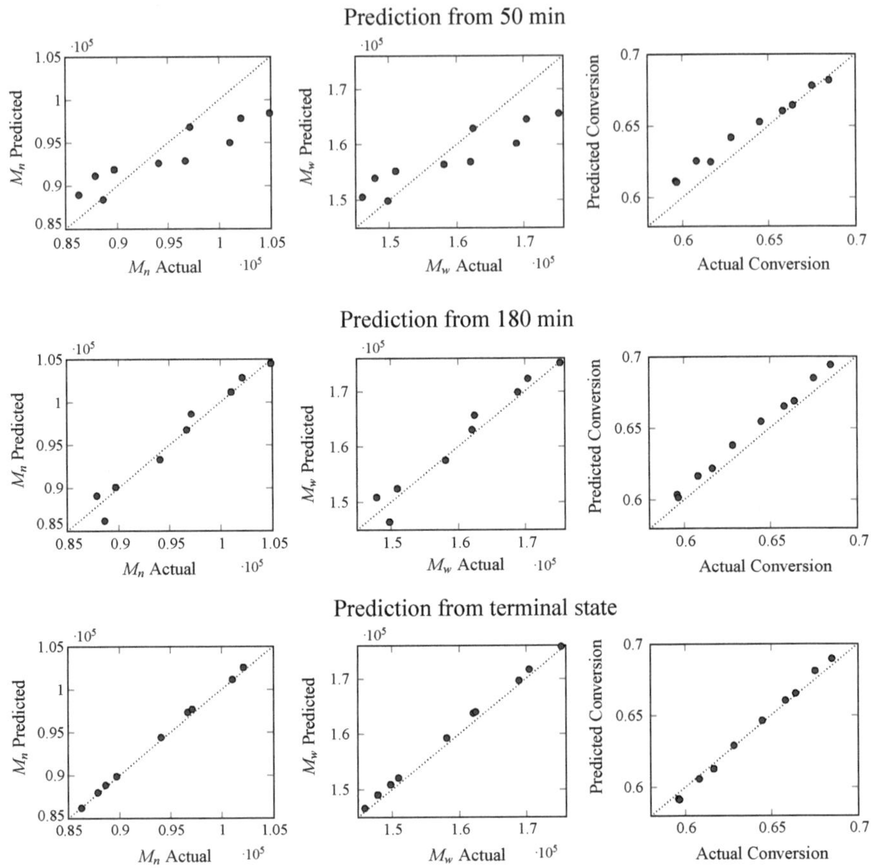

Fig. 13.7 Rows 1 and 2—Overall validation of the combined models for predictions starting 50 and 180 min after the beginning of the batch. Row 3—Validation of the quality model for 'true' terminal state. (Note: points on the diagonal represent perfect predictions)

and minimum relative error for each tuning, are shown in Fig. 13.8. Note that, for all but three tunings, the mean absolute percentage error was lower then that for PI trajectory tracking. In 14 of the 20 tunings, the maximum error over the three qualities and ten batches was also reduced. This demonstrates the relatively low sensitivity of the proposed approach to the specific values of tuning parameters.

For the remainder of this simulation study, the tuning parameters presented in Table 13.5 are used. This tuning was selected from a Monte Carlo approach similar to the one described above. Clearly, this kind of Monte Carlo tuning is not practicable for an actual process. However, this approach was used in this chapter in an effort to demonstrate the efficacy for control of models identified using the proposed identification approach. Future work will address tuning using a more explicit and practical procedure.

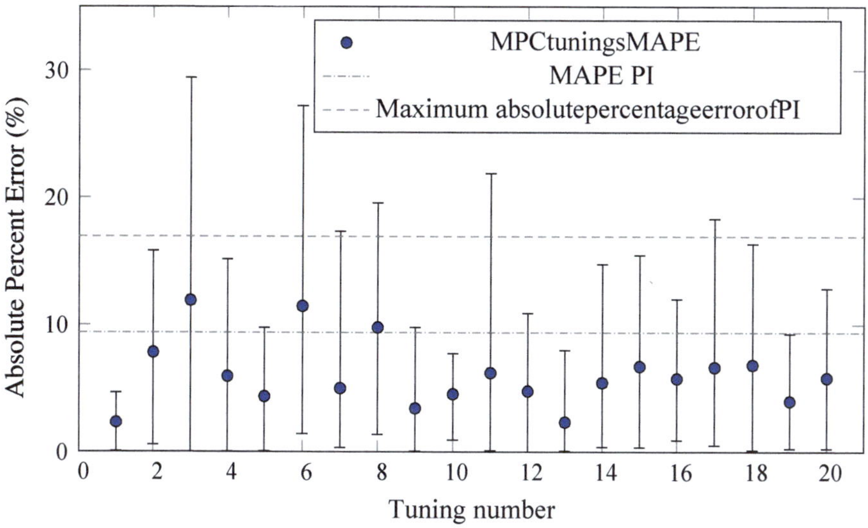

Fig. 13.8 Comparison of 20 random tunings for the proposed controller applied to 10 representative batches. Plotted points are mean absolute percentage errors (MAPE) for all three qualities and all ten batches. Error bars show the maximum and minimum observed absolute percentage errors over the ten batches and three qualities. Similarly, the dot-dashed line shows the MAPE for the same batches under PI control and the dashed line shows the maximum absolute percentage error for PI

Table 13.5 Controller tuning used for the closed-loop studies. Note that all penalties are rescaled so that presented represent relative weight in the objective function (based on magnitude of variables they penalize)

Tuning parameter	Value
Penalty on M_N	10
Penalty on M_W	4
Penalty on conversion	8
Jacket temp. movement penalty	0.5
Observer initial state penalty	diag(0.01)
Observer innovation penalty	diag(200)
Observer state validity penalty	diag(60)
Initial input restriction	50 min
Blackout before monomer addition	25 min
Following monomer addition	15 min
Before batch termination	50 min
Batch end input movement restriction	30 min

13.6.6 Closed-Loop Results

To demonstrate the efficacy of the proposed control structure, the models identified in the previous subsections were applied in the control framework presented in Sect. 13.5. To demonstrate the flexibility of the proposed approach, two case studies were carried out. For both case studies in this section, the same 200 initial conditions, different from the training and validation set, were used. These initial conditions were drawn from a normal distribution centered on the nominal initial batch condition. Table 13.1 shows the nominal initial fundamental state and the standard deviation in each initial condition. The two case studies differ only in the selected quality set-point. Therefore, comparison of these studies demonstrates the efficacy of the proposed approach to drive the process to the desired quality.

Base Case: Nominal Quality Set-Point

In the first case study, the quality set-point for the controller was taken from a nominal batch in the training database. For the nominal batch, the initial fundamental state was exactly the nominal initial fundamental state and trajectory tracking was applied. At the discrete monomer addition, the nominal amount of monomer was added. Because every variable in the nominal batch was forced to the middle of its distribution, the nominal batch quality represents a reasonable expectation for the quality arising from PI trajectory tracking. Therefore, it is a good set-point to use for the proposed controller to provide a comparison to the trajectory tracking approach.

Figure 13.9 shows histograms of the qualities achieved by the proposed VD-SQMPC compared to the trajectory tracking control for the first case study. Clearly,

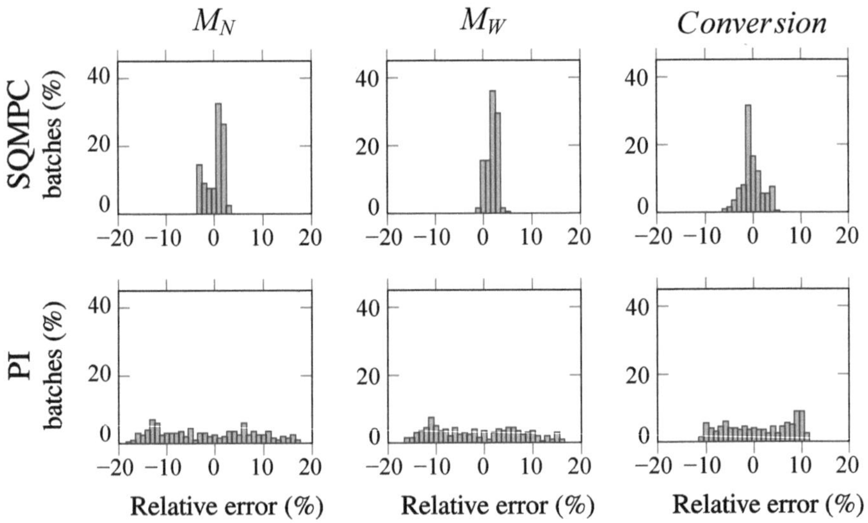

Fig. 13.9 Histograms of relative errors in qualities resulting from PI trajectory tracking and the proposed VD-SQMPC control schemes for the base case

Table 13.6 Mean absolute percentage error (MAPE) in qualities resulting from PI and the proposed VD-SQMPC control schemes

Quality variable	VD-SQMPC (%)	PI (%)
Conversion	1.67	6.13
M_n	1.61	8.66
M_w	1.90	7.87

the proposed control scheme is able to reduce the variance of all three quality variables compared to PI trajectory tracking. Specifically, the mean absolute percentage error (MAPE) values for the base case for both the PI and VD-SQMPC are shown in Table 13.6. Furthermore, in 172 out of the 200 batches, the magnitude of the error was reduced for every one of the three quality measurements when compared with PI control. Figure 13.10 shows a scatter plot matrix of qualities resulting from both PI trajectory tracking and the proposed VD-SQMPC. Note that the VD-SQMPC rejects disturbances in initial condition and, as a result, significantly tightens the cluster of qualities obtained.

Figure 13.11 shows the input trajectories for five representative batches from the proposed VD-SQMPC approach. PI trajectories for the same batches are shown in light gray for reference. Interestingly, the shape of the input trajectory proscribed by the VD-SQMPC varies substantially between batches. This variation reflects the different actions taken by the proposed controller to reject disturbances in the initial conditions. In addition to selecting an appropriate continuous input, the VD-SQMPC is able to coordinate the approach to the continuous input with the discrete addition of monomer. Figure 13.12 is a histogram of the monomer addition amounts for the batches in the base case. The fairly even distribution of monomer addition amounts indicates that the inverted empirical models are able to appropriately proscribe monomer addition amounts.

Remark 13.15 It is interesting to note the high percentage of batches for which the maximum allowable monomer addition (2.5 kg) was made. Since a large number of batches hit this upper bound, it indicates that there would potentially be an advantage to increasing the limit on added monomer. However, to do this, additional training batches with greater amounts of added monomer would be necessary to expand the range of monomer additions. In principle, this could readily be done by using these new batches to re-identify the model and use for the purpose of closed-loop control.

Figure 13.13 is a histogram of the batch durations selected by the controller. Interestingly, the distribution is centered very near the nominal batch duration of 240 min. This makes intuitive sense since the duration of the reaction has a strong impact on the degree of completion of the reaction and therefore the resulting quality. Note, however, that the controller has selected durations varying by as much as 15 min from the nominal duration in order to account for variations introduced to each batch by disturbances in the initial ingredient concentrations.

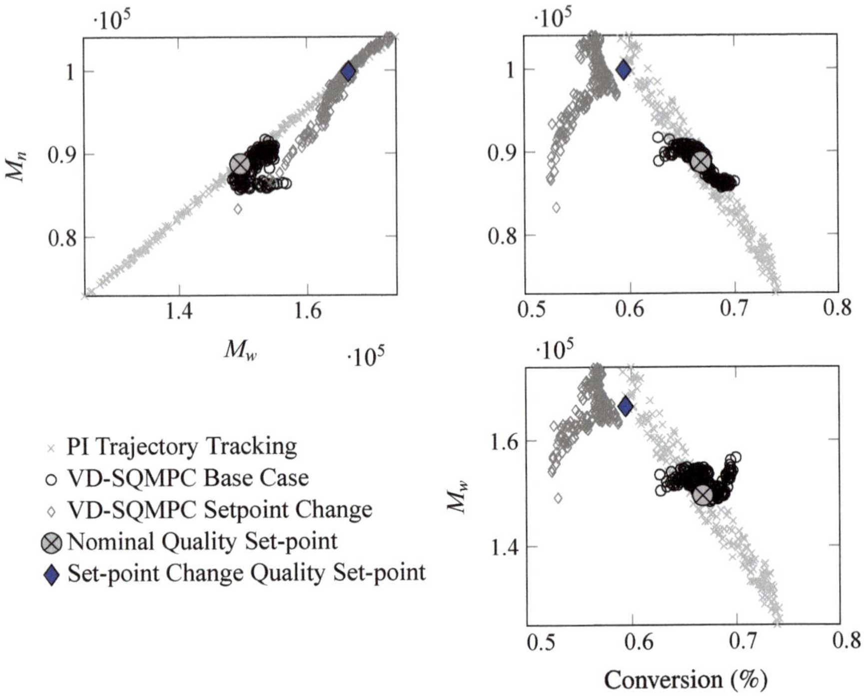

Fig. 13.10 Scatter plot matrix showing quality set-points for both case studies and resulting qualities achieved by PI and VD-SQMPC. Note that all three data sets contain the same initial conditions and quality results are achieved purely through continuous and discrete inputs

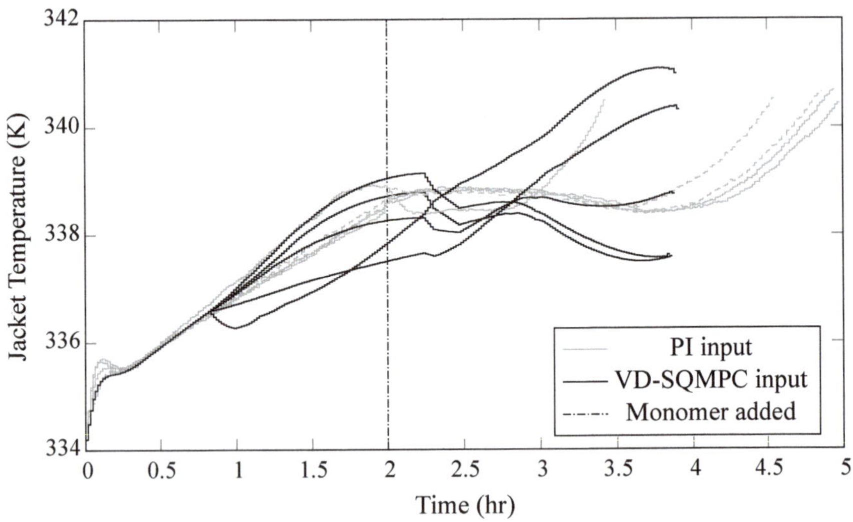

Fig. 13.11 Representative input trajectories for the base case

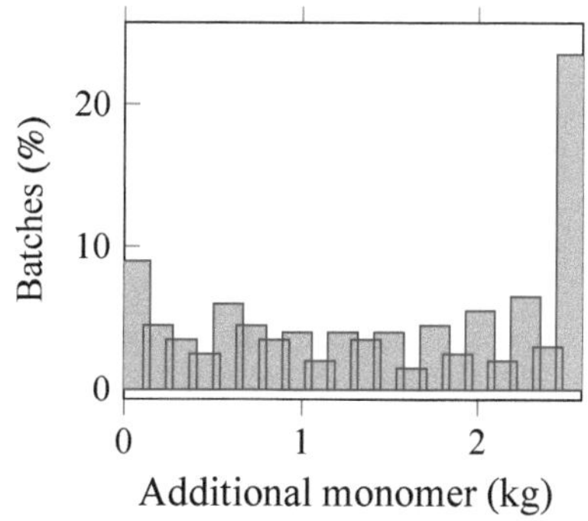

Fig. 13.12 Histogram of monomer added in the base case

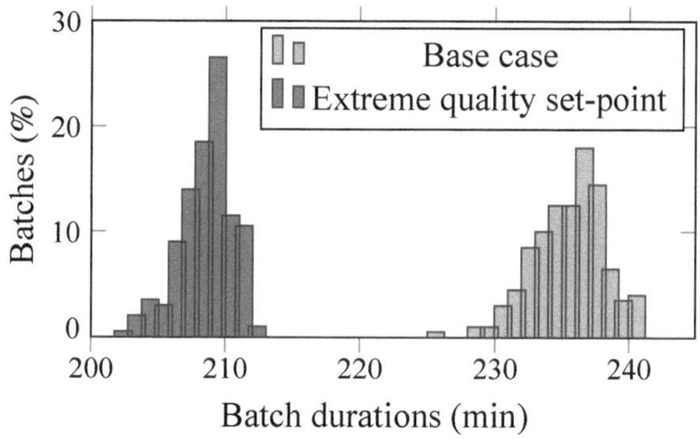

Fig. 13.13 Histograms of batch durations for the base case and the new quality set-point

13.6.7 Case Two: Set-Point Change

In the second case study, the set-point of the controller was changed. Specifically, a quality from an extreme outliers of the qualities obtained by the PI control scheme in the first case study was selected as a new target. Figure 13.14 is a histogram of the resulting relative error in qualities obtained in this case. Note that, in contrast to the previous case study, there is no meaningful way to compare these results to the PI control scheme. That is because there is no way to directly specify a quality set-point under the PI control scheme. Instead, to make a quality change in the trajectory tracking approach, a new temperature trajectory expected to result in the desired quality would have to be determined. This difficulty is one of the principle shortcomings of trajectory tracking control.

Figure 13.10 also shows the resulting qualities for this second case. Clearly, the proposed control scheme is able to drive batches closer to the desired quality. That said, in both Figs. 13.10 and 13.14, it is apparent that the quality control attained in this case is not nearly as tight as that from the base case. The decrease of performance in this case is directly related to the empirical modeling process. Note that models for this second case are not re-identified, instead models from the base case are applied at the far limit of their validity range. This kind of extrapolation from the training

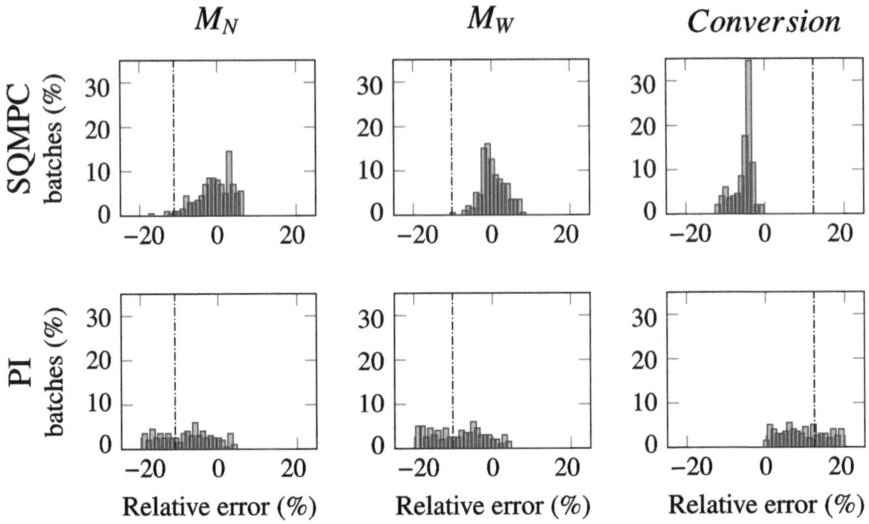

Fig. 13.14 Histograms of the qualities resulting from the proposed controller with a new quality set-point. (The dashed line is the set-point from the base case)

data naturally degrades the predictive ability of the models and therefore their ability to provide good quality control.

Figure 13.13 shows (in dark gray) the histogram of batch durations used in this second case study. Clearly, one of the key ways that the controller is able to meet the new quality set-point is by appropriately shortening the duration of the batch. This capability to recognize and act on the influence of batch durations is one of the strongest advantages of the proposed method.

Remark 13.16 Ideally, one would like to be able to improve the performance for each subsequent batch after a quality set-point change. Intuitively, this should be achievable because each batch operated with the new set-point provides added information about operations in the new state-space region. To achieve this, one option is to add an adaptation step to retrain the models after each batch is complete using the new observations from that batch. In practice, this kind of adaptation needs to be approached carefully because there may be a risk of reducing the predictive ability of the models by diluting the valuable insights from identification batches with too much closed-loop data. As such, methods to successfully adapt the models between batches are an area of future research.

13.7 Conclusions

In this chapter, results from [1] were generalized to encompass two key features of batch processes, namely discrete inputs were incorporated and batch duration was included as a control decision variable. We approached these problems in two steps. First, a new model structure and a corresponding identification method for that structure are proposed to relate process measurements, continuous inputs, and discrete inputs to product quality. Next, a model predictive control scheme was designed for direct quality control. Through a simulation example, the efficacy of this approach was demonstrated. The results of this simulation study clearly demonstrate the advantage of including batch duration as a decision variable.

References

1. Corbett, B., Mhaskar, P.: Subspace identification for data-driven modeling and quality control of batch processes. AIChE J. **62**, 1581–1601 (2016)
2. Moonen, M., Demoor, B., Vandenberghe, L., Vandewalle, J.: Online and off-line identification of linear state-space models. Int. J. Control **49**, 219–232 (1989)
3. Van Overschee, P., De Moor, B.: A unifying theorem for three subspace system identification algorithms. Automatica **31**, 1853–1864 (1995). 10th IFAC Symposium on System Identification, Copenhagen, Denmark, 04–06 July 1994
4. Muske, K., Rawlings, J., Lee, J.: Receding horizon recursive state estimation. In: 1993 Proceedings of the American Control Conference, vol. 1–3, pp. 900–904 (1993)
5. Flores-Cerrillo, J., MacGregor, J.F.: Control of batch product quality by trajectory manipulation using latent variable models. J. Process. Control **14**, 539–553 (2004)
6. Ekpo, E.E., Mujtaba, I.M.: Evaluation of neural networks-based controllers in batch polymerisation of methyl methacrylate. Neurocomputing **71**, 1401–1412 (2008)
7. Fan, S., Gretton-Watson, S., Steinke, J., Alpay, E.: Polymerisation of methyl methacrylate in a pilot-scale tubular reactor: modelling and experimental studies. Chem. Eng. Sci. **58**, 2479–2490 (2003)
8. Rho, H., Huh, Y., Rhee, H.: Application of adaptive model-predictive control to a batch MMA polymerization reactor. Chem. Eng. Sci. **53**, 3729–3739 (1998)
9. Cheng, J., Fang, M., Wang, Y.: Closed-loop subspace identification algorithms for 2-D separable-in-denominator systems. In: IFAC Symposium on System Identification, pp. 217–222 (2015)

Chapter 14
Subspace Identification Based Modeling and Control of Batch Particulate Processes

14.1 Introduction

Particulates production has undergone phenomenal growth in the past few decades. They are used for manufacturing of numerous high-value industrial products such as crystals, polymers. The products are often manufactured using batch processes to provide the flexibility of achieving different products by changing recipes. The physio-chemical and mechanical properties of the end product are known to be linked to the particle size distribution (PSD). For instance, in the case of emulsion polymerization, PSD strongly impacts the polymers' rheology (viscosity), adhesion, drying characteristics etc. [1]. In addition to PSD, certain product qualities are also impacted by variables such as solids content [1] or molecular weight distribution (MWD) [2].

The subspace identification-based batch control approach presented in the previous chapters [3] dealt generally with lumped parameter systems and the applications of data-driven approaches for PBE systems have been limited. Some variation of PLS approach have been proposed for distributed parameter processes. In one approach, [4] a PLS model is used in conjunction with the predictions obtained from PBEs for batch to batch optimization. In another result [5], a similar strategy is presented where only a PLS model is used for predicting the final PSD. As with other latent variable approaches, a limitation with the use of PLS models for batch processes is the need to align the batch lengths. In summary, there exists a lack of results for data-driven subspace modeling of batch particulate process that can readily handle nonuniform batch durations at the modeling and control stage.

Motivated by these considerations, this chapter presents a subspace identification-based modeling and control approach for batch particulate processes. The rest of the chapter is organized as follows: Sect. 14.2 presents a seeded batch crystallizer as a motivating example. Section 14.3.1 presents the subspace identification approach for batch particulate processes that does not require the training batches to be of uniform length. Section 14.3.2 presents the validation approach and clarifies the necessity of an appropriate state estimator. The proposed model predictive control formulation is

P. Mhaskar et al., *Modeling and Control of Batch Processes*,
Advances in Industrial Control, https://doi.org/10.1007/978-3-030-04140-3_14

then presented in Sect. 14.3.3. In Sect. 14.4, an application of the proposed approach to the seeded batch crystallization example is presented with robustness evaluation in the presence of time-varying parameters discussed in Sect. 14.5. Finally, concluding remarks are made in Sect. 14.6.

14.2 Preliminaries

For a particulate process, the state of the system includes the particle size distribution (PSD), and the writing of appropriate balances results in so-called population balance equations (PBEs)-based first-principles models for such processes. These PBEs describe the time evolution of properties of particulate entities such as particles, droplets, when interacting with each other and/or with their environment (usually continuous phase), and result in a set of equation comprising partial and ordinary nonlinear integro-differential equations.

For the purpose of control, a number of studies have taken the approach of developing a first-principles model based on these principles, estimating parameters (typically of a reduced-order model) and using it for closed-loop control [6, 7]. For instances where the development of such a model is possible, it is expected that a good first principles-based model will have better predictive ability and thus better control performance. The focus in the present manuscript is on scenarios where the model is required to be identified from past data, without first developing the governing balances.

14.2.1 Motivating Example: A Seeded Batch Crystallizer

In this section, we briefly describe the simulation test bed. To this end, consider the seeded batch cooling crystallizer [6, 8], used in the manufacturing of potassium sulfate crystals. A first-principles model describing the evolution of the crystal size distribution, $n(r, t)$, under the joint effects of nucleation and crystal growth yields two ordinary differential equations describing the evolution of the solute concentration, C, and reactor temperature, T, and a partial differential equation of the following form:

$$\begin{aligned}
&\frac{\partial n(r,t)}{\partial t} + G(t)\frac{\partial n(r,t)}{\partial r} = 0, \quad n(0,t) = \frac{B(t)}{G(t)} \\
&\frac{dC}{dt} = -3\rho k_v G(t)\mu_2(t) \\
&\frac{dT}{dt} = -\frac{UA}{MC_p}(T - T_j) - \frac{\Delta H}{C_p}3\rho k_v G(t)\mu_2(t)
\end{aligned} \tag{14.1}$$

where ρ is the density of crystals, k_v is the volumetric shape factor, U is the overall heat transfer coefficient, A is the total heat transfer surface area, M is the mass of solvent in the crystallizer, T_j is the jacket temperature, ΔH is the heat of reaction, C_p is the heat capacity of the solution, and μ_2 is the second moment of the PSD given by

$$\mu_2 = \int_0^{\infty} r^2 n(r,t)dr \tag{14.2}$$

The nucleation rate, $B(t)$, and the growth rate, $G(t)$, are given by

$$\begin{aligned} B(t) &= k_b e^{-E_b/RT}\left(\frac{C - C_s(T)}{C_s(T)}\right)^b \mu_3, \\ G(t) &= k_g e^{-E_g/RT}\left(\frac{C - C_s(T)}{C_s(T)}\right)^g \end{aligned} \tag{14.3}$$

where E_b is the nucleation activation energy, E_g is the growth activation energy, b and g are exponents relating nucleation rate and growth rate to supersaturation, C_s is the saturation concentration of the solute (Eq. (14.5)), and μ_3 is the third moment of the PSD given by

$$\mu_3 = \int_0^{\infty} r^3 n(r,t)dr \tag{14.4}$$

Equation (14.5) defines the saturation and metastable concentrations corresponding to the solution temperature T. The solution concentration is required to be between these two concentrations; i.e., $C_s \le C \le C_m$ must hold during the whole batch run.

$$\begin{aligned} C_s(T) &= 6.29 \times 10^{-2} + 2.46 \times 10^{-3} T - 7.14 \times 10^{-6} T^2 \\ C_m(T) &= 7.76 \times 10^{-2} + 2.46 \times 10^{-3} T - 8.10 \times 10^{-6} T^2 \end{aligned} \tag{14.5}$$

Values of the process parameters are given in Table 14.1. To generate data from this virtual test bed, a second-order accurate finite difference method with 1000 discretization points is used. Figure 14.1 shows the evolution of the reactor temperature, T, the solution concentration, C, and the PSD under a linear cooling strategy, where the jacket temperature, T_j, is cooled down linearly from 50 to 30 °C (see Fig. 14.1a).

Fines, or small crystals often formed by nucleation, pose difficulties in downstream processing equipment and affect product quality and process economics. This issue becomes more pronounced in case of a seeded batch crystallizer as the final products primarily grow from the seeds rather than from crystals formed by nucleation. Another downside of excessive fines is that it may require longer batch run time to achieve the desired crystal size distribution. For the present application, crystals smaller than 300 μm (denoted by r_g) are recognized as fines. From a crystal size distribution, one can therefore compute two moments corresponding to fines, and desired crystals, respectively. To be precise, the ith moments can be defined as

Table 14.1 Parameter values for the seeded batch cooling crystallizer of Eqs. (14.1)–(14.3)

Parameter	Value	Unit	Parameter	Value	Unit
b	1.45		g	1.5	
k_b	285.0	$1/(s\ \mu m^3)$	k_g	1.44×10^8	$\mu m/s$
E_b/R	7517.0	K	E_g/R	4859.0	K
ΔH	44.5	kJ/kg	C_p	3.8	$kJ/K \cdot kg$
M	27.0	kg	ρ	2.66×10^{-12}	$g/\mu m^3$
k_v	1.5		t_f	30	min
$T(0)$	50	°C	$C(0)$	0.1742	g/g
UA	0.8	$kJ \cdot s \cdot K$			

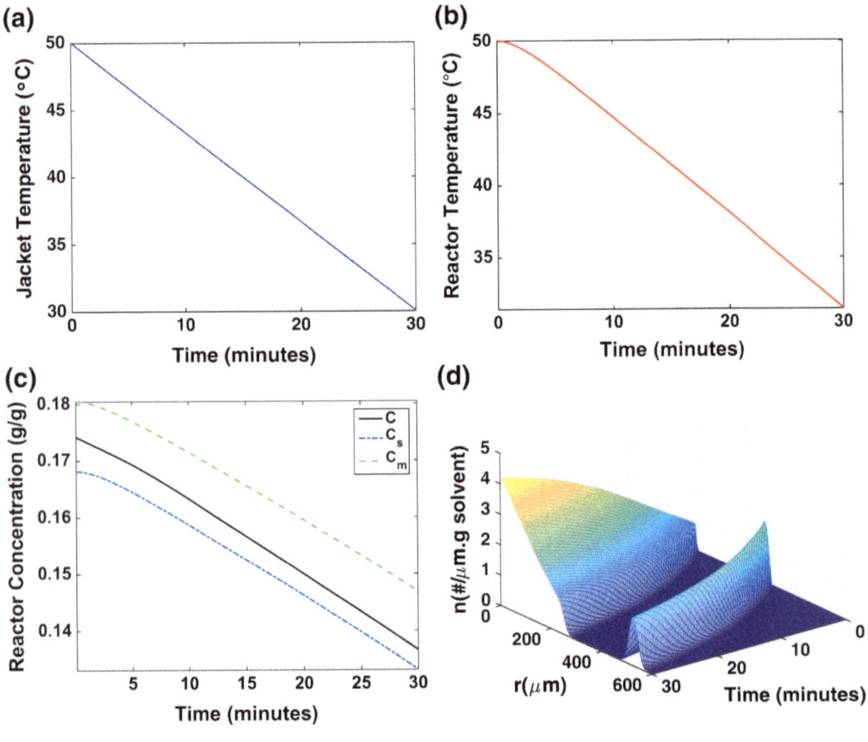

Fig. 14.1 Linear cooling control: **a** jacket temperature profile, **b** reactor temperature profile, **c** reactor concentration profile, and **d** the evolution of crystal size distribution

$$\mu_i^n = \int_0^{r_g} r^i n(r,t)dr$$
$$\mu_i^s = \int_{r_g}^{\infty} r^i n(r,t)dr, \quad i = 0, 1, 2, 3. \tag{14.6}$$

where the superscript, n, stands for nucleation, and s stands for seed. It should be noted here that although in the present study the two modes happen to be distinct (non-overlapping) for the training batches, the proposed approach holds for processes with possibly overlapping modes. In essence, $\mu_3^n[t_f]$ and $\mu_3^s[t_f]$ quantify the total volume of fines (undesired) and the volume of crystals grown from seeds (desired) at the termination of the batch (denoted by time t_f) respectively. The control objective, therefore, is to minimize the volume of fines in the final product along with achieving a desired crystal growth while respecting constraints on the manipulated input and concentration.

14.3 Proposed Modeling and Control Approach

The first step in the proposed approach is to identify a linear state-space model of the batch particulate process using past batch database and identification batches. This is achieved by adapting the subspace identification algorithm for batch processes (also adapted for an application in the context of lumped parameter systems [3]). A key step in subspace identification algorithm is the arrangement of data in Hankel matrices. In case of continuous processes, the training data sets are collected from identification experiments done around a desired steady-state condition and can readily be incorporated in the Hankel matrices for subspace identification of the model. However, in the case of batch processes, a meaningful steady-state is not available around which a model can be identified. Instead, a model for transient dynamics of the process is required over entire range of the process for different initial conditions. More importantly, typically data from multiple batches is available and this nature of the data requires specific adaptation of the database, and subspace identification method for continuous process cannot be directly implemented for batch processes.

14.3.1 Model Identification

Consider the output measurements of a batch process denoted as $\mathbf{y}^{(b)}[k]$, where k is the sampling instant since the batch initiation and b denotes the batch index. The output Hankel submatrix for a batch b is the same as a 'standard' Hankel matrix, given by:

$$\mathbf{Y}_{1|i}^{(b)} = \begin{bmatrix} \mathbf{y}^{(b)}[1] & \mathbf{y}^{(b)}[2] & \cdots & \mathbf{y}^{(b)}\left[j^{(b)}\right] \\ \vdots & \vdots & & \vdots \\ \mathbf{y}^{(b)}[i] & \mathbf{y}^{(b)}[i+1] & \cdots & \mathbf{y}^{(b)}\left[i+j^{(b)}-1\right] \end{bmatrix} \quad \forall\, b = 1, \ldots, nb \qquad (14.7)$$

where nb is the number of batches used for identification. The use of the above Hankel matrix by itself would not enable the use of data available from multiple batches.

At the same time, a naive concatenation of the data across various batches would also lead to erroneous conclusion. More specifically, such a naive concatenation would suggest that the end-point of one batch is the beginning of another, which clearly is not the case. The key, therefore, is to build an appropriate Hankel-like matrix for batch processes in a way that enables use of data across batches, without (incorrectly) assuming that the end-point of one batch is the beginning of another. This is achieved by horizontally concatenating the individual matrices, from each batch, to form a single pseudo-Hankel matrix for both input and output data. Thus, the pseudo-Hankel matrix takes the form:

$$\mathbf{Y}_{1|i} = \left[\mathbf{Y}_{1|i}^{(1)}\ \mathbf{Y}_{1|i}^{(2)}\ \cdots\ \mathbf{Y}_{1|i}^{(nb)}\right] \tag{14.8}$$

Similarly, pseudo-Hankel matrices for input data are formed. This approach for handling multiple batches nicely satisfies the requirements for subspace identification. A distinctive feature of this approach is that it does not impose any constraint on the length of different batches being equal.

Construction of appropriate pseudo-Hankel matrices for input and output enables determination of state trajectories using any of the wide variety of subspace identification algorithms available in the literature (such as the deterministic algorithm [9] used in this chapter). A consequence of concatenation of the Hankel matrices is that the identified state trajectories will be comprised of concatenated state estimates for each training batch. Mathematically, this can be represented as:

$$\hat{\mathbf{X}}_{i+1}^{(b)} = \left[\hat{\mathbf{x}}^{(b)}[i+1]\cdots\hat{\mathbf{x}}^{(b)}\left[i+j^{(b)}\right]\right]\ \forall\, b = 1,\ldots,nb \tag{14.9}$$

$$\hat{\mathbf{X}}_{i+1} = \left[\hat{\mathbf{X}}_{i+1}^{(1)}\ \hat{\mathbf{X}}_{i+1}^{(2)}\ \cdots\ \hat{\mathbf{X}}_{i+1}^{(nb)}\right] \tag{14.10}$$

where nb is the total number of batches used for identification. Once this state trajectory matrix is obtained, the system matrices can be estimated easily using methods like ordinary least squares as shown below:

$$\mathbf{Y}_{reg}^{(b)} = \begin{bmatrix}\hat{\mathbf{x}}^{(b)}[i+2]\cdots & \hat{\mathbf{x}}^{(b)}\left[i+j^{(b)}\right]\\ \mathbf{y}^{(b)}[i+1]\cdots & \mathbf{y}^{(b)}\left[i+j^{(b)}-1\right]\end{bmatrix} \tag{14.11}$$

$$\mathbf{X}_{reg}^{(b)} = \begin{bmatrix}\hat{\mathbf{x}}^{(b)}[i+1]\cdots & \hat{\mathbf{x}}^{(b)}\left[i+j^{(b)}-1\right]\\ \mathbf{u}^{(b)}[i+1]\cdots & \mathbf{u}^{(b)}\left[i+j^{(b)}-1\right]\end{bmatrix} \tag{14.12}$$

$$\left[\mathbf{Y}_{reg}^{(1)}\cdots\mathbf{Y}_{reg}^{(nb)}\right] = \begin{bmatrix}\mathbf{A}\ \mathbf{B}\\ \mathbf{C}\ \mathbf{D}\end{bmatrix}\left[\mathbf{X}_{reg}^{(1)}\cdots\mathbf{X}_{reg}^{(nb)}\right] \tag{14.13}$$

Thus, **A**, **B**, **C**, and **D** are the desired state-space model matrices.

Remark 14.1 It is important to recognize that the subspace identification problem using data from multiple batches is qualitatively different from the existing subspace identification applications. The formation of the pseudo-Hankel matrices is critical in explicitly dealing with the batch nature of the problem, and in ensuring that consistent state estimates are obtained for data from multiple batches. In particular, let us consider the scenario where the traditional subspace identification approaches are used, and for each batch, a corresponding state trajectory is independently obtained, and then, a least squares problem is solved to determine the system matrices. The inherent problem with such an approach would be that when state estimates are obtained for a particular batch, they are on a certain basis, and this basis may be different for each batch. Thus, when state trajectories from different batches are used to determine a common system dynamics, it will inevitably lead to a mismatch. In contrast, in the proposed method, the formation of the pseudo-Hankel matrix and determination of the state trajectories for each batch *by solving a single problem* lead to the state trajectories being identified on the same basis. Thus, when the system dynamics are finally determined, they are consistent across data obtained from the multiple batches.

14.3.2 Model Validation

A model which captures the dynamics of the process reasonably well can be incorporated in a predictive control framework to achieve desired product specifications. This constitutes the second step of the approach. Good model predictions (using state-space models) are dictated primarily by two components: initial state estimate and the system matrices. A good estimate of the system matrices from the training data is obtained utilizing the subspace identification algorithm, which also involves determining the initial state of the system for each batch (based on the entire batch behavior). When implementing such an approach online (i.e., for a new batch), the batch data is unavailable when a new batch starts, and thus, the application of this approach *cannot be done* without using an appropriate state estimation technique. Just as importantly, model validation should not be done using the estimator throughout the batch duration. To judge the ability of the identified model to predict the dynamic behavior, the state estimator should be taken out of the loop after the outputs have converged. Beyond that point, the identified model should be utilized to predict the variable trajectory for the rest of the batch and compared with the observed values. Herein lies the utility of the identified model in determining the appropriate control trajectory for the remaining batch duration, starting from some time into the batch.

To quantify the prediction errors in the validation, once the model predictions are obtained, mean absolute scaled error (MASE) [10] is computed for all the variables. MASE can be computed using the following expression.

$$\text{MASE} = \frac{\sum_{t=1}^{t_p} |e_t|}{\frac{t_p}{t_p-1}\sum_{t=2}^{t_p} |Y_t - Y_{t-1}|} \tag{14.14}$$

where e_t is the prediction error and Y_t is the measured value of a variable at any sampling instant t. t_p denotes the prediction horizon over which MASE is being evaluated.

In the present chapter, a Luenberger observer is used to illustrate the results. However, the method is not restricted to this choice of estimator/observer. Any other estimator, such as Kalman filter, or moving horizon estimator (MHE) can be readily used. Thus, to validate the model, for the initial part of the batch, a standard Luenberger observer of the following form is used to determine good estimates of the states (based on the convergence of the measured output):

$$\hat{\mathbf{x}}[k+1] = \mathbf{A}\hat{\mathbf{x}}[k] + \mathbf{B}\mathbf{u}[k] + \mathbf{L}(\mathbf{y}[k] - \hat{\mathbf{y}}[k]) \tag{14.15}$$

where $\mathbf{L}$ is the observer gain and is design to ensure that $(\mathbf{A} - \mathbf{LC})$ is stable. The poles of the observer are placed appropriately in the unit circle. The initial state estimate could be chosen as zero or initialized at the average of states in the training data set. After the observer has converged, the identified model can be deployed for predicting the rest of the batch profile (for a candidate input profile) both for the purpose of validation and ultimately for feedback control.

14.3.3 *Model Predictive Controller Design*

For the case of batch processes, while the structure of the controller is similar to the standard MPC design, the control objective being different from that of operating around the steady state must be adequately accounted for in the controller implementation. Thus, the MPC formulation takes the following form:

$$\begin{aligned} \min_{\mathbf{u}} \quad & f(y) \\ s.t. \quad & u_{min} \leq u \leq u_{max}, \\ & \alpha[k] \leq f_2(y[k]) \leq \beta[k], \\ & f_3(y[t_f]) \geq \gamma \\ & \mathbf{x}[k+1] = \mathbf{A}\mathbf{x}[k] + \mathbf{B}\mathbf{u}[k] \\ & \mathbf{y}[k] = \mathbf{C}\mathbf{x}[k] + \mathbf{D}\mathbf{u}[k] \end{aligned} \tag{14.16}$$

where $f(y)$ represents the objective function which, depending on the application, could be 'stage cost' or simply depend on the terminal output $y[t_f]$, u_{min} and u_{max} denote the bounds for the manipulated input. The constraint on $f_2(y[k])$ denotes the output path constraints, while the constraint on $f_3(y[t_f])$ denotes the terminal quality constraint. The optimization problem is solved based on the predictions obtained from the identified (LTI) state-space model, utilizing the current state estimate provided by the state estimator. If the constraints and objective can also be described by linear functions, utilization of the subspace model enables converting the optimization problem to a linear/quadratic program that can be readily implemented online.

Remark 14.2 The ability to represent (or reasonably approximate) the dynamics described by PBEs through a linear time-invariant (LTI) state-space model stems from two factors; the first is the ability to approximate PBE models through finite moments models, and the second is the ability to approximate nonlinear dynamics with (possibly higher order) linear time systems for a finite duration. The approximation by an LTI system is enabled by the fact that the linear representation is not obtained by discretization of a nonlinear system around some nominal operating conditions. Instead, a sufficiently high-order representation is estimated to explain the dynamics of the process over a finite duration.

14.4 Applications to the Motivating Example

In this section, first the identification of a state-space model for the batch crystallization process using the subspace identification method is presented. This is followed by the application of the predictive control formulation along with the comparison of the results against conventional approaches.

14.4.1 Model Identification

For the purpose of model identification, historical data from 40 different batches, along with 9 identification experiments, was assumed to be available (see Table 14.2 for details). In these batches, a PI controller tracked the temperature trajectory that yielded the best results under linear cooling (out of 50 batches, also used later for the purpose of comparison with PI and MPC). A saturater is used in series with the PI controller to restrict the input moves between 50 and 30 °C and the gradient within 2 °C/min. The parameters of the PI controller were: controller gain, $K_c = 0.1$, and reset time, $T_i = 1$. The nominal seed is taken to be a parabolic distribution (Fig. 14.2), from 250 to 300 μm, and the maximum density of initial seed distribution, which is 2/μm · g solvent, occurs at 275 μm, i.e.,

$$n(r,0) = \begin{cases} 0 & r < 250\,\mu\text{m} \\ 0.0032(300-r)(r-250) & 250\,\mu\text{m} \le r \le 300\,\mu\text{m} \\ 0 & r > 300\,\mu\text{m} \end{cases} \tag{14.17}$$

To reflect variability across batches, for each batch, the initial conditions for reactor temperature, reactor concentration, and seed distribution vary (randomly) with a zero-mean Gaussian distribution around their nominal initial values as shown in Table 14.1.

For the 49 simulated batches of 30 min duration each, measurements were collected (and feedback implemented) every ($\Delta_s =$) 3 s. As summarized in Table 14.2, the database consisted of PI controlled batches where in the last 9 batches, the tracking trajectory (reactor temperature) was also superimposed with a pseudorandom binary

Table 14.2 Training database summary

Input policy	Number of batches
PI trajectory tracking	40
PI trajectory tracking superimposed with PRBS	9
Total batches	49

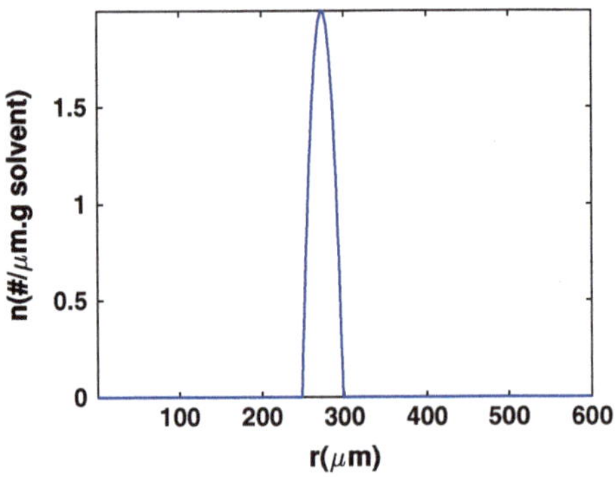

Fig. 14.2 Nominal seed distribution

signal (PRBS). The principle moments calculations are assumed to be available online through online measurements of PSD. Note that in practice, PSD measurements can be obtained using measurement techniques such as laser light scattering method. Further, the concentration and reactor temperature are also assumed to be measured in real time. The measurements of the reactor temperature, reactor concentration, and crystal size distribution are corrupted by measurement noise drawn from a random Gaussian distribution with zero-mean and a standard deviation of 0.001 (appropriately scaled back to each variable by multiplying with the true value of that variable). A moving average filter, of window size 5, was implemented to filter out the measurement noise from the measurements before using them for the training and implementation of the controllers (the same filtering was utilized when implementing the classical controllers as well).

The output vector for the crystallizer thus consists of nine measurements, namely reactor temperature, reactor concentration, saturation concentration, metastable concentration, μ_1, μ_2, μ_3, μ_3^n, and μ_3^s. The input and output data from the training batch database were used to identify a linear state-space model of the process. A parsimonious order of the system was selected based on its prediction accuracy for validation data set. Thus, a 46th-order state-space model of the following form was identified:

$$\begin{aligned} \mathbf{x}[k+1] &= \mathbf{A}\mathbf{x}[k] + \mathbf{B}\mathbf{u}[k] \\ \mathbf{y}[k] &= \mathbf{C}\mathbf{x}[k] + \mathbf{D}\mathbf{u}[k] \end{aligned} \tag{14.18}$$

where $\mathbf{y} = [T\ C\ C_s\ C_m\ \mu_1\ \mu_2\ \mu_3\ \mu_3^n\ \mu_3^s]'$ and $\mathbf{u} = T_j$.

Remark 14.3 In the present chapter, the model order is selected based on cross-validation results. In general, selecting a higher number of states reduces the error in the training data set but may result in increase of prediction errors (for the validation data set) as the higher order of the model effectively attempts to capture the effect of the noise present in the data. Therefore, the order was chosen such that further increase in the model order did not result in any significant reduction in predictions errors in the validation data set.

To validate the identified model, six validation batches, different from the training batches, were assumed to be available. The validation batches were generated using PI trajectory tracking superimposed with PRBS. For validation tests on a new batch, once the observer has converged to a state estimate such that the predicted outputs match the observed outputs (this happens by 12.5 min for the application), the state estimate is used to predict the process output for the remainder of the batch using the remainder of the input sequence. This predicted output is compared to the measured output trajectory to gauge the quality of the estimated model. Results for one of the validations are presented in Fig. 14.3 with the MASE values presented in Table 14.3. The validation results demonstrate the viability of using the developed model for feedback control within an MPC framework.

To further emphasize the merits of the proposed modeling approach, the model predictions were also obtained using the conventional subspace identification algorithm. To represent a conventional subspace identification approach, a single batch data was used for training, while another batch is used for validation. For this, the system identification toolbox in MATLAB was used to identify a deterministic state-space model using CVA algorithm. A model order of 25 was chosen (higher model order did not lead to improvements). The validation results are shown in Fig. 14.4, with the MASE values presented in Table 14.3. As can be observed, the predictions obtained are overall worse. A possible reason for this is that data from a single batch may not exhibit the range of process dynamics that the process typically goes through and thus may not be able to predict the dynamics of a new batch. This reiterates the necessity of including multiple batches for identification of process dynamics. A naive implementation of the conventional subspace identification was also carried out (where all the batches are simply stacked one after the other, see Remark 14.1 for further discussion). This leads to even worse results and are is presented for the sake of brevity. Further comparisons with the conventional subspace identification approach are not pursued further.

Remark 14.4 Note that in the present chapter, online PSD measurements are assumed. It is, in general, possible to develop models with fewer measurements, as long as the key variables of interest remain observable via the measured variables (see [3] for a demonstration). Another direction of future work will address the problem where PSD measurements are available intermittently, while the other variables are available more frequently.

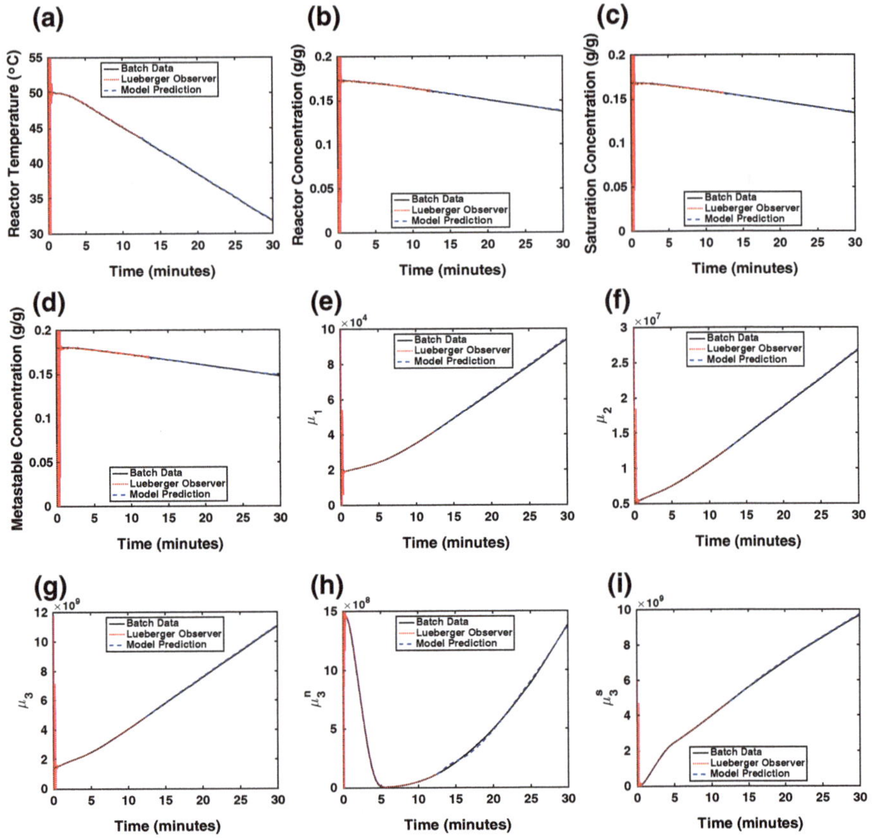

Fig. 14.3 Validation results for identification: **a** reactor temperature, **b** reactor concentration, **c** saturation concentration, **d** metastable concentration, **e** μ_1, **f** μ_2, **g** μ_3, **h** μ_3^n, and **i** μ_3^s

Table 14.3 MASE for validation batch using proposed approach (Fig. 14.3) and conventional subspace identification (Fig. 14.4)

	T	C	C_s	C_m	μ_1	μ_2	μ_3	μ_3^n	μ_3^s
Figure 14.3	0.7175	4.5885	4.5585	5.8187	2.4133	1.9012	1.7803	2.7412	2.3150
Figure 14.4	8.8×10^5	7.9×10^5	1.0×10^6	1.0×10^6	8.2×10^4	12.2	3.6	13.3	4.1

14.4.2 Predictive Control of Batch Particulate Processes

This section presents the application of the predictive controller described in Sect. 14.3.3. In particular, at a sampling instance l, where the current estimate of the states generated by the Luenberger observer is denoted by $\hat{\mathbf{x}}[l-1]$, and the last control action implemented on the process is denoted by $\mathbf{u}[\mathbf{l}]$ the optimal trajectory from the current instance to the end of the batch is computed by solving the following optimization problem:

$$
\begin{aligned}
\min_{\mathbf{U}_f}\ & y_8[l_f - l] \\
s.t.\ & T_{j,min} \le u_f[k] \le T_{j,max}, \quad \forall\ 0 \le k \le l_f - l \\
& y_3[k] + \epsilon_1 \le y_2[k] \le y_4[k] - \epsilon_2, \quad \forall\ k \le H_c \\
& |u_f[0] - u[l-1]| \le \delta, \\
& |u_f[k] - u_f[k-1]| \le \delta,\ \forall\ 1 \le k \le l_f - l \\
& y_9[l_f - l] \ge \gamma + \epsilon_3, \\
& \mathbf{x}[0] = \hat{\mathbf{x}}[l] \\
& \mathbf{x}[k+1] = \mathbf{A}\mathbf{x}[k] + \mathbf{B}\mathbf{u}_f[k] \\
& \mathbf{y}[k] = \mathbf{C}\mathbf{x}[k] + \mathbf{D}\mathbf{u}_f[k]\ \forall\ 0 \le k \le l_f - l
\end{aligned}
\tag{14.19}
$$

where $\mathbf{U}_f = [\mathbf{u}_f[0], \mathbf{u}_f[1], \ldots, \mathbf{u}_f[l_f - l]]$ (the remaining input profile) is the decision variable, $l_f = t_f/\Delta_s$ and y_i is the ith element of the output vector y, predicted using the identified model as discussed before. $T_{j,min}$ and $T_{j,max}$ are the constraints on the manipulated variable, u, and set to 30 and 50 °C in the simulation. y_3 and y_4 are the constraints on the solution concentration. The constant δ, chosen to be 2 °C/min, is the maximum gradient of the jacket temperature. Note that this rate of change constraint is implemented in a way that ensures both that the first piece of the control action respects this constraint with respect to the last implemented control action and that the successive control actions respect this constraint as well. γ, chosen as 8.3301×10^9, denotes the lower bound on the total volume of the crystals growing from the seeds. This constraint on the terminal y_9 represents a desirable quality of the final product and any batch where this constraint is not met is deemed a 'failed' batch. Finally, the last two equations represent the identified discrete time LTI model of the process.

To account for the plant model mismatch, tuning parameters ϵ_1, ϵ_2, ϵ_3, and H_c are used to tighten the constraints when implementing the predictive controller. In this chapter, ϵ_1, ϵ_2, and ϵ_3 were chosen as $0.0035\ g/g$, $0.0035\ g/g$, and 0.4×10^9, respectively. The saturation constraints are implemented only 25 steps ahead and shrunk appropriately as the batch nears completion. Thus, the value of H_c is as follows:

$$
H_c = \begin{cases} 25 & l_f - l \ge 25 \\ l_f - l & l_f - l < 25 \end{cases}
\tag{14.20}
$$

The first piece of the control action is implemented on the process, and the optimization repeated at the next sample time.

Exploiting the fact that the identified model is linear, the MPC optimization problem is formulated as a linear program as discussed below. Consider the estimated subspace state-space model in Eq. (14.18). For the input sequence $\mathbf{U}_f^p =$

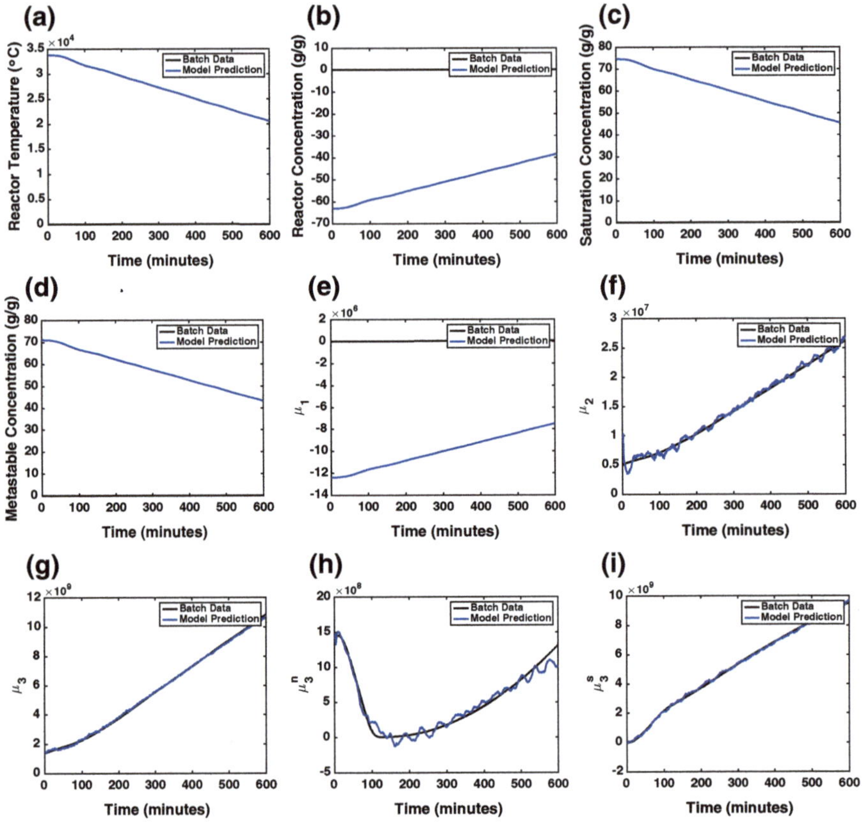

Fig. 14.4 Validation results for identification using conventional subspace identification: **a** reactor temperature, **b** reactor concentration, **c** saturation concentration, **d** metastable concentration, **e** μ_1, **f** μ_2, **g** μ_3, **h** μ_3^n, and **i** μ_3^s

$[\mathbf{u}_f[0], \mathbf{u}_f[1], \ldots, \mathbf{u}_f[p]]$ and initial state as $\mathbf{x}[0]$, the output variable p time steps ahead can be written as

$$\mathbf{Y[p]} = \mathbf{CA}^p\mathbf{x}[l] + \phi_p\mathbf{U}_f^p \tag{14.21}$$

where

$$\phi_p = \left[\mathbf{CA}^{p-1}\mathbf{B}\ \mathbf{CA}^{p-2}\mathbf{B}\ \ldots\ \mathbf{CB}\ \mathbf{D}\right] \tag{14.22}$$

The y_ith measurement, p steps ahead in the future, can be written using Eqs. (14.21) and (14.23) as $Y_i[p] = \mathbf{L}_i\mathbf{Y}[p]$, where

$$\mathbf{L}_i \triangleq \left[\mathbf{0}_{\mathbf{1}\times(\mathbf{i}-\mathbf{1})}\ 1\ \mathbf{0}_{\mathbf{1}\times(\mathbf{9}-\mathbf{i})}\right]\ \ \forall\, i = 1\ldots 9 \tag{14.23}$$

The terminal constraint on the seeded crystals in the final product can thus be written as:

$$\begin{aligned} &\mathbf{L}_9\mathbf{Y}[l_f - l] \geq \gamma \\ \implies &- \mathbf{L}_9\phi_{l_f-l}\mathbf{U}_f^{l_f-l} \leq -\gamma + \mathbf{L}_9\mathbf{C}\mathbf{A}^{l_f-l}x[0] \end{aligned} \tag{14.24}$$

Constraints on the input move is formulated as follows

$$\left| \begin{bmatrix} \mathbf{u}[l] - \mathbf{u}_f[0] \\ \mathbf{u}_f[0] - \mathbf{u}_f[1] \\ \vdots \\ \mathbf{u}_f[l_f - l - 1] - \mathbf{u}_f[l_f - l] \end{bmatrix} \right| \leq \Delta \tag{14.25}$$

$$\implies \begin{bmatrix} 1 & -1 & 0 & \cdots & 0 \\ 0 & 1 & -1 & \cdots & 0 \\ & \ddots & \ddots & \ddots & \\ 0 & \cdots & 0 & 1 & -1 \\ -1 & 1 & 0 & \cdots & 0 \\ 0 & -1 & 1 & \cdots & 0 \\ & \ddots & \ddots & \ddots & \\ 0 & \cdots & 0 & -1 & 1 \end{bmatrix} \begin{bmatrix} u[l-1] \\ \mathbf{U}_f^{l_f-l} \end{bmatrix} \leq \Delta \tag{14.26}$$

Further, the constraints on the concentration, $y_3 \leq y_2$ can be formulated as

$$\begin{bmatrix} \mathbf{L}_3 - \mathbf{L}_2 \\ \vdots \\ \mathbf{L}_3 - \mathbf{L}_2 \end{bmatrix} \circ \Pi\mathbf{U}_{(H_c-1)} \leq - \begin{bmatrix} \mathbf{L}_3 - \mathbf{L}_2 \\ \vdots \\ \mathbf{L}_3 - \mathbf{L}_2 \end{bmatrix} \circ \begin{bmatrix} \mathbf{C} \\ \mathbf{CA} \\ \vdots \\ \mathbf{CA}^{(H_c-1)} \end{bmatrix} \mathbf{x}[0] \tag{14.27}$$

where Π is a Toeplitz matrix given by

$$\Pi = \begin{bmatrix} \mathbf{D} & \mathbf{0} & \mathbf{0} & \cdots & \mathbf{0} \\ \mathbf{CB} & \mathbf{D} & \mathbf{0} & \cdots & \mathbf{0} \\ \mathbf{CAB} & \mathbf{CB} & \mathbf{D} & \cdots & \mathbf{0} \\ & \ddots & \ddots & \ddots & \\ \mathbf{CA}^{(H_c-2)}\mathbf{B} & \cdots & \mathbf{CAB} & \mathbf{CB} & \mathbf{D} \end{bmatrix} \tag{14.28}$$

Similarly, $y_2 \leq y_4$ can be written as

$$\begin{bmatrix} \mathbf{L}_2 - \mathbf{L}_4 \\ \vdots \\ \mathbf{L}_2 - \mathbf{L}_4 \end{bmatrix} \circ \Pi \mathbf{U}_{(H_c-1)} \leq - \begin{bmatrix} \mathbf{L}_2 - \mathbf{L}_4 \\ \vdots \\ \mathbf{L}_2 - \mathbf{L}_4 \end{bmatrix} \circ \begin{bmatrix} \mathbf{C} \\ \mathbf{CA} \\ \vdots \\ \mathbf{CA}^{(H_c-1)} \end{bmatrix} \mathbf{x}[0] \tag{14.29}$$

Here, $\circ$ denotes the Hadamard product. The transformation of the control problem into a linear problem allows fast solutions and enables real-time implementation of the predictive controller.

Remark 14.5 In the proposed MPC formulation, as stated previously, the measurements fed to the MPC are first filtered with a moving horizon filter to mitigate the effects of measurement noise. The use of a filter represents standard practice and in the present case enables a ‘fair’ comparison between the PI and the proposed MPC since both controllers are provided the same filtered measurement for feedback control. A comparison with a first-principles-based approach is not pursued here because, as already stated, if a good first-principles model is already available, its use will naturally lead to better control performance. The comparison is thus restricted to implementations that do not require the development of a first principles model. In particular, the focus in the present manuscript is on addressing the fact that the batch processes is a particulate process (defined inherently through the evolution of particle size distributions) and to demonstrate the robustness of the approach to variability.

Remark 14.6 In this chapter, the objective was to minimize the number of fines in the product at end of the batch. Depending on the type of batch crystallizer, the objective can vary and can be readily accommodated in the proposed formulation. For instance, the quality of crystals could be dependent more explicitly on the shape of the CSD. An example of a desired shape could be a requirement of narrow distribution of crystals. To handle these, the proposed model would have to be extended to capture more of the state variables and connect it with an appropriately built ‘quality’ model. This remains a subject of future work for distributed parameter systems.

Closed-Loop Simulations Results

The subspace state-space model-based linear MPC strategy, proposed in this chapter, is applied to 50 different batches to examine the effectiveness of the proposed approach. The method is also compared with two other classical control strategies, namely open-loop linear cooling and PI controller-based trajectory tracking. It should be noted here that these batches are different from the ones used for identification with the initial conditions generated in a similar fashion.

Table 14.4 summarizes the value of μ_3^n obtained for 50 different batches under three different control strategies, i.e., open-loop with linear cooling, trajectory tracking with PI controller, and a data-driven MPC with the formulation specified in Eq. (14.19). Figure 14.5 compares the variability of the terminal μ_3^n achieved under different control schemes. It is evident from the results that the proposed MPC scheme

Table 14.4 Comparison of control strategies for 50 batches

Control method	$\bar{\mu}_3^n$	$\min(\mu_3^n)$	Number of failed batches
Linear cooling	1.5446×10^9	1.2525×10^9	0
PI	1.4709×10^9	1.1863×10^9	0
MPC	1.0802×10^9	9.4983×10^8	0

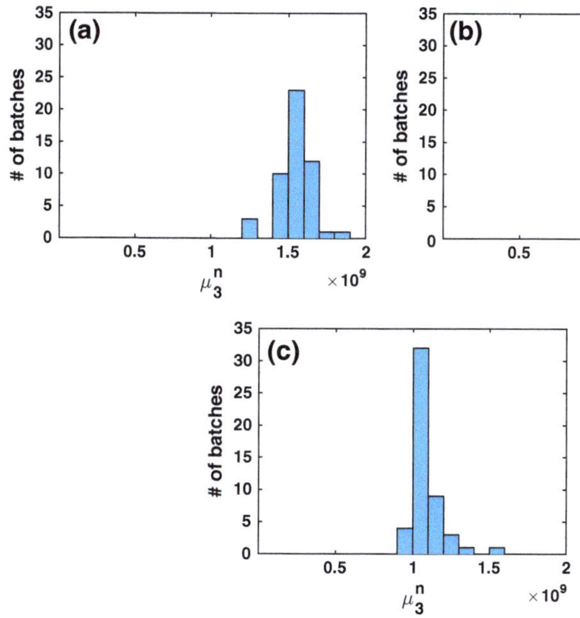

Fig. 14.5 Distribution of μ_3^n for process under **a** linear cooling, **b** PI, and **c** MPC

lowered the volume of fines on an average by over 27 and 30% compared to PI and linear cooling strategy respectively, while satisfying the process constraints in all the batches. Further, the variability in volume of fines in the final product is least for MPC compared to other control schemes.

Figure 14.6 shows the closed-loop profile for the batch for which the MPC yields the best performance. Note that when implementing MPC, the initial 10 min of operation is still under the PI control. During this time, the Luenberger observer estimates the current state of the process. Then, the controller is switched to MPC till the end of the batch. It can be observed from Fig. 14.6 that in the MPC, input trajectory deviates from the PI trajectory after 10 min to drive the process to the optimal product specifications.

To further probe the efficiency of the proposed method, the control schemes are compared for the case where the constraint on the terminal μ_3^s (value of γ) is further increased by 1.5×10^9 from the previous case to reflect the requirement of increase in

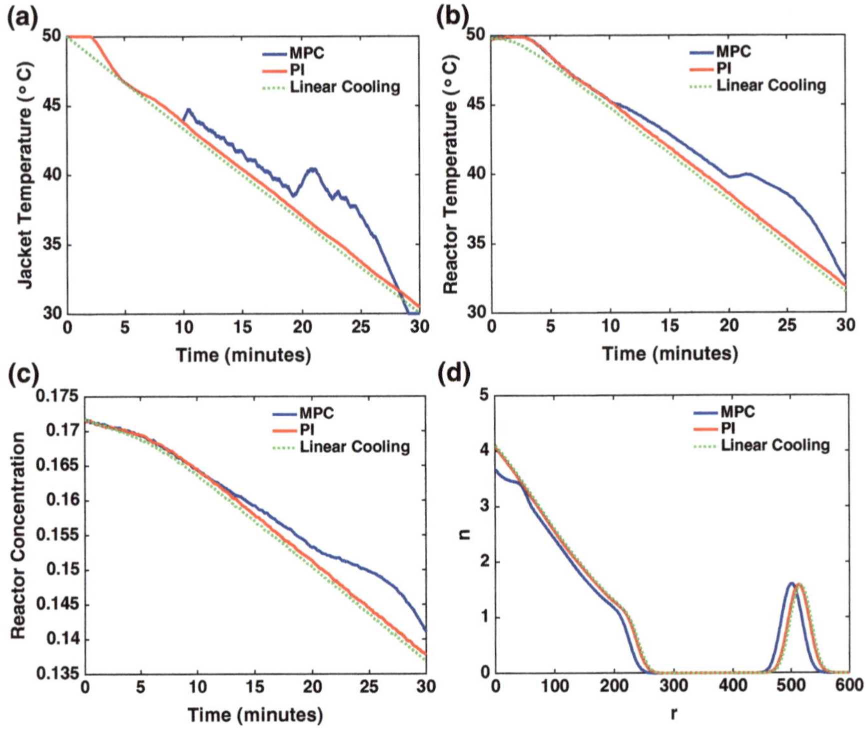

Fig. 14.6 Results corresponding to best MPC batch scenario: **a** input profiles, **b** reactor temperature, **c** reactor concentration, and **d** final CSD

the product quality. For the set of test batches, 24 and 62% batches violate constraints under linear cooling and PI, respectively, and as such do not produce the desired product. In contrast, only 4% of the batches result in loss of product under the MPC.

14.5 Robustness Evaluation in the Presence of Time-Varying Parameters

To reflect process variability, the crystallization process was simulated with b varying randomly with a standard deviation of 0.001 around its nominal value and U decreasing by 0.5% (as a consequence of fouling) in every successive batch.

14.5.1 Model Identification

The process with varying parameters was simulated for 20 batches, among which 15 batches were used for identification and rest for validation. A state-space model of order 31 was identified using subspace identification for batch as discussed before.

14.5.2 Closed-Loop Results

The state-space model identified using the subspace identification was deployed within the previously explained MPC design to control 50 different batches with parameters b and U varying as before. It should be noted that none of these batches were used in the identification batches. The performance of the proposed framework, compared against PI and linear cooling methodologies, are summarized in Table 14.5. As can be noted, the proposed approach resulted in a performance improvement of 21.34 and 7.84% in comparison to PI and linear cooling, respectively. It is important to recognize that the closed-loop performance under the PI controller happens to be poorer than the one under linear cooling. This is not because of poor tuning of the PI controller. The PI controller continues to do a good job of tracking the reference trajectory. It just so happens that with the change in process dynamics due to parameter variation, tracking the same reference trajectory for the temperature ends up resulting in production of more fine particles compared to the nominal linear cooling strategy. The other key points illustrated in these simulations are not so much as to how the MPC continues to do better than the other approaches, but the fact that the MPC continues to generate on-spec product, with improved quality. The temperature and concentration profiles along with the terminal CSD for the scenario with best MPC performance are illustrated in Fig. 14.7. In summary, these results illustrate the robustness of the approach to effectively utilize the available historical batch database to efficiently control a process with varying parameters.

Remark 14.7 Latent variable-based models that are inherently linear time-varying models linearized around a mean trajectory are another excellent tool to model and control batch processes and have found widespread applications [11, 12]. These approaches, because of the necessity to mean center the trajectories, require the identification of some alignment variable (or for the batches to be of the same length). In these approaches, nonlinearity is captured by building as many (as required) models (the time-varying matrices or loadings). In the proposed method, nonlinearity is captured instead by utilizing a sufficiently higher-order LTI model. A comparison of the proposed approach with the PLS-based approaches [3] shows that the proposed method compares favorably to the PLS-based approach. The comparison, therefore, is not repeated in the current study.

Table 14.5 Comparison of control strategies for 50 batches in the presence of time-varying parameters

Control method	$\bar{\mu}_3^n$	$\min(\mu_3^n)$	Number of failed batches
Linear cooling	1.3136×10^9	1.1695×10^9	0
PI	1.5391×10^9	1.0450×10^9	0
MPC	1.2106×10^9	8.9552×10^8	0

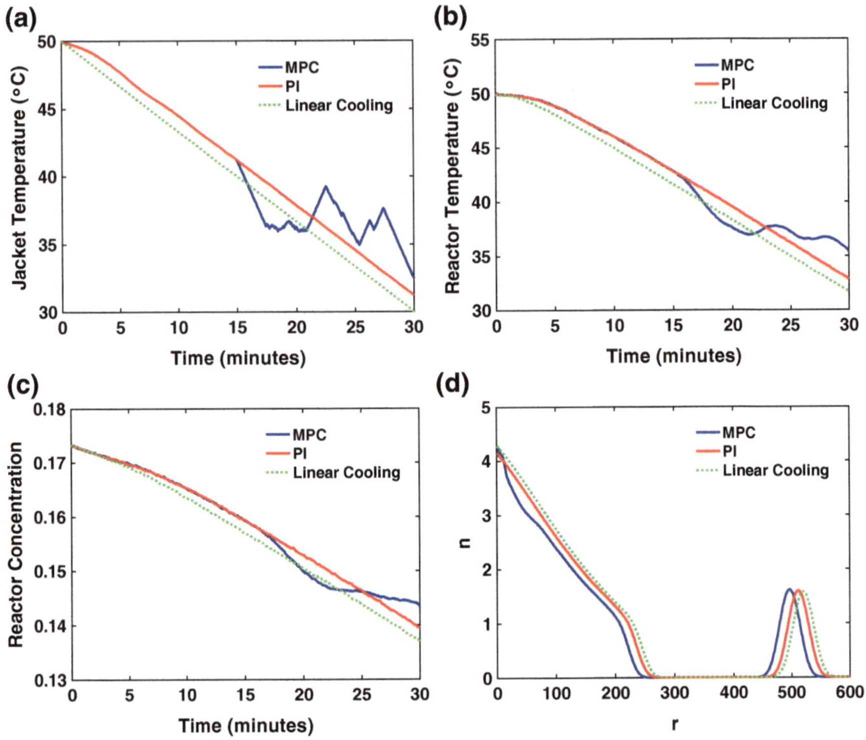

Fig. 14.7 Results corresponding to best MPC batch scenario in the presence of time-varying parameters: **a** input profiles, **b** reactor temperature, **c** reactor concentration, and **d** final CSD

14.6 Conclusions

In this chapter, a novel data-driven model identification and predictive control approach for particulate processes are proposed. Subspace identification method adapted to accommodate batch data structure is used to identify a linear time-invariant state-space model of the process. The identified model is then deployed within MPC to achieve a particle size distribution with desired characteristics. The simulation results show the merits of the proposed approach in terms of improved quality, the ability to achieve a different product specification, and robustness to time-varying process parameters.

References

1. Immanuel, C.D., Wang, Y., Bianco, N.: Feedback controllability assessment and control of particle size distribution in emulsion polymerisation. Chem. Eng. Sci. **63**, 1205–1216 (2008)

2. Alhamad, B., Romagnoli, J.A., Gomes, V.G.: On-line multi-variable predictive control of molar mass and particle size distributions in free-radical emulsion copolymerization. Chem. Eng. Sci. **60**, 6596–6606 (2005)
3. Corbett, B., Mhaskar, P.: Subspace identification for data-driven modeling and quality control of batch processes. AIChE J. **62**, 1581–1601 (2016)
4. Doyle, F.J., Harrison, C.A., Crowley, T.J.: Hybrid model-based approach to batch-to-batch control of particle size distribution in emulsion polymerization. Comput. Chem. Eng. **27**, 1153–1163 (2003)
5. Park, M., Dokucu, M.: Regulation of the emulsion particle size distribution to an optimal trajectory using partial least squares model-based predictive control. Ind. Eng. 7227–7237 (2004)
6. Rawlings, J.B., Miller, S.M., Witkowski, W.R.: Model identification and control of solution crystallization processes: a review. Ind. Eng. Chem. Res. **32**, 1275–1296 (1993)
7. Miller, S.M., Rawlings, J.B.: Model identification and control strategies for batch cooling crystallizers. AIChE J. **40**, 1312–1327 (1994)
8. Shi, D., El-Farra, N.H., Li, M., Mhaskar, P., Christofides, P.D.: Predictive control of particle size distribution in particulate processes. Chem. Eng. Sci. **61**, 268–281 (2006)
9. Moonen, M., Moor, B.D., Vandenberghe, L., Vandewalle, J.: On- and off-line identification of linear state-space models. Int. J. Control **49**, 219–232 (1989)
10. Hyndman, R.J., Koehler, A.B.: Another look at measures of forecast accuracy. Int. J. Forecast. **22**, 679–688 (2006)
11. Yabuki, Y., MacGregor, J.F.: Product quality control in semibatch reactors using midcourse correction policies. Ind. Eng. Chem. Res. **36**, 1268–1275 (1997)
12. Kourti, T., Lee, J., MacGregor, J.F.: Experiences with industrial applications of projection methods for multivariate statistical process control. Comput. Chem. Eng. **20**, S745–S750 (1996)

Chapter 15
Subspace-Based Model Identification of a Hydrogen Plant Startup Dynamics

15.1 Introduction

Hydrogen is an indispensable chemical component extensively used in petroleum and chemical industries. An economic way to commercially produce hydrogen is to utilize steam methane reforming (SMR) [1]. SMR process operation (like most other operations) undergoes scheduled startup and shutdown procedures which is sometimes initiated due to the need for regular maintenance and sometimes by unforeseen circumstances. Startups/shutdowns are implemented using standard operating procedures (SOPs). These procedures present various constraints and challenges such as constraints on the reformer exit temperature and firebox pressure [2]. Further, a typical startup involves a series of operations such as ramping various flows, introducing certain flows, along with making discrete decisions such as starting up certain parts of the unit. Thus, the startup time is not fixed and varies based on the decisions taken at various stages. Startups and shutdowns are fairly resource-intensive operations and stand to gain from optimizing the startup procedure via operating procedure synthesis (OPS).

A key requirement for OPS is a good model for the startup process. There exist several modeling approaches that are well suited for small-scale processes (see [3] for an excellent review) but difficult to implement to large chemical units. Out of these approaches, dynamic simulation-based strategies are most prominent. In these approaches, dynamic simulations are used to select the best startup scenarios. The detailed models allow a good description of the startup dynamics and are valuable from a process analysis standpoint. One of the key contributions of the present chapter is the development of a detailed first-principles model for the hydrogen unit startup, with appropriate adaption of the plant startup SOP. Note however that such detailed first-principles models pose computational challenges when directly embedded in optimization problems, and thus, existing SOP synthesis approaches utilize heuristics in an attempt to determine optimal profiles (e.g., see [4–7]). Performance of a more rigorous optimization necessitates the use of simpler models that can capture the startup/shutdown dynamics, and yet be amenable for online implementation. One

P. Mhaskar et al., *Modeling and Control of Batch Processes*,
Advances in Industrial Control, https://doi.org/10.1007/978-3-030-04140-3_15

of the contributions of the present chapter is the recognition of the startup/shutdown process as a batch-like process that opens up the possibility of adapting data-driven batch modeling techniques to identify models that capture the startup process dynamics reasonably well and yet are not too computationally complex.

While a variety of data-driven modeling techniques exist for modeling and control of batch processes, not all of these are suitable for the problem at hand. One approach that has found wide variety of applications is partial least squares (PLS), which models the process in a projected latent space [8]. PLS models are in essence time-varying linear models, linearized around mean past trajectories, and thus require the batches to be of the same length, or to find an appropriate alignment variable. As indicated earlier, the inherent variabilities of the startup procedure do not result in startups of equal length.

In one recently proposed approach [9], the models were based on the 'current measurements' of the process instead of the 'time.' These developments were followed by contributions in the area of integration of these data-driven models with model predictive control formulations [9, 10]. More recently, a subspace identification-based batch control approach was proposed in [11] where an LTI state-space model of the batch process is identified, and does not require the training batches to be of equal length. The problem of optimizing hydrogen startups thus stands to gain from these recent results. Note that while the approach in [11] has subsequently been adapted to detailed simulations of an electric arc furnace [12] and particle size distribution (PSD) control [13], it has not been applied yet on a process of the present complexity, using real data, and to model startups.

Motivated by these considerations, in this chapter we address the problem of determining a data-driven model for the startup of a hydrogen plant. The chapter is organized as follows. First, necessary background on hydrogen production process is presented in Sect. 15.2. Section 15.3 presents the modifications necessary to adapt a detailed first-principles model in UniSim Design to enable simulating the startup and shutdown phase. Although the primary focus of the proposed approach is to identify reduced-order data-driven model of the process, the development of a detailed first-principles model serves two key purposes. Firstly, it enables the generation of startup and shutdown (simulated) data of the process which can be used to test/develop the data-driven modeling approach. Secondly, a detailed simulator capable of mimicking the startup/shutdown can serve as a good analysis tool for control practitioners. Before implementing subspace identification methods to identify a reduced-order linear time-invariant model of the process, the suitability of a linear model to explain the dynamics of a finite duration nonlinear process is illustrated through simulations in Sect. 15.4.1. Then, an LTI data-driven model of the process using subspace identification-based method is identified in Sect. 15.4.2. Simulation results illustrate the prediction capabilities of the identified model validated against the detailed first-principles simulation model. Next, the approach is implemented on data collected from a Praxair hydrogen unit, and validated in Sect. 15.4.3. Finally, a few concluding remarks are presented in Sect. 15.5.

15.2 Preliminaries

This section presents an overview of the hydrogen production process and subspace identification approaches.

15.2.1 Process Description

Hydrogen is commercially manufactured by steam methane reforming, where natural gas (NG) and superheated steam are fed to a chemical reactor called reformer, which consists of catalyst tubes filled with nickel reforming catalyst as depicted in Fig. 15.1. Natural gas is first purified by removing any liquid that may have condensed due to low ambient temperature. It is then split into two streams. Most of the natural gas is compressed as the feed to the reformer. A small amount is used as fuel to provide heat for the reformer. The steam is produced from water by a steam system in the heat recovery block. The NG feed stream is heated using downstream process heat and further processed by removing any unfavorable compounds to the reformer catalyst. Under normal conditions, the pressure of the NG fuel is controlled by an upstream valve, and the flow rate of the NG fuel is controlled by a downstream valve. The majority of the hydrogen is produced in the reformer through the following chemical reactions:

$$C_x H_y + x H_2 O \rightleftarrows x C O + \left(x + \frac{y}{2}\right) H_2 \tag{15.1}$$

$$C O + H_2 O \rightleftarrows C O_2 + H_2 \tag{15.2}$$

Reaction (15.1) is known as reforming, and reaction (15.2) is also called shift conversion. Both the reactions are reversible. The overall reaction is endothermic. Reformer exit temperature (RET) is an important process variable for this process and is expected to be kept at a specific value by heating the reformer. This heat is provided by burning the off-gas from the pressure swing adsorber (PSA) and NG fuel

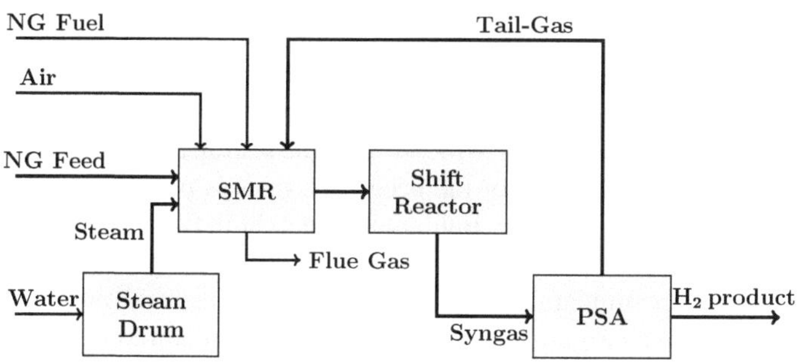

Fig. 15.1 A schematic of the hydrogen production process

stream. The NG fuel stream is used to regulate the reformer exit temperature under nominal operating scenarios. Moreover, the temperature is affected by the flow rates of the combustion air and the superheated steam.

The reformer effluent process gas passes through a reactor, where additional hydrogen is produced by shifting most of the carbon monoxide in the process gas to carbon dioxide and hydrogen through reaction (15.2). The reactor effluent stream then passes through the heat recovery and is sent to the PSA, where the hydrogen is produced. The PSA process is based on the physical adsorption phenomena. High volatile compounds with low polarity such as hydrogen are practically nonadsorbable compared to water, nitrogen, carbon monoxide, ammonia, methane, sulfur compounds, and hydrocarbons. Most of the impurities in the gas can be selectively adsorbed, resulting in high purity hydrogen. During normal operation, the off-gas out of the PSA is used as the primary fuel that provides heat for the reformer.

The combustion heat resulting from the ignition of air and fuel in burners heats the reformer tubes. A fan is used to supply air to the burners and another one draws the combustion products, which are termed flue gas, out of the reformer firebox. The firebox pressure should not reach its lower and upper limits for safety. If the pressure is too low, the fire can be extinguished. If it is too high, it may impose safety hazards to facility and personnel. The pressure is controlled by adjusting the position of the suction louvers of Fan 2. As the louvers open, the fan draws more flow, resulting in a lower pressure. Conversely, as the louvers close, the fan draws less flow to increase the pressure.

The focus of the present chapter is the modeling of the startup process. A typical startup involves ramping the natural gas fuel to reach a sufficient reformer exit temperature, introduction of natural gas feed and steam, regulation of nitrogen flow in the reactor tubes. It also requires making discrete decisions such as starting up the PSA, recirculating the tail gas from the PSA, and ramping up to full capacity in several stages. Specific details of the startup procedure are omitted due to confidentiality reasons.

15.3 Detailed First-Principles Model Development

To serve as a test bed, a UniSim Design model of the hydrogen production unit capable of simulating the plant startup and shutdown dynamics was first developed. To this end, a UniSim Design model capturing the nominal production phase provided by Praxair Inc. was suitably modified to incorporate the startup and shutdown simulation capabilities. A brief description of the necessary steps is provided here to serve as guidelines to adapt other dynamic simulation models built for simulating the nominal continuous operation (around some steady-state equilibrium point) of the plant.

To enable startup simulation, the flowsheet requires several modifications. For instance, the control configurations from the P&ID are required to be incorporated into the model. Apart from these, several dummy controllers are required to mimic the manual operations by the operator during the course of startup/shutdown. In

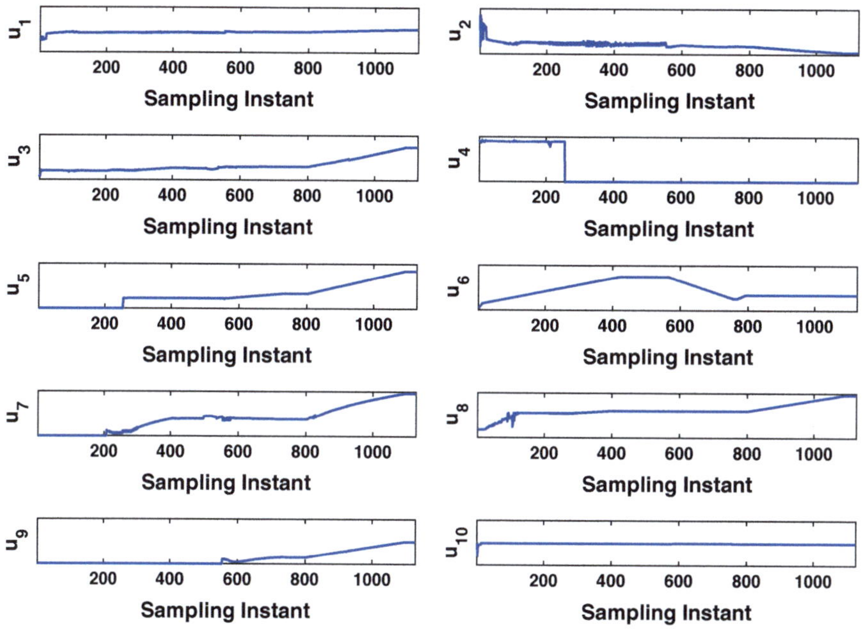

Fig. 15.2 Simulated startup: manipulated variables profile

shutdown phase simulation, it is not possible to drive the model to a complete shutdown state due to the limitations of simulator. Therefore, the flows were reduced to a nominal low value which can simulate the process without running into numerical issues. This minimal state is treated as shutdown state in the simulation. The startup procedure is then implemented on this attained shutdown state. Note that the goal of the first-principles model development in this chapter was only to capture the key startup and shutdown trends (validated against real startup data). This model can be then used as a test bed for the application of the proposed identification approach.

Note that the SOP available from plant operations was not amenable for direct implementation in a simulated environment. Therefore, a 'minimalistic' SOP implementable in the UniSim Design model was developed along the lines of real plant SOP. Data from several startups is generated by implementing this adapted SOP. For implementing the modified SOP in UniSim Design, any of the following two approaches can be used. Either, the startup strategy is defined within UniSim Design itself using Event Schedulers (as done in the present work) or using a MATLAB–UniSim interface to drive the process using MATLAB. However, for implementing a closed-loop control strategy the latter approach might be more suited and will be the subject of future work. One such simulated startup manipulated and output profiles are illustrated in Figs. 15.2 and 15.3 (showing the scaled manipulated inputs and measured outputs, respectively). While specific details of the inputs and outputs remain confidential, the inputs include various gas flow rates and the outputs include temperatures and pressures.

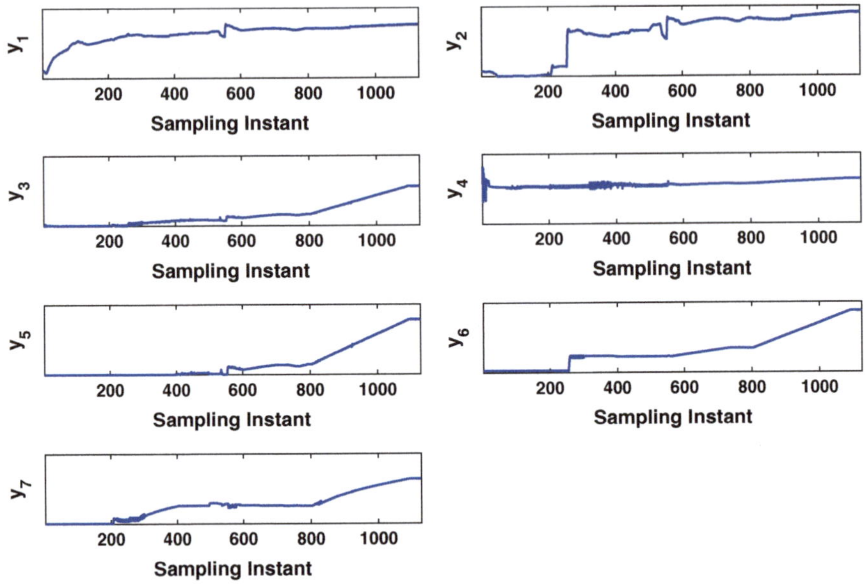

Fig. 15.3 Simulated startup: output variables profile

15.4 Identification of State-Space Model

We recognize that the startup phase can be treated as a batch process. Thus, consider the output measurements of a startup batch denoted as $\mathbf{y}^{(b)}[k]$, where k is the sampling instant since the startup initiation and b denotes the startup index. The constructions of the matrices and model identification follow the procedure outlined in Sect. 14.3.1.

One of the key 'tuning parameters' in identifying the model is the number of subspace states. In general, a higher number of subspace states result in better fit in the training set, but could likely lead to over-fitting, and poor prediction for a fresh startup. This is avoided by choosing the tuning parameter on the basis of cross validation, that is, utilizing a startup that was not part of the training set. The validation of the model (and eventual use as a model within a feedback control tool) however also requires the use of a state estimator. This is because for a new batch, only the initial outputs are available as measurements, and not the value of the subspace states for that particular batch. If the subspace model is integrated from an arbitrary initial (subspace states) condition, it is unlikely that the model will be able to predict the dynamics of the new batch. An integral feature of the subspace identification approach for batch processes, therefore, is to use a state estimator for the initial duration of the batch to first estimate the subspace state trajectory for the initial duration of the batch. It is important to note that the potential use of the model in a closed-loop setting relies on its ability to predict future outputs for candidate input moves. The state estimator, however, can only be utilized to determine the current state estimates. After the initial duration, therefore, prediction of the outputs for the rest of the batch (without using the state estimator) is used for validation of the identified model.

In the present chapter, a Luenberger observer is used to illustrate the approach. However, the method is not restricted to this choice of estimator/observer. Any other estimator, such as Kalman filter, or moving horizon estimator (MHE) can be readily used. Thus, to validate the model, for the initial part of the batch, a standard Luenberger observer of the following form is used to first determine good estimates of the states (based on the convergence of the measured output):

$$\hat{\mathbf{x}}[k+1] = \mathbf{A}\hat{\mathbf{x}}[k] + \mathbf{B}\mathbf{u}[k] + \mathbf{L}(\mathbf{y}[k] - \hat{\mathbf{y}}[k]) \tag{15.3}$$

where $\mathbf{L}$ is the observer gain, picked to ensure that $(\mathbf{A} - \mathbf{LC})$ is stable, k is the current sampling instant, $\hat{\mathbf{x}}$ is the state estimate, $\mathbf{y}$ denotes the measurement, and $\hat{\mathbf{y}}$ is the output prediction obtained using the estimated model. The poles of the observer are placed appropriately in the unit circle.

Another decision that must be made when dealing with a fresh startup is determining the initial condition for the subspace model. One reasonable way of picking the initial state estimates is, for instance, the average of states in the training data set. The observer is initialized at $k = 0$ and is updated with measurements until $k = l$, such that outputs converge. Then for $k = l + 1$ onward, the identified LTI model is used to predict forward for validation purposes.

Once the model predictions are obtained, the prediction errors in the process outputs are quantified using mean absolute scaled error (MASE) [14]. The specific expression for computing MASE is given as follows

$$\text{MASE} = \frac{\sum_{t=1}^{T} |e_t|}{\frac{T}{T-1}\sum_{t=2}^{T} |Y_t - Y_{t-1}|} \tag{15.4}$$

where e_t is the prediction error and Y_t is the measured value of a variable at any sampling instant t.

Remark 15.1 One could argue that it should be possible to carry out subspace identification individually on each training batch using existing identification approaches to determine state trajectories for each batch, and then identify a single dynamic model from the resulting state trajectories. However, doing so would necessitate reconciling the different state projections realized for each training batch, which is simply not possible. One of the principal advantages of the proposed approach is that by combining multiple batches in the pseudo-Hankel matrix a common subspace basis is selected for all training batches. As a result, the state trajectories identified for each training batch are consistent.

Remark 15.2 As mentioned earlier, the proposed approach readily permits the use of batches of varying durations. However, it should be noted that, in this approach, each observation has equal weight. Therefore, a batch of greater duration will have a correspondingly greater influence in the identified model. Practically however, this result is appropriate since longer batches contain more information about the dynamics of the system. This additional information is a direct result of the greater number of measurements and inputs perturbations available in longer duration batches. This has the implication that longer duration batches may play a greater role in guaranteeing persistent excitation as later discussed in Conjecture 1.

15.4.1 Illustrative Simulations for Batch Subspace Identification Methodology

In this section, the capabilities of the proposed data-driven modeling methodology are illustrated and discussed with the aid of simulations. In particular, the objective is to illustrate improved prediction ability by using higher number of states in the identified model. To this end, consider a switched system described as follows

$$x[k+1] = \begin{cases} 0.7x[k] + 0.5u[k] & \text{if } x[k] \leq 0 \\ 0.9x[k] + 0.5u[k] & \text{if } 0 < x[k] < 0.4 \\ 0.4x[k] + 0.5u[k] & \text{if } x[k] > 0.4 \end{cases} \tag{15.5}$$

$$y[k] = x[k] \tag{15.6}$$

The state of this single state process is assumed to be measured and treated as process output. A set of five batches were generated for this process. The first four batches were used for training, while the last batch is used for validation. A PRBS input signal was designed such that the for two of the batches, the process starts in mode 2 and goes to mode 3 and stays there, while for the other three batches, the process goes from mode 2 to mode 3 and switches batch to mode 2 for a short duration. Two models are identified, one with a single state and another with two states. The validation results and the corresponding error evolution are shown in Figs. 15.4 and 15.5, respectively.

These simulations are presented to illustrate two key points. The first is that using higher number of states allows the model identification procedure to better capture the dynamics of the process in different regions, by the higher flexibility in choosing the corresponding state trajectories. Thus, it is expected that the model using say two subspace states should perform better than a subspace model with a single state. The second key point is the use of the observer, which as the process evolves and transits from one operation region to another, allows the model to 'learn' the current dynamics of the process better. Thus, in Fig. 15.5, compared to the error evolution of the prediction using the state estimates at the 20th time instance, the error is significantly lower for the prediction using the state estimate at the 60 time instant.

Remark 15.3 For identification of a process model, a key requirement is to have a rich data set. In continuous processes, operating around a nominal steady state, sufficient excitation of the input signal is in general difficult, especially for closed-loop data.

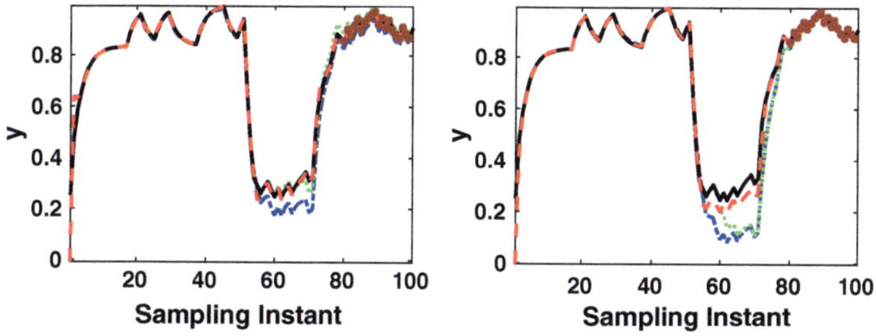

(a) Two states model (b) One state model

Fig. 15.4 A Luenberger observer (red) is used to estimate the initial condition of the identified state-space model till 20, 60, and 80th sampling instant, and then, the identified model predictions are shown by blue dashed lines, green dotted lines, and brown circles, respectively. Black lines denote the plant output. The simulation illustrates the the prediction error goes down as the updated state estimate is used

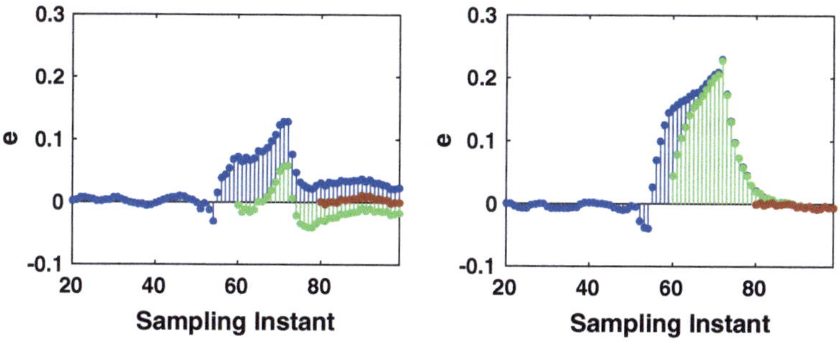

(a) Two states model (b) One state model

Fig. 15.5 Residual plot from 20, 60 and 80th sampling instants onward

However, in a process like startup/shutdown, this excitation occurs naturally since the process is never really at steady state, and the variation in feedstocks (and even operating procedure) provides variations across batches. The proposed approach, by recognizing startup/shutdown process as a batch-like process and using the data sets across different startup/shutdown scenarios, exploits the richness in the data to identify a reasonable model of the process, as demonstrated in this chapter.

To understand the benefits of using multiple batches for identification, we recognize that we are identifying a linear model of an appropriate order from the data. The order of the model that can be uniquely identified from the data depends on the 'excitation' of the input signal. To conjecture that data from multiple batches makes the identification data set richer, we first invoke a definition of persistence of excitation from classical system identification literature [15, 16]:

Definition 15.1 A quasi-stationary signal $u(t)$, with the spectrum $\Phi_u(\omega)$, is said to be persistently exciting of order n if, for all filters of the form

$$M_n(q) = m_1 q^{-1} + \cdots + m_n q^{-n} \tag{15.7}$$

the relation

$$|M_n(e^{iw})|^2 \Phi_u(\omega) \equiv 0 \tag{15.8}$$

implies that $M_n(e^{iw}) \equiv 0$

A frequency domain interpretation of this definition is that $u(t)$ is persistently exciting of order n, if $\Phi_u(\omega)$ is non-zero for at least n points in the interval $-\pi < \omega \leq \pi$ [15].

Building on this frequency domain interpretation, consider p different batch data sets with inputs and outputs denoted as $\mathbf{u}_1, \ldots, \mathbf{u}_p$ and $\mathbf{y}_1, \ldots, \mathbf{y}_p$, respectively. Also, the input spectrum is denoted as $\Phi_1, \ldots, \Phi_p$. Then, we conjecture that the use of multiple batches improves the identifiability of a model by essentially improving its order of excitation in the inputs. Formally, this is stated as follows:

Conjecture 1 *For batches $B_1, \ldots, B_p$, let the set of frequencies different from zero in the input spectrum for each batch be denoted as $\Psi_1, \ldots, \Psi_p$, and then, the overall persistence of excitation in the input for the multiple batches can be given by the number of frequencies in the set Ψ^b, where*

$$\Psi^b = \Psi_1 \cup \cdots \cup \Psi_p \tag{15.9}$$

that is, number of unique frequencies in all the batches that are different from zero.

Thus, including multiple batches implies that the order of the model that can be identified will be at least as much as that identified using a single batch. We next present a modified definition of persistent excitation for batch data sets.

Definition 15.2 The inputs from multiple batches are said to be *persistently exciting* if Ψ^b contains all the frequencies in the interval $-\pi < \omega \leq \pi$.

15.4.2 Model Identification and Validation for Simulated Data

A database of five historical startup batches and one identification batch corresponding to step test around the nominal operating phase for different inputs was used as training data sets for the subspace identification algorithm. The database included batches ranging from 1560 to 6840 sampling instant duration. They consist of ten manipulated variables and seven outputs as illustrated in Figs. 15.2 and 15.3. The variability among different startups is captured by suitably altering the various event triggers in the startup recipe. All the startups thus are of different lengths, i.e., have a different startup time. A subspace state-space model of order 52 was estimated (see

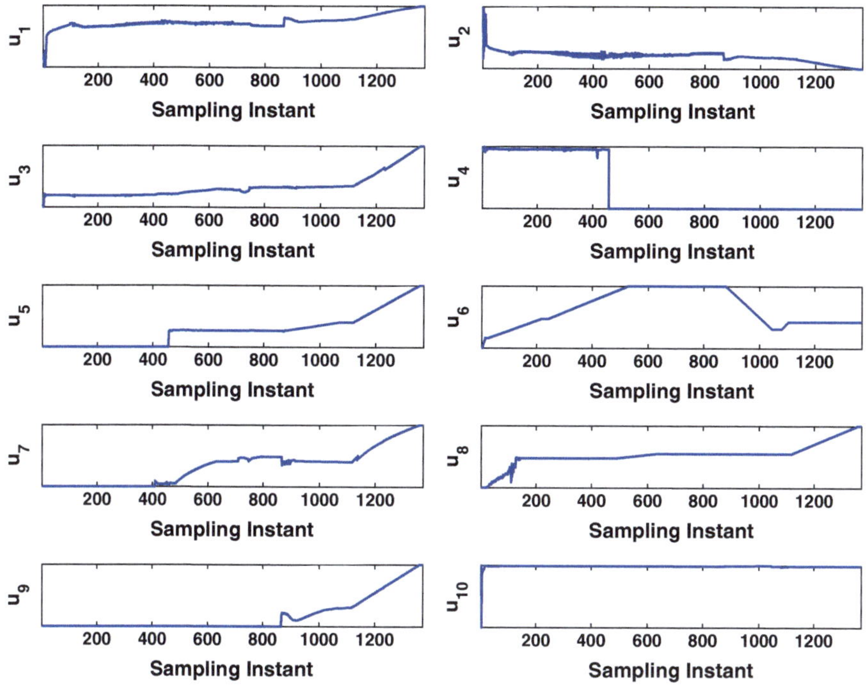

Fig. 15.6 Manipulated variables for the validation batch

Remark 15.4 for a discussion on this). Note that it was verified that the input signal for the simulated batches was persistently exciting, resulting in the ability of the data to identify the appropriate underlying model.

The identified model was validated on another startup batch, and the results are presented in Fig. 15.7. For validation, Luenberger observer was utilized until outputs converge (this happens at 440th sampling instant). Then, the estimated LTI model is used to make the output predictions for the remainder of batch using the remainder of the input profile and the current state estimate as shown in Fig. 15.6. As can be seen from Fig. 15.7, the model predictions match the simulated data (the prediction errors in the validation batch are quantified in Table 15.1). In essence, the identified LTI model is able to capture the complex nonlinear dynamics sufficiently well.

Remark 15.4 Note that the high order of the model is a direct consequence of the requirement to capture process nonlinearity of a complex system. It is important to recognize the distinction (and similarity) between models obtained using latent variable methods. While not directly applicable to the present application, a latent variable model [8] for batch processes typically includes far fewer latent variables or states. However, the model parameters are not necessarily any less. The latent variable models include 'loadings' corresponding to each sampling instance. Thus, the reduced number of latent variables is offset by the number of loadings (matrices

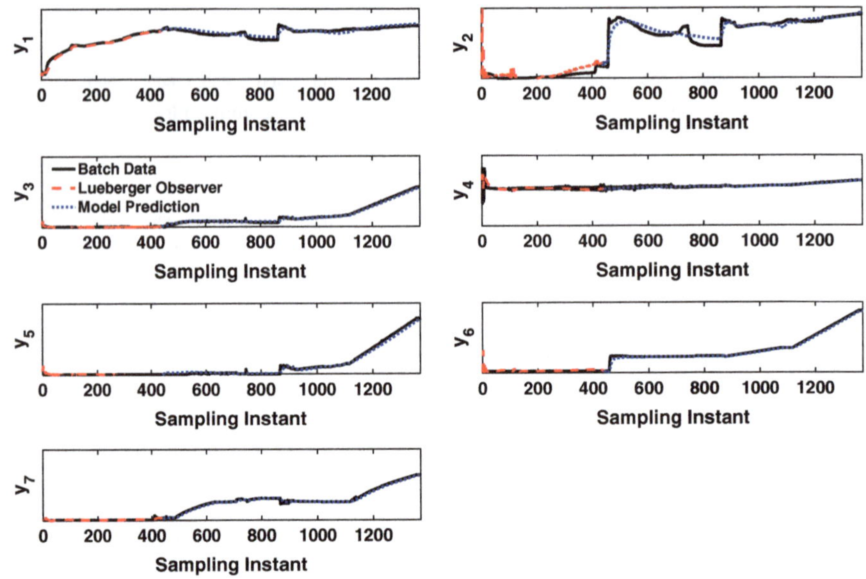

Fig. 15.7 Validation results

Table 15.1 MASE for validation results in Fig. 15.7

Outputs	y_1	y_2	y_3	y_4	y_5	y_6	y_7
MASE	12.88	10.19	5.82	0.97	7.43	9.64	4.85

of the linear time-varying model) that need to be determined. One of the key benefits of the proposed approach is that it enables capturing process nonlinearity, without any higher model complexity than the latent variable approaches. Further, the use of higher number of states did not result in any numerical issues in the present work.

15.4.3 Model Identification and Validation for Plant Data

The proposed approach was next utilized to model data from a hydrogen unit startup. In this work, four data sets corresponding to cold startup of a hydrogen unit of Praxair Inc. were used to model the startup dynamics while the model was validated using another startup from the same plant. The data set consisted of ten manipulated and eight output variables. A model of order 46 was estimated. It should be noted that all the startups were of different duration. For the validation batch, input profiles and the outputs along with predictions are shown in Figs. 15.8 and 15.9, respectively. Some of the manipulated and output variables in the plant historian are different from the ones used for the simulated study. This is because not all measurements are stored in the historian and thus cannot be used for model estimation.

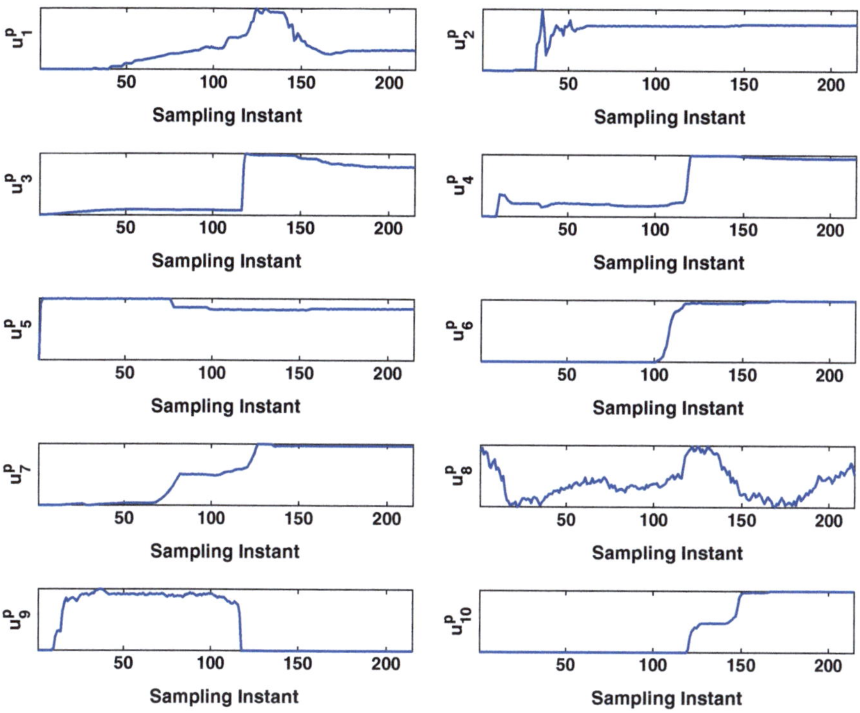

Fig. 15.8 Input profiles for the validation startup batch from plant

For the validation data set, the Luenberger observer was utilized until outputs converge. Then, the estimated LTI model is used to make the output predictions for the remainder of batch using the inputs as shown in Fig. 15.9. The dashed lines show the outputs from the Luenberger observer. Note that the estimated output predictions staying reasonably close to the observed measurements simply show that the observer is well tuned. The Luenberger observer outputs alone are not sufficient from a closed-loop implementation perspective. The key for closed-loop implementation is the ability of the model to predict future output behavior for a candidate input profile from a certain point into the batch. These are shown by the dashed green and solid blue lines in Fig. 15.9, where the predictions are made starting from sampling instant 50 and 180, respectively. It can be seen that the open-loop predictions are reasonably good even starting from 50th sample (see Table 15.2 for a quantitative error measure) and improve considerably for the case where the predictions are made from 180th sampling instant onward. The identified model is thus an excellent candidate for closed-loop control and optimization.

Remark 15.5 Note that during the startup, various streams were turned on at different points for the training and validation data. This however does not necessitate building different models. The very fact that all these stream flow rates are inputs that are part of the modeling and validation data sets enables capturing the effect of these inputs

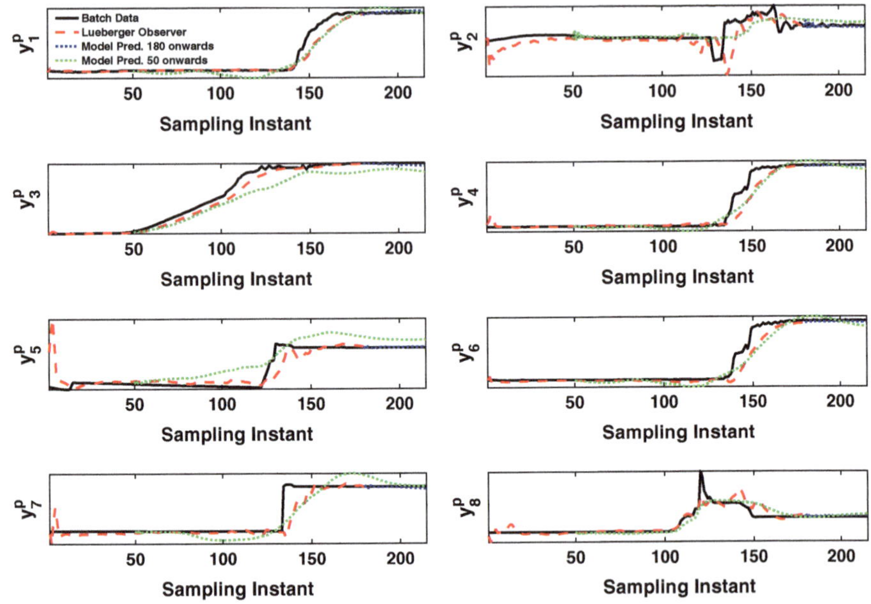

Fig. 15.9 Output profiles prediction for plant data, with open-loop prediction starting from 50 (dotted lines) to 180 (solid blue line) sampling instance. The red dashed line shows outputs from the Luenberger observer

Table 15.2 MASE for validation results in Fig. 15.9 for predictions starting from 50th sampling instant onward

Outputs	y_1^p	y_2^p	y_3^p	y_4^p	y_5^p	y_6^p	y_7^p	y_8^p
MASE	11.35	4.59	15.46	8.21	30.25	9.89	23.76	4.59

on the process dynamics. In essence, the ability of the model to capture the dynamics follows from the fact that varying the inputs to an LTI system does not make the system time varying.

Remark 15.6 The initial duration, over which the outputs converge, serves as an important adaptive mechanism in the proposed approach and is consistent with latent variable approaches where the scores for a particular batch can only be reasonably well estimated after some data from the new batch has been collected. In particular, the initial period where the observer 'estimates' the state for the new batch provides the proposed approach with the ability to 'learn' of the variations in the new batch, including variation in feedstock, and to account for the differences in the dynamics that result from that.

15.5 Conclusion

This chapter presents a data-driven model of the hydrogen production unit. To this end, a detailed first-principles model of the entire plant was developed in UniSim Design, capable of simulating the startup and shutdown phase. Several simulated startups were used as training database for the identification of reduced-order data-driven model. An LTI data-driven model of the process using subspace identification-based methods was estimated along with a simulation-based study to demonstrate the suitability of linear models to approximate nonlinear dynamics. Simulation results demonstrate the prediction capabilities of the identified model and its suitability for deployment in an optimization framework.

References

1. Kroschwitz, J., Seidel, A.: Kirk-Othmer Encyclopedia of Chemical Technology. Encyclopedia of Chemical Technology, vol. 20. Wiley, New York (2006)
2. Du, M., Mhaskar, P., Zhu, Y., Flores-Cerrillo, J.: Safe-parking of a hydrogen production unit. Ind. Eng. Chem. Res. **53**(19), 8147–8154 (2014)
3. Batres, R., Soutter, J., Asprey, S.P., Chung, P.: Operating procedure synthesis: science or art. Knowl. Eng. Rev. **17**(3), S0269888902000498 (2002)
4. Scenna, N., Ruiz, C., Benz, S.: Dynamic simulation of start-up procedures of reactive distillation columns. Comput. Chem. Eng. **22**, S719–S722 (1998)
5. Yang, X., Xu, Q., Li, K., Sagar, C.D.: Dynamic simulation and optimization for the start-up operation of an ethylene oxide plant. Ind. Eng. Chem. Res. **49**(9), 4360–4371 (2010)
6. Zhao, Y., Zhang, J., Qiu, T., Zhao, J., Xu, Q.: Flare minimization during start-ups of an integrated cryogenic separation system via dynamic simulation. Ind. Eng. Chem. Res. **53**(4), 1553–1562 (2014)
7. Reepmeyer, F., Repke, J.-U., Wozny, G.: Time optimal start-up strategies for reactive distillation columns. Chem. Eng. Sci. **59**(20), 4339–4347 (2004)
8. Flores-Cerrillo, J., MacGregor, J.F.: Control of particle size distributions in emulsion semibatch polymerization using mid-course correction policies. Ind. Eng. Chem. Res. **41**(7), 1805–1814 (2002)
9. Aumi, S., Corbett, B., Mhaskar, P., Clarke-Pringle, T.: Data-based modeling and control of nylon-6, 6 batch polymerization. IEEE Trans. Control Syst. Technol. **21**, 94–106 (2013)
10. Aumi, S., Mhaskar, P.: Integrating data-based modeling and nonlinear control tools for batch process control. AIChE J. **58**, 2105–2119 (2012)
11. Corbett, B., Mhaskar, P.: Subspace identification for data-driven modeling and quality control of batch processes. AIChE J. **62**, 1581–1601 (2016)
12. Rashid, M.M., Mhaskar, P., Swartz, C.L.: Multi-rate modeling and economic model predictive control of the electric arc furnace. J. Process Control **40**, 50–61 (2016)
13. Garg, A., Mhaskar, P.: Subspace identification based modeling and control of batch particulate processes. Ind. Eng. Chem. Res. 56(26), 7491–7502 (2017)
14. Hyndman, R.J., Koehler, A.B.: Another look at measures of forecast accuracy. Int. J. Forecast. **22**, 679–688 (2006)
15. Ljung, L.: System Identification: Theory for the User. Pearson Education, Boston (1998)
16. Zhu, Y.: Multivariable System Identification For Process Control. Elsevier Science, New York (2001)

MIX
Papier aus verantwortungsvollen Quellen
Paper from responsible sources
FSC® C105338

If you have any concerns about our products,
you can contact us on
ProductSafety@springernature.com

In case Publisher is established outside the EU,
the EU authorized representative is:
Springer Nature Customer Service Center GmbH
Europaplatz 3, 69115 Heidelberg, Germany

Printed by Libri Plureos GmbH
in Hamburg, Germany